T0361791

# Metamorphosis in
# Fish

# Metamorphosis in Fish

*Editors*

**Sylvie Dufour**

Research Unit BOREA "Biology of Aquatic Organisms and Ecosystems"
Muséum National d'Histoire Naturelle
Paris Cedex
France

**Karine Rousseau**

Research Unit BOREA "Biology of Aquatic Organisms and Ecosystems"
Muséum National d'Histoire Naturelle
Paris Cedex
France

**B.G. Kapoor**

Formerly Professor of Zoology
Jodhpur University
India

CRC Press
Taylor & Francis Group
an **Informa** business
www.taylorandfrancisgroup.com

6000 Broken Sound Parkway, NW
Suite 300, Boca Raton, FL 33487
711 Third Avenue
New York, NY 10017
2 Park Square, Milton Park
Abingdon, Oxon OX14 4RN, UK

Science Publishers
Jersey, British Isles
Enfield, New Hampshire

Published by Science Publishers, an imprint of Edenbridge Ltd.
- St. Helier, Jersey, British Channel Islands
- P.O. Box 699, Enfield, NH 03748, USA

E-mail: _info@scipub.net_                     Website: _www.scipub.net_

*Marketed and distributed by:*

CRC Press
Taylor & Francis Group
an **Informa** business
www.taylorandfrancisgroup.com

6000 Broken Sound Parkway, NW
Suite 300, Boca Raton, FL 33487
711 Third Avenue
New York, NY 10017
2 Park Square, Milton Park
Abingdon, Oxon OX14 4RN, UK

Copyright reserved © 2012

ISBN 978-1-57808-713-6

**Cover Illustrations:** Reproduced by kind courtesy of:
Richard G. Manzon for the figures of larval & juvenile Lampreys
Keisuke Yamano/Hideki Tanaka [Figure 8 from Chapter 3]
Yasuo Inui and Satoshi Miwa for the figure of Flatfish eye migration

```
          Library of Congress Cataloging-in-Publication Data
Metamorphosis in fish / editors, Sylvie Dufour, Karine Rous-
seau, B.G.
Kapoor.
       p. cm.
  Includes bibliographical references and index.
  ISBN 978-1-57808-713-6 (hardcover)
1.  Fishes--Metamorphosis.  I. Dufour, Sylvie. II. Rousseau,
Karine. III.
Kapoor, B. G.
  QL615.M48 2011
  571.1'7--dc23
                                                    2011019798
```

# Contents

# List of Contributors

**Aroua Salima**
Research Unit BOREA "Biology of Aquatic Organisms and Ecosystems" Muséum National d'Histoire Naturelle, CNRS 7208, IRD 207, UPMC, 7 rue Cuvier, CP32, 75231 Paris Cedex 05, France.

**Boeuf Gilles**
Laboratoire Arago, Research Team "Models in cell and evolutionary biology", University Pierre and Marie Curie- Paris 6/CNRS, Banyuls-sur-mer and Muséum national d'Histoire naturelle, 57 rue Cuvier, 75231 Paris Cedex 05, Paris, France
E-mail: boeuf@mnhn.fr

**Dufour Sylvie**
Research Unit BOREA "Biology of Aquatic Organisms and Ecosystems" Muséum national d'Histoire naturelle, CNRS 7208, IRD 207, UPMC, 7 rue Cuvier, CP32, 75231 Paris Cedex 05, France.
E-mail: dufour@mnhn.fr

**Ellien Céline**
Research Unit BOREA "Biology of Aquatic Organisms and Ecosystems" Muséum national d'Histoire naturelle, CNRS 7208, IRD 207, UPMC, 47 rue Cuvier, CP26, 75231 Paris Cedex 05, France.
E-mail: ellien@mnhn.fr

**Inui Yasuo**
Tamaki, Mie, 519-0414, Japan.
E-mail: inuiyj@yahoo.co.jp

**Keith Philippe**
Research Unit BOREA "Biology of Aquatic Organisms and Ecosystems" Muséum national d'Histoire naturelle, CNRS 7208, IRD 207, UPMC, 47 rue Cuvier, CP26, 75231 Paris Cedex 05, France.
E-mail: keith@mnhn.fr

**Lord Clara**
The University of Tokyo, Atmosphere and Ocean Research Institute, Division of Marine Life Science, 5-1-5 Kashiwanoha, Kashiwa, Chiba, 277-8564, Japan.

**Manzon Richard G.**
Department of Biology, University of Regina, Regina, Saskatchewan S4S0A2 Canada.
E-mail: richard.manzon@uregina.ca

**Martin Patrick**
Conservatoire National du Saumon Sauvage, 43 300 Chanteuges, France.
Email: pmartin@cnss.fr

**Miwa Satoshi**
Inland Station, National Research Institute of Aquaculture, Tamaki, Mie 519-0423, Japan.
E-mail: miwasat@affrc.go.jp

**Rousseau Karine**
Research Unit BOREA "Biology of Aquatic Organisms and Ecosystems" Muséum national d'Histoire naturelle, CNRS 7208, IRD 207, UPMC, 7 rue Cuvier, CP32, 75231 Paris Cedex 05, France.
E-mail: rousse@mnhn.fr

**Taillebois Laura**
Research Unit BOREA "Biology of Aquatic Organisms and Ecosystems" Muséum national d'Histoire naturelle, CNRS 7208, IRD 207, UPMC, 47 rue Cuvier, CP26, 75231 Paris Cedex 05, France.
E-mail: taillebois@mnhn.fr

**Yamano Keisuke**
National Research Institute of Aquaculture, Fisheries Research Agency, Minamiise, Mie 516-0193, Japan.
E-mail: yamano@fra.affrc.go.jp
Tel: 81-599-66-1830

**Youson John H.**
Department of Biological Sciences, University of Toronto Scarborough, Toronto, Ontario M1C1A4 Canada.
E-mail: youson@utsc.utoronto.ca or jhy@rogers.com

# CHAPTER 1

# Introduction to Fish First and Secondary Metamorphoses

*Karine Rousseau*[1,a,]* and *Sylvie Dufour*[1,b]

## 1.1 Metamorphosis

The term metamorphosis is commonly used in metazoa to define remarkable developmental body changes accompanied by a drastic shift in habitat or behavior. Such events have been described in various groups such as cnidaria, insects, crustacean, molluscs, tunicates and vertebrates (amphibians and some fishes).

Back to basics, metamorphosis comes from the Greek *meta*- « change » and *morphe* « form », indicating that remarkable morphological change should be one major criteria to define metamorphosis. The problem thereafter is how to define a morphological change; does it need to be drastic so that the premetamorphic and the postmetamorphic individuals do not look alike or could it be any anatomical remodelling? Moreover, in animal as in vegetal kingdom, the need to change form comes from a change of modes of life or of habitats.

In 2006, a symposium, entitled "Metamorphosis: A multi-kingdom approach" was presented at the annual meeting of the Society for Integrative and Comparative Biology (SICB). Afterwards, several participants from different backgrounds presented each of their conceptions of metamorphosis in one review paper (Bishop et al., 2006). As they suggested, this review could be presented as a reference tool, as it shows so well how difficult the task is. Different criteria for metamorphosis were defined: « habitat

[1]Research Unit BOREA "Biology of Aquatic Organisms and Ecosystems" Muséum National d'Histoire Naturelle, CNRS 7208, IRD 207, UPMC, 7 rue Cuvier, CP32, 75231 Paris Cedex 05, France.
[a]E-mail: rousse@mnhn.fr
[b]E-mail: dufour@mnhn.fr
*Corresponding author

shift, major morphological change, change in adaptive landscape, rapidity, change in feeding mode, post-embryonic event, usually pre-reproductive to reproductive transition, hormone-regulated transition », but none of them was found to be common to all 14 authors' definitions. As you will notice in this book, change of habitat and feeding mode, in addition to acute morphological changes and hormonally controlled processes, are common features to all fish metamorphoses described here.

## 1.2 Metamorphosis in Fishes

### 1.2.1 Phylogeny

Fish is a term commonly used to name various groups of aquatic vertebrates, which inhabit a variety of fresh and seawater environments including extreme ones. As vertebrate species they all belong to chordate deuterostomian phylum. However, they do not represent a monophyletic group and are classified in different major groups including agnatha (lampreys), chondrichthyes (sharks, rays) and osteichthyes (sarcopterygii = lungfish and actinopterygii = chondrostei and teleostei) (Fig. 1). Teleosts, with more than 25,000 species, represent the largest group, accounting for more than half of vertebrate species.

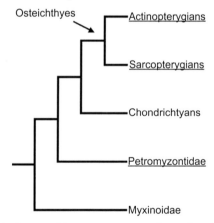

**Figure 1.** Craniate phylogeny. The groups presenting a larval metamorphosis are underlined.

### 1.2.2 Different developmental strategies in fishes

Four major developmental periods can be recognized in fish: embryonic, larval, juvenile and adult (Youson, 1988). Youson (1988) proposed three pathways of ontogeny between the embryo and the adult: direct development from the embryonic period leading to the juvenile and adult period (many examples such as goldfish);—a non-classical metamorphosis occurring during the juvenile period (example: the secondary metamorphosis, smoltification, in salmons);—a typical indirect development involving a larval metamorphosis (also called first or "true" metamorphosis) leading to the juvenile period (example: flounder and eel).

### 1.2.3 Two types of metamorphosis in fishes

Pr. Youson, in Bishop et al. (2006), wrote that "Fish with a metamorphosis are lampreys, eels, and flatfishes (for example flounder and sole). From my point of view, the majority of fishes do not have a true metamorphosis. I do not consider parr-smolt transformation in salmonids as a metamorphosis". It is true that parr-smolt transformation (smoltification) in salmonids or silvering in eels do not correspond to a drastic change of form, as you will see in Chapters 6 and 7. However, major physiological and behavioural changes occurring during these transformations are necessary for the fish to reach its next habitat and survive in the new environment.

Therefore, in this book, we will consider two types of metamorphoses in fishes (Fig. 1 and 2). First or larval metamorphosis is a true metamorphosis and can be observed in agnatha (lampreys; Chapter 2), elopomorphs (eels; Chapter 3) and pleuronectiforms (flatfishes; Chapter 4). Secondary metamorphosis occurs in juveniles of some diadromic migratory teleosts and compared to larval metamorphosis, involves less drastic morphological changes. This is the case of smoltification in salmons (Chapter 6) and silvering in eels (Chapter 7).

## 1.3 Endocrine Control of Metamorphosis

### 1.3.1 The reference of the amphibian larval metamorphosis

In vertebrates, the most described metamorphosis is the transformation in amphibians of the aquatic larva (tadpole) into the terrestrial juvenile. The role of the thyroid gland in this metamorphosis was first demonstrated by Gudernatsch as soon as 1912. When feeding tadpoles with thyroid gland extracts, he observed that their transformation to frogs was accelerated. Then, Allen (1916) was able to completely prevent metamorphosis by thyroidectomy. The pair of thyroid glands is first detectable after embryogenesis when the tadpole begins to feed. Under the control of pituitary thyrotropin (TSH), these glands produce thyroid hormones (TH: thyroxine, T4 and triiodothyronine, T3), which act on target organs via specific receptors. While TRH (thyrotropin-releasing hormone) is the brain neuropeptide controlling TSH in mammals and adult amphibians, it is CRH (corticotropin-releasing hormone), which is responsible for the activation

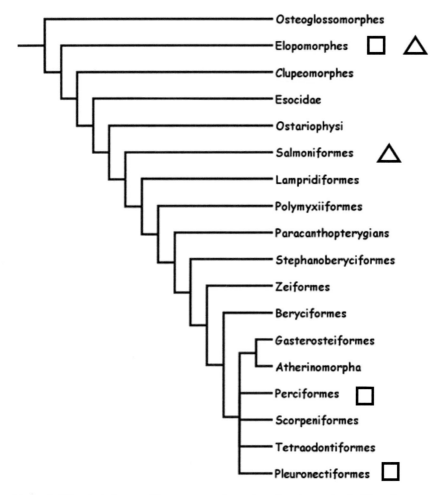

**Figure 2.** Teleost phylogeny. Groups presenting a larval metamorphosis (□) and groups presenting a secondary metamorphosis (△).

of TSH production at the time of amphibian metamorphosis (Denver 1999) (Fig. 3). As the thyrotropic axis is activated, a series of sequential morphological transformations occur. An early change is the growth and differential of the limbs, which in the absence of hormone, still form but will not progress beyond the bud stage. The final morphological change, tail resorption, occurs when the level of TH is highest at the climax of metamorphosis (for reviews: Kanamori and Brown, 1996; Tata, 2006).

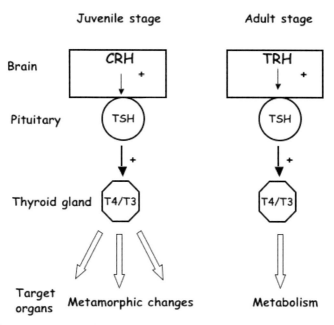

**Figure 3.** Thyrotropic axis in amphibians. Note the change in the brain stimulatory neurohormone between juvenile (CRH=corticotropin-releasing hormone) and adult (TRH=thyrotropin-releasing hormone). TSH=thyrotropin; T4=thyroxine; T3=triiodothyronine.

## 1.3.2 Thyroid-hormone induced metamorphosis in other vertebrates and chordates

This triggering effect of THs on metamorphosis was afterwards found also in teleosts, (elopomorphs, Chapter 3 and pleuronectiforms, Chapter 4). Concerning secondary metamorphosis, smoltification in salmonids is also partially controlled by changes in THs (Chapter 6). This suggests that TH-induced metamorphosis may be a common regulatory mechanism among vertebrates (Fig. 4). However, as highlighted in Chapter 2, in the lamprey, a petromyzontidae, larval metamorphosis is driven in contrast by a drop of TH, and not by an increase of TH levels (Chapter 2).

Recent data in amphioxus, a cephalochordate, showed that iodothyronines induced metamorphosis by binding to a receptor homologous to vertebrate thyroid hormone receptors (Paris et al. 2008). These findings suggested an ancestral origin of thyroactive compound-induced metamorphosis in chordates (Denver, 2008) and supported a definition of metamorphosis based on its hormonal control by TH-like

Neuroendocrine axes

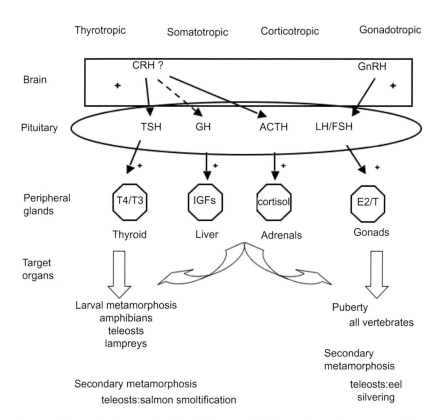

**Figure 4.** Neuroendocrine axes involved in the control of first and secondary metamorphoses in vertebrates. Larval/first metamorphosis is mainly controlled by the thyrotropic axis in amphibians and teleosts (positively), and lampreys (negatively). Secondary metamorphoses (salmonid smoltification and eel silvering) are triggered by different neuroendocrine axes, with smoltification being controlled by thyrotropic and somatotropic axes and eel silvering by gonadotropic axis, both with some synergism of corticotropic axis. CRH=corticotropin-releasing hormone; GnRH=gonadotropin-releasing hormone; TSH=thyrotropin; T4=thyroxine; T3=triiodothyronine; GH=growth hormone; IGF=insulin-like growth factor; ACTH=corticotropin; LH=luteinizing hormone; FSH=follicle-stimulating hormone; E2=estradiol; T=testosterone; 11KT=11-ketotestosterone.

compounds (Paris and Laudet, 2008). However, TH are also the major hormones controlling whole development and metabolism in all vertebrates, including species with a classical direct development. Thus, such a definition of metamorphosis based on its hormonal control is leading to the confusion of the concept of metamorphosis with any other developmental event. For instance, Brown (1997) described TH-induced "zebrafish metamorphosis", a teleost species with a rather typical direct development. In the present book, we choose to use the full definition of metamorphosis, based on remarkable morphological, physiological, behavioral and habitat changes, which accounts for its key-role in the structure of complex life cycles with different ecophases.

### 1.3.3 Metamorphosis, puberty and sex change

If the hormonal control by TH is, from our point of view, not at all sufficient to define metamorphosis in vertebrates, we believe it is a necessary criteria to distinguish metamorphosis from other major late developmental events such as puberty, or sex change in some teleosts (Fig. 4). Puberty in all vertebrate species is defined as the acquisition by the individual of the capacity to reproduce for the first time (Romeo, 2003). It is often accompanied by drastic morphological, physiological, behavioral and even habitat changes, which represent species-specific secondary sexual characters. Furthermore, in some teleost species, adults may change gender, after a certain age or under the effect of specific environmental and social cues (Frisch, 2004). A remarkable remodelling of secondary sexual characters can be observed. All these changes at puberty or at sex change would fit with many criteria of metamorphosis. They are however under the major control of sexual steroids, the production of which is induced by the gonadotropic axis. Based on their hormonal control by sex steroids, such changes have therefore never been considered as a metamorphosis.

Interestingly, the other example of secondary metamorphosis in teleosts is silvering in eels (Chapter 7). Our own results revealed that during eel silvering, gonadotropic axis is activated instead of thyrotropic axis as commonly observed during other metamorphoses. In addition, treatments with sex steroids and especially androgens are able to induce silvering morphological changes. These data lead us to think of eel silvering as a puberty rather than a metamorphosis.

## 1.4 First Metamorphosis in Fishes

Chapter 2 Lamprey metamorphosis. By J. Youson and R. Manzon.

Chapter 3 Metamorphosis of elopomorphs. By K. Yamano.

Chapter 4 Metamorphosis of flatfish (Pleuronectiformes). By Y. Miwa and S. Inui

Chapter 5 Gobiid metamorphosis: a true metamorphosis? By P. Keith et al.

The typical larval metamorphosis, also called true metamorphosis or first metamorphosis in fishes (in opposition to secondary metamorphosis) is classically restricted to lampreys (Chapter 2), and to two teleost goups, elopomorphs (Chapter 3) and pleuronectiformes (Chapter 5).

Lampreys are representative species of agnatha, an ancient group of vertebrates, while teleosts belong to osteichtyans, actinopterygians (for their phylogenetic positions, see Fig. 1). There is no report of larval metamorphosis in any other "fish" representatives such as chondrichtyans (shark and ray) or osteichtyans sarcopterygians (lungfish). This suggests that larval metamorphosis may have been acquired independently in the agnathan lineage and the teleost lineage. The opposite role of TH in the control of larval metamorphosis in lampreys and teleosts further supports this hypothesis.

Elopomorphs and Pleuronectiforms are representative of some of the most ancient and most recent subgroups of teleosts, respectively. This suggests that larval metamorphosis in a biological life cycle has probably been acquired independently in these two groups during teleost evolution. The alternative hypothesis is that larval metamorphosis could have been lost in all the other teleost groups. However, as reviewed in Chapter 5, transformation of oceanic larva into freshwater juvenile in another teleost group, the gobiids (order Perciformes), presents typical characteristics of true metamorphosis, with TH-induced remarkable morphological, physiological, behavioral, feeding and ecophase changes. We strongly suggest to include this teleost group among those presenting a larval metamorphosis, in addition to elopomorphs and pleuronectiforms. As discussed above, this does not mean that we extend the concept of metamorphosis to any TH-induced developmental transition.

The thyrotropic axis possesses a major activator role in the control of larval metamorphosis in representatives of sarcopterygians (amphibians) and actinopterygians (teleosts: elopomorphs, perciform and pleuronectiformes). This suggests that the stimulatory action of this axis may have an ancient origin in a common osteichthyan ancestor to the sarcopterygian and actinopterygian lineages. An alternative hypothesis is that the stimulatory involvement of the thyrotropic axis in the control of larval metamorphosis may have been acquired several times independently during the evolution of osteichthyans (in amphibians and in different groups of teleosts). As the thyroid hormones are thought to be inhibitory in the control of larval metamorphosis in lampreys, a switch in the role of the thyrotropic axis may have occurred after the emergence of petromyzontidae. The two opposite roles of THs in vertebrates (and in fishes) show that, during the evolution of vertebrates, the role of thyroid hormones in the endocrinology of larval metamorphosis may have differed dramatically, being inhibitory in lampreys (representative of a very ancient group of vertebrates, petromyzontidae) and stimulatory in osteichytans.

## 1.5 Secondary Metamorphosis in Fishes

Chapter 6 Salmonid smoltification. By Rousseau et al.

Chapter 7 Eel secondary metamorphosis: silvering. By Rousseau et al.

Secondary metamorphosis presents similar criteria as first/larval metamorphosis, meaning that it involves various morphological, physiological and behavioural modifications that preadapt the animal to life in the next and new environment. However, in contrast to first metamorphosis, it occurs after a juvenile period. Moreover, the morphological modifications involved are less drastic than those observed during the larval metamorphosis. Two examples are found in fishes and restricted to teleosts: smoltification (also called smolting or parr-smolt transformation) in salmon (Chapter 6) and silvering in eel (Chapter 7). They both concern extreme migratory diadromic teleosts and are related to their complex life cycles. These migratory teleosts migrate between freshwater and seawater as part of their life cycle and are capable of withstanding a wide range of salinity. Some fish need to swim at important water depth during their life cycle and thus are able to withstand important variations of pressure. Complex migratory life cycles are related to a strict dependency to special environments for growth or for reproduction. Such life cycles, with ecophases and migrations in various environments, are permitted by complex regulatory systems involving the interplays of internal and external factors. These changes are induced via the occurrence of special post-embryonic developmental events: the metamorphoses.

Concerning endocrine control of these secondary metamorphoses, distinct neuroendocrine axes are shown to be involved. Smoltification is largely controlled by thyroid hormones as are first metamorphoses. Thus, in teleosts, the involvement of the thyrotropic axis is observed not only in the true larval metamorphoses in elopomorphs, pleuronectiformes, and some perciformes, but also in secondary metamorphosis in salmoniformes (smoltification) (Fig. 4). This indicates an additional recruitement of the thyrotropic axis for the induction of metamorphic changes during the evolution of teleosts. However, differently from larval metamorphosis, another major neuroendocrine axis is also compulsory for the completion of smoltification: the somatotropic axis.

In contrast, the secondary metamorphosis of the anguilliformes (silvering) is induced primarily by the gonadotropic axis (Fig. 4), and sex steroids are able to induce peripheral morphological changes, as do thyroid hormones in the previous examples. This let us regard silvering as a pubertal rather than a metamorphic event. The term "prepuberty" was first used by our group, as during eel silvering, puberty is blocked at an early stage and sexual maturation only occurs during the reproductive oceanic migration.

It is thus of great interest to note that while smoltification and silvering share many similarities in term of morphological, physiological and behavioral changes, the endocrinology of these two secondary metamorphoses drastically differs, with the major involvement of different neuroendocrine axes, the thyrotropic/somatotropic ones for smoltification and the gonadotropic one for silvering. This suggests that secondary metamorphoses may have been acquired independently, via different endocrine mechanisms, during teleost evolution. The convergence between some morphological (skin silvering, eye size and pigments), metabolic and behavioural (downstream migration) changes reflects that the control of the same peripheral target organs (skin, eye, muscle, brain…) and target genes is exerted by different hormonal receptors (thyroid hormone receptors in salmon versus androgen receptors in the eel). This discovery suggests an independent recruitment of different endocrine axes for the induction of secondary metamorphoses during teleost evolution.

Most of the data available on the endocrinology of fish metamorphoses concern peripheral hormones. Future studies should aim at investigating the pituitary and brain components of the neuroendocrine axes, as well as the mechanisms of cerebral integration of internal and environmental factors responsible for the timing of the metamorphoses and ecophases during the life cycles.

## Acknowledgements

We thank Jérémy Pasquier (Research Unit BOREA MNHN) for his help in the design of figures.

## References

Allen, B.M. 1916. Extirpation experiments in *Rana pipiens* larva. Science 44: 755–757.

Bishop, C.D., D.F. Erezyilmaz, T. Flatt, C.D. Georgiou, M.G. Hadfield, A. Heyland, J. Hodin, M.W. Jacobs, S.A. Maslakova, A. Pires, A.M. Reitzel, S. Santagata, K. Tanaka and J.H. Youson. 2006. What is metamorphosis? Integrative and Comparative Biology 46: 655–661.

Brown, D.D. 1997. The role of thyroid hormone in zebrafish and axolotl development. Proceeding of the National Academy of Sciences 94: 13011–13016.

Denver, R.J. 1999. Evolution of the Corticotropin-releasing Hormone Signaling System and Its Role in Stress-induced Phenotypic Plasticity. Annals of the New York Academy of Sciences 897: 46–53.

Denver, R.J. 2008. Chordate metamorphosis: ancient control by iodothyronines. Current Biology 18: R567–569.

Frisch, A. 2004. Sex-change and gonadal steroids in sequentially-hermaphroditic teleost fish. Reviews in Fish Biology and Fisheries 14: 481–499.

Gudernatsch, J.F. 1912. Feeding experiments in tadpoles. I. The influence of organs given as food on growth and differentiation. A contribution to the knowledge of organs with internal secretion. Arch. Entwicklunsmech. Org. 35: 457–483.

Kanamori, A. and D.D. Brown. 1996. The analysis of complex developmental programmes: amphibian metamorphoses. Genes to cells 1: 429–435.

Paris, M. and V. Laudet. 2008. The history of a developmental stage: metamorphosis in chordates. Genesis 46: 657–672.

Paris, M., H. Escriva, M. Schubert, F. Brunet, J. Brtko, F. Ciesielski, D. Roecklin, V. Vival-Hannah, E.L. Jamin, J-P Cravedi, T.S. Scanlan, J-P. Renaud, N.D. Holland and V. Laudet. 2008. Amphioxus postembryonic development reveals the homology of chordate metamorphosis. Current Biology 18: 825–830.

Romeo, R.D. 2003. Puberty: a period of both organizational and activational effects of steroid hormones on neurobehavioural development. Journal of Neuroendocrinology 15: 1185–1192.

Tata, J.R. 2006. Amphibian metamorphosis as a model for the developmental actions of thyroid hormone. Molecular and Cellular Endocrinology 246: 10–20.

Youson, J.H. 1988. First metamorphosis. In: *Fish Physiology* Vol XI, W.S. Hoar and D.J. Randall (Eds.). Academic Press, San Diego pp. 135–196.

# CHAPTER 2

# Lamprey Metamorphosis

*John H. Youson[1],* * and *Richard G. Manzon[2]*

## 2.1 Introduction

Lampreys are one of two extant members of the ancient group of jawless vertebrates, the Agnatha, whose ancestry can be traced back to at least 550 million years ago to the armoured ostracoderms. The other extant agnathan is the strictly marine group, the hagfishes, and there is recent fossil evidence from southern China that lamprey- and hagfish-like agnathans coexisted (Shu et al., 1999) in the lower Cambrian period. Identifiable lamprey fossils appear in mostly marine deposits from the upper Carboniferous period and morphological evidence suggests that lamprey evolution has been relatively conservative for the past 300 million years (Potter and Gill, 2003). The taxanomic relationship between the agnathans, the lampreys and hagfishes, has been an on-going controversy. Although extensive morphological and physiological, and some molecular, data suggest that lampreys are more similar to the jawed fishes than they are to hagfishes, other molecular data imply a close relationship, in fact monophyly, of the two extant agnathan groups (see Potter and Gill, 2003; Hardisty, 2006). One way in which hagfishes and lampreys differ is in their development.

In particular, the two groups of extant agnathostomes, the lampreys and hagfishes, differ markedly in their mode of post-hatched development. Whereas immediately post-hatched hagfishes closely resemble the adult and a have direct development into sexually mature adults all in a marine environment, all lamprey species have phenotypically distinct larvae

[1]Department of Biological Sciences, University of Toronto Scarborough, Toronto, Ontario MICIA4 Canada.
E-mail: youson@utsc.utoronto.ca or jhy@rogers.com
[2]Department of Biology, University of Regina, Regina, Saskatchewan S4S0A2 Canada.
E-mail: richard.manzon@uregina.ca
*Corresponding author

(ammocoetes) and adults. Furthermore, the larva are restricted to freshwater habitats (Morris, 1980) and some adult lampreys may occupy both fresh- and salt-water and others are compelled to live in either fresh water or in a marine environment for at least part of their adult life. To accommodate all of the physical and physiological differences between larval and adult lampreys requires a major interval of change between the two periods of the life cycle, i.e., a phase of metamorphosis. Since there is a metamorphosis in the life cycle of lampreys, the post-hatched development of larvae to adults is indirect (Youson, 1988).

## 2.2 Adult Life History Types

There are two adult life history types among the 38, known, species of lampreys, but this number will undoubtedly change in the coming years (for example, see Renaud and Economidis, 2010). Thirty-four of these species reside in the Northern Hemisphere (holoarctic) and there are four species of southern hemisphere lampreys. Twenty of these species, referred to as non-parasitic, are restricted to fresh water during their adult life and they never feed again once metamorphosis commences. The remaining 18 species, parasitic species, feed either on the blood or tissues of host fishes as juveniles in either fresh water or salt water. In some cases a non-parasitic and a parasitic species share a common evolutionary history, likely arising from a common parasitic ancestor. These related species are referred to as "paired" or "satellite" species and there are situations where they cohabit a stream. There are some isolated reports of unusually large-sized, sexually mature non-parasitic individuals that likely must have fed as adults (Manion and Purvis, 1971). Also, there is at least one population of non-parasitic western brook lamprey, *Lampetra richardsoni*, that produces two immediately postmetamorphic morphotypes (R.J. Beamish, 1987). One of these delays sexual maturity for a year and is capable of feeding and the other follows the normal pattern of non-parasitic species and starts sexual maturation shortly after the completion of metamorphosis (Youson and Beamish, 1991). These cases, and other evidence, has led to the belief that there is a certain amount of plasticity in lamprey metamorphosis (Youson, 1999; Youson, 2004). It is commonly accepted that, despite the apparent plasticity towards the timing of sexual maturation, neoteny does not exist among lampreys (Hardisty, 2006).

## 2.3 Life Cycle

The lamprey life cycle begins with external fertilization of eggs by sperm in nest of stones in a freshwater stream. The embryos develop relatively quickly with the prolarvae ready to embed themselves in soft silt and sand within about 17 days post-fertilization (PF). At this time the prolarvae are dependent on their yolk sacs for nutrition but by about 33 days PF the young larvae, 9–10 mm long, begin to feed on detritus, algae, desmids and diatoms. Although the larvae (ammocoetes) likely move within their natal stream their primary habitat is a burrow within the soft sediment. Since they cannot tolerate even dilute seawater, they remain far upstream from any tidal effects. On the other hand, should their natal stream empty into a larger body of freshwater, such as in the Great Lakes or in Lake Champlain of North America, they can be found at the mouth of the stream or even in the substrate of a lake.

The length of the larval period of the lamprey life cycle varies with the species, but it is usually never less than 2 years and seldom higher than 7 years. According to Hardisty (1979), there has been a trend to lengthen larval life and eliminate feeding during the evolution of lampreys. Hence, if one accepts the view that non-parasitic lampreys are the more derived adult life history type, then larvae of non-parasitic species are older than those of parasitic species at the time of the onset of metamorphosis. However, there are documented cases in experimental situations in the wild where larvae of parasitic, *Petromyzon marinus*, have been recorded at ages greater than 10 years (Potter, 1980). Historically, the age of larval lampreys has been determined through modes that appear within length-frequency data from animals collected in a given stream. More recently, annuli on statoliths from the otic capsules have been applied, but data from this new approach is often correlated with animal length or length-frequency data. Also, one has to sacrifice the animal to obtain the statolith data. Although the early year classes are quite clear as modes in the length-frequency data, likely due to rapid growth spurts, it is the older age groups where age distinction is less clear because growth slows as the larva approaches the upper ages. The rates of growth also seem to vary between different populations (i.e., different streams) of the same species due to differences in both habitat and water temperature conditions. The slowing or arresting of growth in the oldest larvae is reflected in an overlap in lengths of animals which are either immediately pre-metamorphic and metamorphosing and those that will wait another year before they reach either of these intervals. Thus, in some species of lampreys it is difficult to be certain which larvae are about to enter metamorphosis. The two most important criteria to consider in selecting immediately pre-metamorphic animals are knowledge of the time of the year when they generally undergo metamorphosis and the length of metamorphosing animals in wild populations. There are also cases where lipogenesis and fat storage is reflected in increased body weight of immediately pre-metamorphic animals so that animal length and weight (collectively referred to as size) can be used as indicators as the season for metamorphosis approaches.

As will be described in detail below, metamorphosis in lampreys involves extensive modification to the internal organs and tissues. The changes to the internal anatomy are of such a magnitude and so widespread that the animal is incapable of feeding until metamorphosis is fully completed; i.e., lamprey metamorphosis is a non-trophic phase of the life cycle. In addition, external changes involve body colouration, a complete remodeling of the mouth from the larval buccal funnel to the adult suctorial disc with teeth and a tongue-like piston, completion of development of the eyes, and alterations to the shape of the branchiopores. It is these external changes that are the basis for the description of stages (see below) of this developmental process. Although there are some subtle differences in the events and processes of metamorphic change between lamprey species, the end result is basically the same. That is, metamorphosis results in a juvenile. In non-parasitic species, juveniles usually have a larger gonad, compared to their parasitic member in a paired species, likely as a consequence of their more protracted interval of larval life (Bird and Potter, 1979b). Despite the fact that they have the same adult internal anatomy as a parasitic species, the juveniles of non-parasitic species almost immediately commence sexual maturation and can be capable of spawning (sexually mature adults) within 5–6 months of the completion of metamorphosis. During this interval of sexual maturation and after reaching sexual maturity, the adults of non-parasitic species move only a short distance within their natal stream. In contrast, juveniles of parasitic species are required to feed before sexual maturation commences and in some cases, such as in anadromous species, they undertake an extensive downstream migration to the ocean. Some parasitic, freshwater species may undergo a short migration to find a host within their natal stream. In the case of the landlocked, *P. marinus*, and other landlocked species, the downstream migration is to a large body of fresh water such as a lake. The length of the parasitic period in adult life is quite variable among the species. Information is most prevalent on the sea lamprey, *P. marinus*, where a rapid-growth, parasitic period of around 1 year for landlocked forms (Bergstedt and Swink, 1995) and around 2 years for the anadromous form (Beamish, 1980) are estimated. The smaller size of parasitic forms that remain in their natal stream, suggest a much shorter feeding period.

Feeding in adult life ceases at the time the animals commence an upstream migration to the spawning grounds. Once again, the length of time for this upstream migration is quite variable even among anadromous forms. For example, although the anadromous *P. marinus* (and also the landlocked form) migrate and spawn between March and September (Beamish, 1980), both *Mordacia mordax* and *Geotria australis* of the southern hemisphere and the northern hemisphere, river lamprey, *Lampetra fluviatilis*, have migration periods of much longer duration; i.e., up to 15 months for *G. australis* (Hardisty, 2006).

## 2.4 Time of Metamorphosis

The period of the year when a given stream population of lampreys commence their metamorphosis is seasonal and synchronized from year to year. This synchrony is likely important to permit the animal to go through the immense changes under conditions which are most favourable and the least threatening to survival. However, there is some intraspecific variability due to both the range of their temperate distribution and to the conditions of water discharge of each stream (Potter, 1980). For the most part, metamorphosis is initiated in July to early August in holarctic lampreys and the equivalent months of January and February in southern hemisphere species. As will be seen below, there is clear evidence, at least in some species, that water temperature is an important cue for initiating lamprey metamorphosis. The commencement of metamorphosis in landlocked and anadromous populations of *P. marinus* in North America seems to be highly synchronized from population to population and from year to year, despite varied conditions (Potter et al., 1978b; Youson, 2003). Data from three successive years of metamorphosis in *G. australis* in Australia shows a similar synchrony (Potter et al., 1980). In contrast, Hardisty (2006) emphasizes the weather-sensitive nature of larval populations of *Lampetra planeri* and *L. fluviatilis* in the British Isles with a resulting higher degree of variability in the occurrence of metamorphosis. Lower stream temperatures tend to result in earlier commencement of metamorphosis in *L. planeri* (Bird and Potter, 1979a) and *G. australis* (Potter et al., 1980). There is also a marked difference in the onset of metamorphosis in closely related forms such as *M. mordax* (~ early March) and *Mordacia praecox* (~ early November), a non-parasitic form likely derived from the former (Potter, 1970).

## 2.5 Stages of Metamorphosis

The reader is referred to a comprehensive account of the historical literature and present status of the methods for determination of the sequence of stages in lamprey metamorphosis (Potter et al., 1982). The present system is based on identical and simultaneous assessment of features in the metamorphosis of *L. planeri* and *L. fluviatilis* (Bird and Potter, 1979a), *P. marinus* (Youson and Potter, 1979), and *G. australis* (Potter et al., 1980). There are five main, external morphological features that are used to identify seven sequential stages in lamprey metamorphosis with the first two features particularly important for recognizing the onset of the process (Fig. 1). These features are highlighted by Potter et al. (1982) as: 1. changes in the appearance, shape, and size of the eye; 2. modification of the buccal funnel (oral hood) and prebranchial region of larva into the adult oral disc and snout; 3. growth of the fins; 4. changes in body colouration; 5. changes in the branchial region

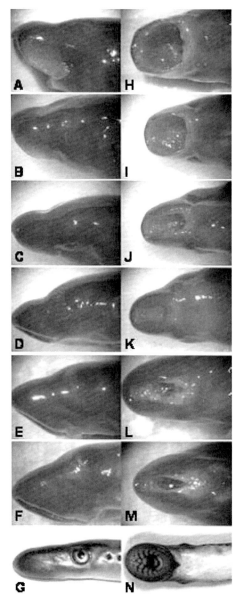

**Figure 1.** Anterior region in lateral (A- G) and ventral (H-N) views of larval *Petromyzon marinus* (A, H) induced to metamorphose following treatment with potassium perchlorate (KClO$_4$) (B-F and I-M) and juvenile *P. marinus* immediately following the completion of spontaneous metamorphosis (G, N). Approximate staging for KClO$_4$-induced metamorphosis is as follows: Stage 1 (B, I); Stage 2 (C, J); Stage 3 (D, K); Stage 4 (E, L); Stage 5/6 (F, M). Photos A-F and H-M kindly provided by Dr. John R. Gosche, University of Nevada.

*Color image of this figure appears in the color plate section at the end of the book.*

including the shape of the branchiopores. It is noteworthy that modification to the larval buccal funnel includes a complex set of changes including the loss of larval cirrhi, the development of teeth and a tongue-like piston, and special laminae, and the appearance of a ridge of fimbriae on the outer surface of the developing oral disc. The sequential development of these features in the oral region are all important for stage identification.

Detailed descriptions of all events in the alteration of the five main features is beyond the scope of this review, but several are worthy of a summary. For instance, the eye which is beneath the skin in larva, first becomes larger and almost oval-shaped patch (stage 1) and darker and more rounded by stage 2. The circular shape persists as the eye further enlarges and starts to show separation of the pupil and iris (stage 3); these features develop further to such an extent that the eye bulges from the body surface (stage 4). By stage 5 eyes with distinct pupil and pale-grey iris protrude further from the surface. Subsequent stages show further eye enlargement and a much more silver iris. The branchiopores show a progressive change from triangular shape in larva and early stages to more oval by stage 4. Body colouration retains the larval brown ventral and dorsal surfaces until stage 4 when a blue-grey sheen is observed. Subsequent stages can be distinguished on the basis of the degree to which the dorsal surface becomes blue-black and the ventral surface a silver sheen. Stage 4 is identified when the transverse and lateral lips of the larval buccal funnel fuse to produce the oral disc, the larval cirrhi are reduced to a few papillae-like projections, and the undeveloped piston is first prominently visible. Teeth are first visible as small points at stage 5 when the piston shows its laminae and the infraoral lamina is discernible. The changes in these and other features also manifest themselves in changing behaviour that can be species specific. For instance, the changes are so advanced by stage 6 in *P. marinus* that the animals may leave their burrows and may be capable of adhering with their suctorial disc (Youson and Potter, 1979).

Since the animal length does not change appreciably in most lamprey species, in addition to the five main visible external changes, measurements of body proportions (% of total length) are also useful to follow during metamorphosis. These types of measurements show reductions in branchial and trunk lengths and an increase in prebranchial length. Although tail length remains relatively constant, the length of the disc increases, particularly in species such as *P. marinus* (Potter et al., 1978b). A comparison of three species showed a common increase in eye diameter and height of both the first and second dorsal fin (Potter et al., 1982).

Following the reports cited in the previous paragraph, the seven stages of lamprey metamorphosis (1, earliest to 7 final stage) have been universally adopted. Staging is now available for the southern brook lamprey, *Ichthyomyzon gagei* (Beamish and Thomas, 1984), the far eastern (sand) brook lamprey, *Lampetra reissneri* (Tsuneki and Ouji, 1984), the mountain brook lamprey, *Ichthyomyzon greeleyi* (Beamish and Austin, 1985), the American brook lamprey, *Lampetra appendix* (Holmes et al., 1999), and the Pacific lamprey, *Lampetra tridentata* (McGree et al., 2008).

## 2.6 Preparation for Metamorphosis

### 2.6.1 Age

This subject received some treatment in the section on the lamprey life cycle and the length of larval life. It is important, however, to emphasize that larval age is an important parameter in lamprey metamorphosis. There are no reports of larvae entering metamorphosis one year after hatching and even cases of two-year old larvae entering metamorphosis are relatively rare. Larvae of non-parasitic species metamorphose at much longer lengths than parasitic species and likely are at least 5 years of age. As mentioned above, the most widely used method of determining age in larvae is through defining length modes via the use of length-frequency data from field collections. Age is unquestionably an important parameter in dictating the end of the larval growth period, but there is a great deal of variability, even within the same stream population, of lengths (age?) of metamorphosing animals. The most extensive studies of age at metamorphosis have been reported on the landlocked sea lamprey, *P. marinus*, in the watershed of the Great Lakes of North America. These studies are part of the ongoing challenge to eliminate or reduce the predation of parasitic-phase, adult *P.marinus* on sport and commercial fishes. Although there are some newer methodologies directed at post-metamorphic periods of the life cycle (Twohey et al., 2003; Docker et al., 2003), the primary target has been the larval phase before they can enter metamorphosis or at least before the immediately post-metamorphic animals, the juveniles, can start their downstream migration to a site where they will commence their parasitic feeding. The control method is the use of the lampricide, 3-trifluoromethyl-4-nitrophenol (TFM), on larval populations on a 3-4 year cycle. This treatment cycle period was initially determined from length-frequency data. However, it can now be assumed that, with the exception of a few survivors of the TFM treatment, called residuals, the age (length) at metamorphosis can be determined from the numbers of years, post-treatment, that metamorphosing animals first appear. Some stream populations are treated after three years, whereas others are again subjected to the lampricide in the fourth year post-treatment. Post-treatment surveys have revealed, as have general larval population surveys, that not all animals of the same length (age) enter metamorphosis. The study by Hollett (1998) used a mark and recapture protocol and age through statolith analysis to conclude that metamorphosing animals are often one year older than non-metamorphosing animals of the same length. The study also showed that in some post-treatment streams, the third year of larval life is an immediately pre-metamorphic year when growth in length may be stable but the animals prepare themselves physiologically for metamorphosis. These results suggest that either growth rates, as reflected in length, differ within individuals of the same cohort or that there must be other factors that are important to preparation of the animals for metamorphosis. As will be discussed below, water temperature is an important cue for metamorphosis, however, field studies have shown that larvae in warm water creeks do not metamorphose at a younger age than larvae from colder water creeks (Hollett, 1998).

## 2.6.2 Mass, length, and condition factor

All studies to date, on a variety of species, have some record of the weight (mass) of metamorphosing animals relative to that of animals of similar length that did not enter metamorphosis. Although there is usually some higher mass in animals undergoing metamorphosis in most species, the most dramatic and potentially significant, increase is observed in *P. marinus*. Extensive laboratory studies have determined that animals at least 120 mm in length and 3.0 g in mass are likely to enter metamorphosis (Holmes and Youson, 1994, 1997, 1998; Holmes et al., 1994). As a result, reference is made to the size (length and mass) of metamorphosing animals. The mass has been correlated with the deposition of fat in strategic locations within the body and organs of animals about to enter metamorphosis (Youson et al., 1979). The nephric fold, fat column, muscle septa, and subcutaneous adipose tissue are primary sites (Fig. 2). This deposition is a consequence of lipogenesis that occurs in pre-metamorphic animals where lipids reach a threshold level of 14% of wet body mass (Lowe et al., 1973; O'Boyle and Beamish, 1977). Attempts at correlating length and mass through calculation of a condition factor (CF, where $CF = M/L^3 \times 10^6$; M is mass [g] and L is total length [mm]) have shown some promising results. This CF has been useful in selecting animals about to enter metamorphosis in *P. marinus* (Fig. 3). Results from laboratory studies (Youson et al., 1993; Holmes and Youson, 1994) have provided for a prediction in the spring for metamorphosing animals with a CF of 1.50 for landlocked *P. marinus* of at least 120 mm long and 3.0 g, but it seems that the anadromous *P. marinus* may metamorphose at a smaller size (110 mm and 2.0 g).

The mass, length, and CF criteria have been applied in attempts to select immediately pre-metamorphic animals in populations under the conditions of their natural habitat (Hollett, 1998; Henson et al., 2003). Larvae which fit the above three criteria were fitted with coded wire tags in the summer and fall prior to the summer that they would be expected to

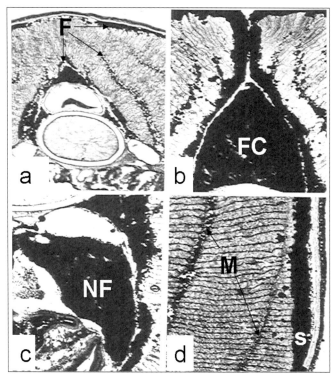

**Figure 2.** Transverse sections through various regions of the branchial (a) trunk (b, c, d) regions of immediately pre-metamorphic larvae of *P. marinus* stained with Sudan Black B for fat. a. Various regions above the notochord and spinal cord stain positively for fat (F). X25. b. The fat column (FC) and the surrounding muscle layers possess abundant fat. X540. c. The nephric fold (NF) is a primary site for fat deposition. X460. d. Fat is present in the myosepta (M) and as a thick layer in subcutaneous tissue (S). X550. (Modified from Youson et al., 1979).

show signs of metamorphosis. The stream populations represented a good cross section of the types of habitat occupied by larvae of the landlocked *P. marinus*. The three criteria were correct predictors for 30 to 73% of the recaptured larvae in six study creeks, but it was suggested that specific criteria for each creek may be more reliable (Hollett, 1998). A second field study supported this observation of stream variability where mass alone could predict metamorphosis in two streams but, depending on the method used, sometimes length and CF were better predictors in one stream over the other (Henson et al., 2003). The results are mixed, but for the most part there is little doubt that mass should be considered in selecting pre-metamorphic larval *P. marinus*. Whether this is cost effective in the sea lamprey control program is still in question. The most recent study to provide a predictive model for metamorphosis in this species concluded that a model that used criteria such as animal length in the fall, animal density, stream lamprey

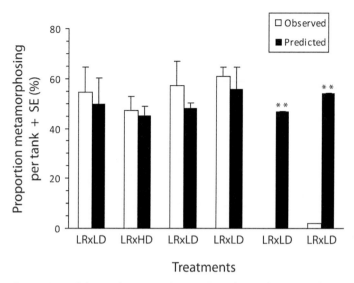

**Figure 3.** Comparison of observed metamorphosis in *P. marinus* and metamorphosis predicted using a condition factor criterion of 1.45 or greater in the fall. Prediction bars topped with asterisks (**) differ significantly from the corresponding observed bars. Treatment combinations consist of low-temperature (LR), high temperature (HR), and cold temperature (CT) regimes and of low (LD) and high (HD) larval densities. See the text for further detail and the significance of these data. (From Holmes and Youson, 1997).

production, and stream latitude and longitude was a 20% more effective than any other model (Treble et al., 2008).

In contrast to the above, a study of metamorphosis in the non-parasitic species, *Lampetra appendix*, showed that the size and condition factor parameters of *P. marinus* do not apply as easily to this species. Firstly, the larvae of *L. appendix*, that often cohabit with *P. marinus*, metamorphose at lengths of 155 to 189 mm and masses of 5.40 to 9.74 g. The CF of the earliest metamorphosing animals was 1.25. Statolith aging of *L. appendix* indicates that they are at least 1 year older than *P. marinus* when they commence metamorphosis (Beamish and Medland, 1988). The large size of this species, rather than lipid reserves as in *P. marinus*, might reflect both the preparation for the non-trophism of metamorphosis and that they will commence sexual maturation immediately after completing metamorphosis. The mean lengths and mass, and the CF of immediately pre-metamorphic individuals of the much smaller, parasitic, southern hemisphere lamprey, *G. australis*, were ~90 mm, 0.95 g, and 1.35, respectively (Potter et al., 1980, 1982). The decline in mass and CF during the metamorphosis of *G. australis* suggests a similar pattern of utilization of fat stores during metamorphosis, without the cost of any major change in length, that most closely resembles the situation in *P. marinus* (Potter et al., 1978b, 1982).

## 2.6.3 Physiological preparation

It is now clear that larval lampreys have to be prepared for the non-trophic phase of metamorphosis. Although not all aspects of the hormonal and metabolic involvement in this preparation have been identified, the role of the thyroid axis and lipid metabolism has received the most attention. Much of what we know about these two parameters of metamorphic preparation arises from studies of the sea lamprey, *P. marinus*. These two areas will receive in depth treatment in this section with data from this species. Similarity and differences between *P. marinus* and other species will be highlighted.

### 2.6.3.1 Lipid metabolism and fat storage

As mentioned above, larva of immediately pre-metamorphic *P. marinus* are heavier than animals of similar length that are unlikely to enter metamorphosis in that same season. Both morphological and biochemical evidence have shown that this added weight can be directly correlated with the deposition of fat (Lowe et al., 1973; O'Boyle and Beamish, 1977; Youson et al., 1979; Kao et al., 1997a,1997b). Immediately pre-metamorphic *P. marinus* and the earliest intervals of metamorphosis show lipogenesis as reflected in marked deposits of fat in sites such as the nephric folds of the kidney (Fig. 2) and increased activity of the enzymes, acetyl-CoA carboxylase and diacylglycerol acyltransferase (Kao et al., 1997a) compared to animals that will not metamorphose (Table 1). These lipid stores are primarily in the form of triacylglycerol (Sheridan and Kao, 1998) and lipolysis will then follow in later stages of metamorphosis with increases in triacylglycerol lipase activity in the several sites of lipid deposition (Kao et al., 1997a). Therefore, there are two phases of lipid metabolism in sea lamprey metamorphosis, lipogenesis

**Table 1.** Summary of the features of lipid metabolism in the liver and kidney of *Petromyzon marinus* during spontaneous and induced metamorphosis and following the blocking of $KClO_4$-inudced metamorphosis with exogenous thyroid hormones (TH-blocked). From Youson, 2003.

|  | Spontaneous | | $KClO_4$-induced | | TH-blocked | |
|---|---|---|---|---|---|---|
|  | liver | kidney | liver | kidney | liver | kidney |
| total lipid | ↓ | ↓ | ↓ | ↓ | ↑ | ↑ |
| lipolysis | ↑ | ↑ | ↑ | ↑ | ↓ | ↓ |
| lipogenesis |  |  |  |  |  |  |
| ACC | ↓ | ↓ | ↓ | n/d | ↑ | n/d |
| DGAT | ↓ | n/d | ↓ | ↓ | n/d | ↑ |

ACC, acetyl-CoA carboxylase; DGAT, diacylglycerol acyltransferase; ↑ increase; ↓ decrease; n/d not determined.

and lipolysis (Fig. 4) with the latter occurring later and reflected in a decline in mass and condition factor of late stages of metamorphosis in both *P. marinus* (Potter et al., 1978a) and *G. australis* (Potter et al., 1980). Changes in the fatty acid compositions of triacylglycerols and phospholipids have also been examined in metamorphosis of *G. australis* (Bird and Potter, 1983).

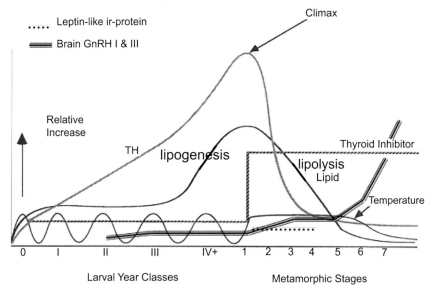

**Figure 4.** Diagrammatic representation of the factors involved in metamorphosis of the sea lamprey, *P. marinus*. There is a gradual increase in the serum level of TH over the IV+ larval year classes but metamorphosis (stages 1–7) is marked by a sharp decline in both thyroid hormones (TH), $T_4$ and $T_3$, at stage 1. The decline in TH occurs when stream temperatures arise from winter lows and when body fat is highest as result of lipogenesis in older larva and early metamorphic stages. Note the decline in body lipid is associated with lipolysis when stored fat is used by the non-feeding animals. There is some evidence to suggest that a leptin-like protein from the fat-storing tissues may be a trigger or be influenced by other factors in metamorphosis. Concentrations of gonadotropin-releasing hormones (GnRH-I and -III) in the brain increase at the time that serum TH levels decline. Since TH declines, a putative inhibitor of activity in the thyroid gland is indicated in the scheme. (From Youson, 2007, modified from Youson, 1994).

### 2.6.3.2 Endocrine

As discussed in detail below, thyroid hormones (TH) play a role in lamprey development and metamorphosis. However, the characteristic decline observed early in lamprey metamorphosis contrasts with the rise in TH levels during the metamorphoses of all other vertebrates studied to date. In addition to their potential role in the metamorphic process, it has been suggested that elevated TH levels in larvae might function as a juvenilizing hormone preventing metamorphosis, as in juvenile hormone in insects, and at the same time prepare the larval lamprey for the metamorphic event (Fig. 4; Youson, 1997). Serum 3, 5, 3′, 5′-tetraiodothyronine (thyroxine; $T_4$) and 3, 5, 3′-triiodothyronine ($T_3$) concentrations in larvae of pre-metamorphic size (year class 4 [IV]), are significantly greater than those of year class 2 (II) or 3 (III) larvae (Youson et al., 1994). Moreover, serum $T_3$ concentrations progressively increase in each successive year class (Youson et al., 1994). The concentrations of both $T_4$ and $T_3$ also increase significantly between young and immediately pre-metamorphic larvae of *L. appendix* (Holmes et al., 1999). Thus, it is likely that the gradual rise in serum TH concentrations and their peak in immediately pre-metamorphic larvae are in preparation for the impending metamorphosis (Fig. 4). Consistent with this idea are observations that serum $T_4$ concentrations increase seasonally between March and June, the months prior to the initiation of metamorphosis (Wright and Youson, 1980b). Extrapolating these data, the lower incidence of $KClO_4$-induced metamorphosis in smaller size groups (Holmes and Youson, 1993; see below) might be due to the lack of a previous rise and peak in TH levels that are necessary to prepare for metamorphosis. It is conceivable that the rise in serum TH concentration prior to their rapid decline is a critical component of normal metamorphosis. For instance R.G. Manzon et al. (2001) report a correlation between the magnitude of the decline in serum TH levels and the incidence of goitrogen-induced metamorphosis. Elevated TH levels might function to stimulate lipogenesis and the accumulation of lipid reserves which are characteristic of pre-metamorphic larvae. Exogenous TH have been shown to up-regulate a variety of processes associated with lipogenesis, and a suppression of serum TH levels following $KClO_4$ treatment coincides with lipolysis (Kao et al., 1999b; see below). Whether the gradual rise and peak of TH levels in larvae is simply a consequence of the increased capacity of the larval endostyle to produce TH and the lack of central hypothalamic-pituitary regulation or is an actively regulated process remains to be determined. Nonetheless, the existing data strongly suggest that elevated TH levels are important in preparing larval lampreys for metamorphosis.

### 2.6.3.3 Reproductive system

There is no clear evidence that the state of development of the ovary or testes is an important preparatory cue for lamprey metamorphosis. However, it should be emphasized that in the few cases where parasitic and non-parasitic members of paired species have been compared, the gonads of the non-parasitic member are much more advanced in their development (Bird and Potter, 1979a). Since larvae of non-parasitic species metamorphose at a larger length (and likely age) than larvae of parasitic species, it is possible that gonad size and the degree of development may be a factor in determining the time of metamorphosis. On the other hand, the larger gonad could be a consequence of the protraction of larval life in non-parasitic species and the longer period of potential for both somatic and gonadal growth. As will be discussed later, the hypothalamic hormone, gonadotropin-releasing hormones (GnRH-I and -III), become elevated in immediately pre-metamorphic larvae and early stages of metamorphosis of *P. marinus* (Youson and Sower, 2001) and the non-parasitic, *L. appendix* (Youson et al., 2006), but there is no evidence that the target is the gonad (Fig. 4). For instance, there is speculation, based on existing data from extant lampreys, that the interaction between the reproductive and thyroid axes has been a key factor in the development of metamorphosis as a life cycle strategy in lampreys (Youson and Sower, 2001).

## 2.7 Cues for Metamorphosis

There are two primary and essential cues for lamprey metamorphosis that have been recognized to date, and they are not mutually exclusive. These cues involve TH and water temperature. Other critical factors may be a signal from the fat stores (leptin-like molecule) and from the reproductive system, as illustrated through GnRH. In *P. marinus*, there is undoubtedly some involvement of fat within the adipose tissue but whether activity within this tissue is a consequence of a cascade of events or an essential cue is uncertain. These definitive cues and potential cues are summarized in Fig. 4. Other secondary factors may be photoperiod and animal density.

### 2.7.1 Water temperature

That metamorphosis is a seasonal event and restricted to summer and early autumn of the majority of lamprey species in both the Northern and Southern Hemispheres is not just a coincidence. Water temperature is not only a cue for the event but also is likely a critical feature to permit survival of the animals during a time when they are most vulnerable due to the immense changes that occur during their metamorphosis (Potter, 1970). Recently, Hardisty (2006) summarized the exceptions to the summer-fall incidence scenario described above around the viewpoint of altitude and latitude, and hence, water temperature. He described a metamorphosis occurring in early April in a Mexican freshwater lampreys (*Lampetra [Tetrapluerodon] spadicea* and *geminis*) living at high altitude at latitude 20°N, that is, far below the southerly limit of any other holarctic species (Lyons et al., 1994). Other exceptions were found in most northerly (61°N latitude) Arctic lampreys (*Lethenteron camtschaticum*) with June-March metamorphosis and variations (mid July to October) within populations of the Ukranian lamprey (*Eudontomyzon mariae*) that were related to their distribution between 35 and 47°N (Holcík, 1986; Holcík and Renaud, 1986; Renaud, 1986). Potter et al. (1980) showed the month variation of the onset of metamorphosis in *G. australis* from January in southern Tasmania (43°S latitude, 147°E longitude) to early February in southwestern Australia (34°S lat., 116°E long.) There are also examples within species where altitude influences the seasonal onset of metamorphosis (Hardisty, 2006). For an extended discussion and interesting speculation on how altitude, latitude might influence the timing of metamorphosis and water temperature we recommend Hardisty (2006).

Potter (1970) was among the first to show, through laboratory experiments with *M. mordax*, that metamorphosis occurs earlier in larval lampreys exposed to the higher of two water temperatures. This evidence along with the reports of earlier metamorphosis of *L. planeri* and *L. fluviatilis* following an early spring (Bird and Potter, 1979a) and the consistent correlation of metamorphosis in the anadromous sea lamprey (*P. marinus*) with the breakup of ice in the St. John River system in New Brunswick, Canada (Potter and Beamish, 1977), led Potter (1980) to suggest that the onset of metamorphosis is more related to a pattern of change rather than the reaching of an optimum water temperature. If there is a critical temperature to initiate metamorphosis and it varies among populations of the same species, the variation in the time of initiation could be a consequence of differences in their exposure to annual temperature regimes (Potter et al., 1982).

The importance of water temperature to lamprey metamorphosis has been explored through extensive laboratory studies on larvae of the landlocked sea lamprey (*P. marinus*). These studies have been recently reviewed by Youson (2003) and there have been no other studies since this review. The reader is referred to this most recent review and, herein, we will highlight the most salient points coming from the individual investigations. Purvis (1980) conducted field and laboratory studies without intra- and inter-experimental controls to show 75 to 100% metamorphosis of animals housed at 20 to 21°C, 46–76% at 14 to 16°C, and 5 to 10% at 7 to 11°C and concluded that controlled laboratory experiments were needed. The first of these subsequent laboratory experiments (Youson et al., 1993) showed that when animals of uniform size (common masses and lengths) are subjected to water temperatures of either 13 or 21°C on June 10–11 a significantly higher incidence of metamorphosis occurs at the higher temperature in the following month, the normal month for metamorphosis. In this study that was duplicated over two years, there was a much higher incidence of metamorphosis at both temperatures in the second year (overall 76% compared to 11.2% in year 1) when the animals were at least 120 mm long and weight at least 3.0 g. The incidence of metamorphosis was 66% at 13° and 84% at 21°C but development was much slower and even suppressed at the lower temperature in both years of the study. Although this was the first study to control for size, it was concluded that the true test for water temperature as an important cue was to use larvae of immediately pre-metamorphic size (≥120 mm and ≥3.0 g) exposed over a longer period to a water temperature regime that mirrored that in a sea lamprey stream (1 to 21°C) or to the mean temperature of that stream (21°C) at the time of metamorphosis. Such a study was carried out from mid-September to mid-August with ~ 1000 larvae, with the above size criteria, collected from the same stream (Holmes et al., 1994). The result was no metamorphosis at the constant temperature and the predicted number (based on size and condition factor [CF], see above for formula) in the ambient temperature group. The suggestion from this study was that it is not the highest temperature alone that is important to metamorphosis but it is likely that a cool water temperature followed by the spring rise is a critical cue. However, since this study was also exploring the effects of feeding-starvation, animal density, and photoperiod there was need for further long-term investigations to explore the temperature influence without as many variables.

Two investigations were undertaken to study the influence of temperature and the CF of the animals in the fall to the incidence of metamorphosis in *P. marinus* in the following summer (Holmes and Youson, 1994, 1997). In one case larvae with a CF of 1.50 or greater combined with a mass of at least 3.0 g and a length of at least 120 mm were marked with latex dye and predicted to enter metamorphosis the following summer (Holmes and Youson, 1994). Animals of similar size (mass and length) but with a CF less than 1.50 were deemed to be presumptively non-metamorphic and marked with a different color of latex dye. Holding tanks contained a mixture of both presumptively metamorphic and non-metamorphic animals that were subjected to either an ambient (3–21°C) or a constant (21°C) water temperature regime for 9 months. The overall incidence of metamorphosis was 53% in the ambient group and 2% in the constant temperature. While 64% of presumptively metamorphic animals entered metamorphosis in the ambient regime, only 10% of this group entered metamorphosis when kept at the constant temperature. Among the presumptively non-metamorphic group, 50% metamorphosed in the ambient temperature and no metamorphosis occurred in animals kept in a constant temperature regime. These results clearly showed that a more favorable environment is created by the ambient temperature and particularly, the seasonally lower winter temperatures, that place fewer demands on the animals and perhaps even permit anabolic activity, such as lipogenesis. In contrast, the constant temperature creates an over-wintering, catabolic activity that was reflected in a rapid negative change in animal mass. The ambient temperature regime also created the cool winter condition followed by the spring rise in water temperature that was suggested to be a critical cue in an earlier study (Holmes et al., 1994). It is likely that the problem lies with the absence of a change in the temperature, rather than that the constant temperature of 21°C is too high, for even a constant temperature of 15–16°C for 9 months on animals of >130 mm failed to produce metamorphosis (Holmes and Youson, 1994).

The study by Holmes and Youson (1994) showed a high incidence of metamorphosis in animals in the ambient temperature regime that were predicted in the fall to be non-metamorphic based on a CF of <1.50. O'Boyle and Beamish (1977) had suggested that lipogenesis occurs in the fall and winter months and it seems that the ambient temperature regime of the study by Holmes and Youson (1994) created conditions for fall, non-metamorphic animals of appropriate mass and length to prepare themselves for the non-trophic phase of metamorphosis in the following summer. Thus, a lower CF than 1.50 may be more appropriate for suitably sized animals in the fall prior to metamorphosis. A CF of 1.45 in animals of minimum 120 mm and 3.0 g in the fall and the spring rise in temperature resulted in a 56% incidence of metamorphosis (Holmes and Youson, 1997). It is noteworthy that groups of animals subjected to a spring rise in water temperature from the winter low of ~ 4°C to either ~ 13° or ~ 21°C did not vary significantly in the incidence of metamorphosis and accounted for 92% of the observed metamorphosis (Fig. 3). In contrast, animals kept between 8–9°C over the fall (November) to the summer period of metamorphosis (July-August) showed a significantly lower incidence of metamorphosis compared to what was predicted based on their size and CF.

The above studies provide support for the hypothesis (Holmes et al., 1994) that rising water temperature in the spring is an important cue for metamorphosis in *P. marinus* (Fig. 4). Furthermore, the study by Holmes and Youson (1997) showed that the magnitude of the increase in water temperature does not affect the incidence of metamorphosis. However, a subsequent study showed that there is an optimal upper temperature of 21°C and a lower thermal limit of between 9 and 13°C in the spring immediately prior to metamorphosis that are important factors influencing the incidence of metamorphosis (Fig. 5). In a study that compared the incidence of metamorphosis in appropriately sized and CF *P. marinus* kept at water temperatures of either 9, 13, 17, 21, and 25°C starting at the beginning of June, 0% metamorphosis occurred at 9°C, 80% at 21°C, and 58% at 25°C. The incidence of metamorphosis only met expectations, based on size and CF, at 21°C (Holmes and Youson, 1998). Overall, the above results may explain why *P. marinus* is distributed over such a wide thermosphere in both North America (Beamish, 1980) and Europe. All the regions of distribution of this species are subjected to seasonal variation in water temperature of the streams but they do not reach comparable highs nor are the changes of great magnitude in all cases. As shown above, it is the spring rise from the seasonal low that is the important water temperature cue for metamorphosis

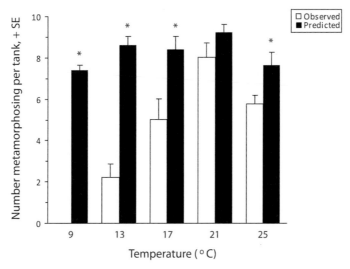

**Figure 5.** A comparison of the number of observed (August, open bars) and predicted (based on size and condition factor in June) of *P. marinus* metamorphosing/tank following treatment in one of five different water temperature regimes. The asterisk (*) above the prediction bar indicates a significant difference (P ≤ 0.05) from observed at each temperature. The data indicate that 21°C is near optimum for metamorphosis, whereas between 9 and 13°C water temperature has an inhibitory effect. (From Holmes and Youson, 1998).

and this feature, given the potential for sexual maturation in post-larval life, is critical for the perpetuation of the species.

### 2.7.2 *Animal density*

It is well established that high density among larval lampreys increases the length of the larval growth phase and reduces growth rates (Mallatt, 1983; Morman, 1987; Murdoch et al., 1992; Rodriguez-Munoz et al., 2003). It might be expected then that metamorphosis might occur earlier (at a younger age) in low-density populations of larva where growth rates may be enhanced relative to a high-density population. This feature of metamorphosis has been shown in *P. marinus* (Purvis, 1979; Morman, 1987) and in the northern brook lamprey, *Ichthyomyzon fossor* (Purvis, 1970) over several consecutive years in wild populations. Growth in length, an indicator of larval age and preparation for metamorphosis, slows as density increases and, therefore, increases the duration of larval life. In contrast, however, a more closely controlled laboratory study (Fig. 3) of 9-month duration on immediately pre-metamorphic larval *P. marinus* (appropriate CF and size) showed no significant difference in the incidence of metamorphosis between high-density (164 larvae/m²) and low-density (66 larvae/m²) populations in three temperature regimes (Holmes and Youson, 1997). In a more recent mark-recapture study in a number of streams in the Great Lakes watershed, larval density was recognized as an important variable in predictive models of determining when *P. marinus* will enter metamorphosis (Treble et al., 2008). The effects of animal density in controlling metamorphosis requires further study, for there has been some suggestion that some chemical or biological agent is released into the water by the larvae that can influence growth (Rodriguez-Munoz et al., 2003). If such is the case, then larval density could prove to be an important factor in regulating metamorphosis.

### 2.7.3 Photoperiod

As described above, there is a high degree of annual synchrony of lamprey metamorphosis with the majority of species undergoing this phase of the life cycle in the summer months. Although water temperature is critical at this time, it is also the annual period when the daily period of light is at optimum. It has long been speculated that metamorphosis in lampreys may be triggered by a change in the length of the periods of light and darkness, i.e., a photoperiod stimulation. Lampreys have a prominent pineal complex, a photoreceptive organ made up of pineal and parapineal glands (Cole and Youson, 1982), that yield melatonin in a circadian fashion. Pinealectomized larval lampreys do not enter metamorphosis (Eddy, 1969; Cole and Youson, 1981). The study by Cole and Youson (1981) used *P. marinus* that were immediately pre-metamorphic, based on CF and size criteria (for review see Youson, 2003). However, since groups of sham-operated animals, control animals, and animals kept in total darkness were not significantly different in the incidence of metamorphosis, the role of photoperiod is in question. Similar studies over a longer period to test for the effects of accumulated light should be undertaken.

The role of photoperiod was also tested in *P. marinus* in a short-term experiment immediately prior to metamorphosis (Youson et al., 1993). There was no difference in the incidence of metamorphosis in groups of larvae kept under artificial 15 h light : 9 h dark cycle or in constant darkness from mid June to the end of July. A similar result was found with appropriately sized (pre-metamorphic) animals kept under either an ambient-light regime or a constant 15 h light : 9 h dark regime from September to the middle of the following August (Holmes et al., 1994). Although the pinealectomy studies suggest some involvement of the pineal complex in lamprey metamorphosis, its role may not be in response to photoperiod changes. There is still a possibility that the pineal complex may be involved in regulating or responding to changes in metabolism and even changes in water temperature or responding to some endogenous factor (Youson, 1994).

## 2.7.4 Endocrine

Historically, research on the endocrine regulation of lamprey metamorphosis has focused largely on the thyroid system and its peripheral regulators, but the recent identification of hypothalamic-releasing and pituitary hormones in lampreys has set the stage for the study of central regulatory mechanisms. Moreover, the importance of lipid accumulation and lipid metabolism prior to and during metamorphosis in *P. marinus*, respectively, suggests that lipolytic and lipogenic hormones such as insulin and somatostatin, and a leptin-like protein may also be regulators of lamprey metamorphosis.

Biologists have sought to understand the role of TH and the thyroid system in lamprey metamorphosis since Gudernatsch (Gudernatsch, 1912) first discovered that thyroid gland extracts induce precocious metamorphosis in anuran amphibians. Most information regarding the endocrine regulators of lamprey metamorphosis has been obtained from studies on the thyroid gland, and the thyroid hormones, T4 and the more biologically active $T_3$. Also included are the peripheral regulators of TH action, such as binding proteins and deiodinase enzymes. It is evident that TH are important for lamprey development and metamorphosis, however, their precise role has yet to be fully elucidated and present evidence is not necessarily consistent with the anuran model. TH modulate, via their interaction with nuclear receptors, most essential gene expression cascades, either directly or indirectly, that are required for anuran metamorphosis and perhaps the metamorphoses of gnathostome fishes (see other chapters in this volume). Included among these anuran gene expression cascades are those associated with cellular proliferation and *de novo* synthesis of adult organs (i.e., limbs and lungs), cellular apoptosis and regression of larval structures (i.e., tail and gills), and remodeling of larval organs into their adult counterparts (i.e., liver, nervous system, intestine) (reviewed in Shi, 2000; Denver et al., 2002; Brown and Cai, 2007).

### 2.7.4.1 Thyroid hormones

The first experiments to examine the role of the thyroid axis in lamprey metamorphosis failed to trigger metamorphosis using approaches which had been successful in anurans. Included among these approaches were immersion in iodine-containing compounds or thyroid extracts (Horton, 1934) and injection of either $T_4$ (Leach, 1946) or pituitary extracts (Young and Bellerby, 1935; Knowles, 1941). Furthermore, despite using *L. planeri* of presumed pre-metamorphic size and age, the thyrotropic fraction of ox pituitary extracts Knowles (1941) failed to induce precocious metamorphosis.

The first indication of TH involvement in lamprey metamorphosis was the paradoxical induction of a partial metamorphosis in larval year classes 1, 2, and 3 of *L. planeri* following immersion in the anti-thyroid agent (goitrogen) potassium perchlorate ($KClO_4$) (Hoheisel and Sterba, 1963) (Fig. 1). The effects of goitrogens on endostyle morphology and iodine metabolism suggested that $KClO_4$ inhibited thyroidal activity (Jones, 1947; Klenner and Schipper, 1954; Clements-merlini, 1962a,1962b), but the relationship to metamorphosis remained unclear. This idea that thyroidal inhibition might be involved in lamprey metamorphosis was later supported by the observations that serum $T_4$ (Wright and Youson, 1977) and $T_3$ (Lintlop and Youson, 1983b) concentrations declined coincident with the first external signs of metamorphosis in *P. marinus*. Subsequent studies on *L. reissneri* claimed a complete metamorphosis following treatment with either $KClO_4$, sodium perchlorate ($NaClO_4$), proplythiouracil (PTU) or thiourea (TU) (Suzuki, 1986,1987,1989). Conversely, precocious metamorphosis was not observed in *G. australis* following 70 days of PTU treatment despite significant decreases in both serum $T_4$ and $T_3$ concentrations (Leatherland et al., 1990). These results led Leatherland et al. (1990) to further suggest that, since the decline in TH occurs after the first external signs of metamorphosis are visible, the decline in TH could not be the stimulus that initiates metamorphosis.

These early studies raised an important question regarding the role of TH in lamprey metamorphosis: Is the decline in serum TH levels required for the initiation of metamorphosis or is it simply a consequence thereof? These two possibilities need not be mutually exclusive, for instance the decline, in fact, might be a consequence of metamorphosis with the resultant lower TH a requirement for metamorphosis. The transformation of the larval endostyle to a follicular thyroid during metamorphosis may be responsible for the observed decline in serum TH levels (Youson, 2007). Unlike all other vertebrates, but more like that seen in some protochordates, TH synthesis in larval lampreys occurs in the subpharyngeal endostyle rather the typical vertebrate thyroid follicles (or tubules). The transformation of the endostyle commences at stage 1–2 of metamorphosis with typical follicles first appearing at stage 4 of metamorphosis (Wright and Youson, 1976,1980a). Although the transforming endostyle contains thyroglobulin and binds iodide throughout metamorphosis, there appears to be a reduced capacity to bind iodide at this time which might equate to a decrease in TH synthetic capacity (Wright and Youson, 1976; Wright et al., 1980). It is possible that the elevated TH levels observed in pre-metamorphic larvae (see above) are required to ensure sufficient reserves for metamorphosis when synthetic capacity is reduced during transformation of the thyroid tissue. To help resolve the puzzle over the importance of the decline in serum TH levels for metamorphosis a series of thyroid ablation and replacement experiments were conducted on larval lamprey.

Partial clarification of the importance of the TH decline arose from work on goitrogen-induced metamorphosis in *P. marinus* and to a lesser extent *L. appendix*. The induction of metamorphosis in *P. marinus* with $KClO_4$ was first demonstrated by Holmes and Youson (Holmes and Youson, 1993). Not only were the larvae in this study induced to metamorphose at a time of year when spontaneous metamorphosis does not naturally occur, but $KClO_4$ treatment superseded the size and CF requirements for metamorphosis. Furthermore, it was discovered that $KClO_4$-induced metamorphosis was size dependent with 98%, 52%, and 22% induction of metamorphosis in larvae > 130 mm, 110–119 mm and 65–95 mm in length, respectively (Holmes and Youson, 1993). When the sera from the lampreys used in this study were analyzed, it was found that $KClO_4$ significantly suppressed $T_3$ concentrations in all size groups and $T_4$ concentrations in the small and intermediate size groups (Youson et al., 1995b). The physiological link to the thyroid and the absolute requirement for a decline in TH levels for induced metamorphosis was confirmed by observations that the $KClO_4$-induced metamorphosis of *P. marinus* larvae could be completely blocked by maintaining serum $T_4$ and $T_3$ concentrations at or above control levels with exposure to either exogenous $T_4$ or $T_3$ (R.G. Manzon and Youson, 1997; R.G. Manzon et al., 1998) (Fig. 6). Moreover, a variety of different goitrogens ($KClO_4$, $NaClO_4$, potassium thiocyanate, and methimazole) were effective at inducing metamorphosis and the incidence of induction was correlated to the magnitude of the decline in serum TH concentrations (R.G. Manzon et al., 2001; Fig. 7). Consistent with earlier work on *G. australis* (Leatherland et al., 1990) and

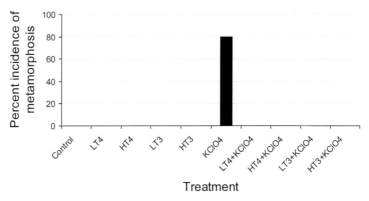

**Figure 6.** Percent metamorphosing *P. marinus* following exposure to a high or low dose of exogenous thyroxine (T4; 0.5 or 1 mg/L) or triiodothyronine (T3; 0.25 or 1 mg/L) in the presence or absence of potassium perchlorate (KClO4; 0.05%) for 16 weeks. 80% of larvae exposed to KClO4 alone were induced to metamorphose. Exogenous T4 and T3 did not induce metamorphosis in any instance and completely blocked KClO4-induced metamorphosis at both treatment doses. Sample for each treatment group is equal 30. (R.G. Manzon et al., 1998).

*L. appendix* (Holmes et al., 1999), PTU did not induce metamorphosis in *P. marinus* despite a suppression of serum TH concentrations (discussed below) (R.G. Manzon et al., 2001). Finally, *in vitro* studies confirmed that $KClO_4$ acts directly on the larval endostyle to inhibit iodide uptake and organification (R.G. Manzon and Youson, 2002; Fig. 8).

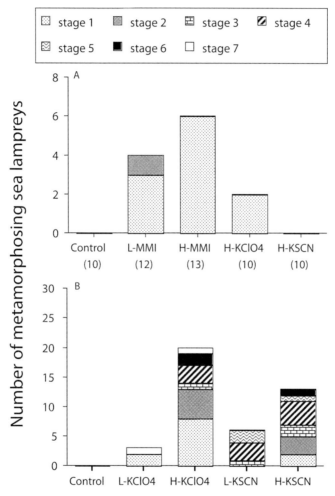

**Figure 7.** A) Number of metamorphosing sea lamprey and stage of metamorphosis in control larvae and following a 6 week exposure to methimazole at low (L-MMI; 0.087 mM) and high doses (H-MMI; 0.87), potassium perchlorate at the high dose (H-KClO4; 0.72 mM), or potassium thiocyanate at the high dose (H-KSCN; 0.51 mM). B) Number of metamorphosing sea lamprey and stage of metamorphosis in control larvae and following a 16 weeks exposure to KClO4 at low (L-KClO4; 0.072) and high (H-KClO4) doses or KSCN at low (L-KSCN; 0.051 mM) and high (H-KSCN) doses. Sample is equal to 30 unless indicated otherwise below the abscissa. (From R.G. Manzon et al., 2001).

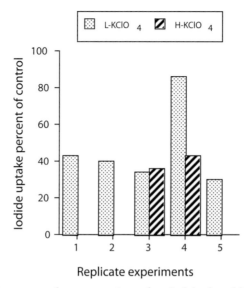

**Figure 8.** [125]I⁻ uptake, expressed as a percentage of control, by larval lamprey endostyles following a 4 hour *in vitro* incubation in the presence of 0.72 nM (L-KClO$_4$) or 3.6 mM (H-KClO$_4$) potassium perchlorate. Endostyles were incubated with 3 or 30 µCi of Na[125]I in the presence or absence (controls) of KClO$_4$. In all instances, [125]I⁻ uptake by KClO$_4$-treated endostyles was significantly lower than controls (ANOVA, $P < 0.05$). (From R.G. Manzon and Youson, 2002).

Collectively, the work discussed above convincingly shows that a decline in serum TH levels is required for induced metamorphosis. When coupled with observations that serum TH levels gradually increase throughout the larval period in preparation for metamorphosis and peak prior to metamorphosis in *P. marinus* (Youson et al., 1994), decrease early in spontaneous metamorphosis in *P. marinus* (Wright and Youson, 1977; Lintlop and Youson, 1983b; Youson et al., 1994), *G. australis* (Leatherland et al., 1990), and *L. appendix* (Holmes et al., 1999), and that exogenous T$_3$ can disrupt spontaneous metamorphosis in *P. marinus* (Youson et al., 1997) it is difficult to discount the importance of a decline in TH levels during lamprey metamorphosis. Although a decline appears to be essential for metamorphosis, it is likely not the only factor required. For instance, PTU suppresses TH levels in three species of lamprey (Leatherland et al., 1990; Holmes et al., 1999; R.G. Manzon et al., 2001), but unlike other goitrogens does not induce metamorphosis. KClO$_4$ cannot supersede the water temperature requirement for metamorphosis and is not able to induce metamorphosis when lampreys are maintained at winter water temperatures (2–3°C) (R.G. Manzon and Youson, 1999). Finally goitrogen-induced metamorphosis is not as synchronized as spontaneous metamorphosis and, to date, there is no evidence that animals that are the product of inducement are capable of feeding, thus, other factors are likely important for the completion of normal metamorphosis (Fig. 1). Although PTU toxicity has not been eliminated as a factor contributing to the absence of metamorphosis following PTU treatment and $KClO_4$ treatment at more moderately cool temperatures remain to be tested, these studies indicate that other factors such as water temperature, lipid reserves (size), and other components of the thyroid system (i.e., peripheral regulators) are important for the initiation and progression of lamprey metamorphosis, at least in *P. marinus*.

The peripheral regulators of the lamprey thyroid system which have received some attention include the serum TH distributor proteins (THDP), TH deiodinases, and TH nuclear receptors (TRs) and their heterodimeric partners the retinoid-X receptors (RXRs). Other components, including the TH transmembrane cellular uptake transporters, the cytosolic TH binding (transport) proteins (CTBPs), and transmembrane receptors, have yet to be investigated in lampreys but should be the focus of future studies. The observations that larval lamprey serum has a tremendous capacity to store and transport TH, and treatment with exogenous TH can result in elevations in serum $T_4$ and $T_3$ concentrations that are 5–20 fold and 3–10 fold greater than control levels, respectively, (R.G. Manzon and Youson, 1997; R.G. Manzon et al., 1998) raises several questions. Is the decline in TH at the onset of metamorphosis due to a combination of a decrease in TH synthetic capacity of the transforming endostyle (discussed above), reduced TH-binding and transport capacity of the serum, increased TH uptake by cells, increased cytosolic TH deiodination, and/or increased TH nuclear binding?

Serum THDP are important since they allow the serum to function as a reservoir for TH and ensure a uniform distribution to all cells and tissues. In the absence of serum THDP, TH will readily partition into the first cells they encounter (Mendel et al., 1987). An investigation of serum THDP in *P. marinus* has identified at least four distinct proteins capable of binding $T_4$ and $T_3$ (Gross and R.G. Manzon, unpublished). Included among these THDP are the albumin-like proteins AS and SDS-1, the glycolipoprotein band CB-III and an unidentified protein band designated Spot-5 (Gross and R.G. Manzon, unpublished). TH binding studies indicate that these THDP are developmentally regulated: AS is the dominant THDP in larvae and throughout metamorphosis; AS is replaced by SDS-1 and CB-III beginning at stage 7 of metamorphosis; SDS-1 and CB-III are the primary TH-binding proteins in juveniles and upstream-migrant adults (Gross and R.G. Manzon, unpublished). This shift from AS to SDS-1 and CB-III using TH-binding as a metric is consistent with other reports on these proteins using immunometric methodologies (Filosa et al., 1982, 1986). Despite the change in the type and number of THDP in lamprey sera during the life cycle, there did not appear to be any dramatic change in the overall total of TH binding capacity of

*P. marinus* sera throughout development, thus, it is very unlikely that the decline in serum TH concentrations during metamorphosis is a result of a decrease in serum transport or binding capacity (Gross and R.G. Manzon, unpublished). The cDNA for a fifth THDP, transthyretin (TTR) was isolated from *L. appendix* and *P. marinus* and its mRNA transcript was shown to be developmentally regulated in *P. marinus* liver and widely expressed in other tissues (R.G. Manzon et al., 2007). TTR is a dominant serum THDP in other fish, anurans, birds and mammals; however, it was not identified as a THDP in lamprey serum (Gross and R.G. Manzon, unpublished). Collectively, these studies are among the first data on THDP in lampreys.

Peripheral deiodination of TH appears to play a particularly important role in the regulation of TH homeostasis in fishes, including "primitive" fishes (Youson, 2007), and in many instances may be a more significant regulator of thyroid status than the hypothalamic-pituitary axis (Eales and Brown, 1993). Three classes of deiodinases designated type 1, 2 and 3 (D1, D2, and D3, respectively) have been identified in most vertebrates studied to date. D2 is an outer ring deiodinase (ORD) responsible for the conversion of $T_4$ to the more biologically active $T_3$ and, thus, is an activating enzyme. Conversely, D3 is an inner ring deiodinase (IRD) that inactivates both $T_4$ and $T_3$ via their conversion to reverse $T_3$ ($rT_3$) and $T_2$ (diiodothyronine), respectively. Mammalian D1 is capable of both IRD and ORD reactions (see Bianco et al., 2002).

TH deiodinases may contribute to the precipitous decline in serum TH levels during lamprey metamorphosis (Eales et al., 2000). Contrary to other vertebrates, the intestine and not the liver is the primary site of $T_4$ ORD in *P. marinus* larvae, but lower activities can be detected in the liver, kidney and muscle. This finding is consistent with the hypothesis that the primary route of entry of TH into the circulation of lampreys is by intestinal absorption (see Eales, 1997). The larval lamprey endostyle produces not only TH, but also large quantities of mucopolysaccarides that aid in the trapping of nutrients during filter feeding; both of these compounds are secreted into the pharynx and transported through the digestive tract where absorption takes place (see Barrington, 1972; Eales, 1997; Youson, 2007). Only $T_4$ ORD activity was detected in larval intestines, while parasitic-phase juveniles and upstream-migrant adults have $T_4$ ORD, $T_4$ IRD and $T_3$ IRD activities (Eales et al., 1997). $T_4$ ORD and $T_4$ IRD (i.e., TH activation and inactivation reactions, respectively) show a reciprocal relationship throughout lamprey development. Intestinal $T_4$ ORD activity is low in *P. marinus* larvae, increases in pre-metamorphic larvae, peaks at stage 1 of metamorphosis (Eales et al., 2000) and then declines to very low levels that are sustained through to the upstream-migrant period. In contrast, $T_4$ IRD activity levels are at or below the limits of detection in larvae and during the first two stages of metamorphosis, but increase significantly at stage 3 and reach peak levels at stage 7 of metamorphosis. Following the completion of metamorphosis, IRD activity falls to moderate levels in parasitic-phase juveniles and upstream-migrant adults (Eales et al., 2000). The sharp decline in $T_4$ ORD and a surge in $T_3$ IRD activity may contribute to the decline in serum TH levels that is observed at the start of metamorphosis and, thus, may act to further minimize the level of active hormone available throughout the metamorphic process. The cDNA for *P. marinus* D2 has recently been cloned and real-time PCR expression data for the intestine, liver and kidney are consistent with $T_4$ ORD activity data (R.G. Manzon, unpublished data).

Collectively, all data pertaining to the lamprey thyroid axis, including serum TH concentrations during spontaneous metamorphosis, various ablation and replacement experiments, exogenous TH treatments, and both THDP and TH deiodinase data, are consistent with the hypothesis that a suppression of the thyroid system is important for the normal progression of metamorphosis in lampreys. Although these data are very convincing and strongly support the need for a decline in serum TH concentrations during lamprey metamorphosis, they do not represent the complete picture. A close examination of serum TH levels demonstrates that despite the decline in serum TH concentrations during spontaneous metamorphosis, these relatively low levels might be comparable to peak levels during anuran metamorphosis. In larval lamprey serum, $T_4$ and $T_3$ concentrations peak at values in the range of 1550–3153 µg/dl and 7800 ng/dl, respectively (Lintlop and Youson, 1983a; Youson et al., 1994). By stage 3 of metamorphosis, TH concentrations drop to 965 µg/dl and 860 ng/dl for $T_4$ and $T_3$, respectively, and are at their lowest levels of 98 µg/dl of $T_4$ and 61 ng/dl of $T_3$ at stage 6 of metamorphosis. Relative to larval lamprey, the TH concentrations measured during metamorphosis are very low, however, they are comparable to the peak $T_3$ concentrations of 75–150 ng/dl as measured during metamorphic climax in *R. catesbeiana* and *X. laevis* (White and Nicoll, 1981; Tata et al., 1993).

One view is that the decline in serum TH levels during spontaneous lamprey metamorphosis is a result of a decrease in TH synthesis and secretion in conjunction with an increase in cellular uptake and nuclear binding at sites where TH act to regulate morphogenesis. Thus, despite the decrease in serum concentrations, the function of TH in lamprey metamorphosis might be fundamentally similar at the level of the gene to the situation in teleost and anuran metamorphoses. There are data that suggest that the nuclear binding of $T_3$ by hepatocytes increases during *P. marinus* metamorphosis (Lintlop and Youson, 1983a). Using *in vitro* binding assays, hepatocyte nuclei where shown to bind $T_3$ with high affinity ($K_d$ 2.69 X10$^{-10}$ M) and their binding capacity increased slightly from 1.89 pg $T_3$/µg DNA in larvae to peak values of 2.62 pg $T_3$/µg DNA at metamorphic stage 5 and 6 and then decreased in parasitic adults and upstream migrants to levels more typical of other vertebrates (0.78 and 0.12 pg $T_3$/µg DNA, respectively) (Lintlop and Youson, 1983a). These data must be interpreted with some caution as the increase in binding capacity at metamorphosis was not statistically significant, sample size was limited, and the data are derived from isolated nuclei and, thus, do not account for cellular uptake and cytosolic or nuclear transport mechanisms. However, they do suggest that $T_3$ may exert some influence on liver morphogenesis. It will be interesting to determine if similar increases in nuclear binding capacity are observed in other tissues during metamorphosis and whether cellular uptake and transport mechanism are likewise elevated during metamorphosis.

TH action on vertebrate development is predominately mediated by interaction with nuclear receptors that are TH-regulated transcription factors and are, therefore, directly responsible for modulation of the associated gene expression cascades. To elaborate on the hepatocyte nuclear binding data of Lintlop and Youson (1983b), information on lamprey TRs and RXRs was sought. Two TR (PmTR1 and PmTR2) and three RXR (PmRXR 1, PmRXR 2, and PmRXR 3) cDNAs were cloned from *P. marinus* (L.A. Manzon, 2006). Interestingly, although these receptors are highly conserved, phylogenetic analyses suggest that the lamprey TRs diverged from the gnathostome lineage prior to the TRα and TRβ split, hence their designation as PmTR1 and PmTR2 (Escriva et al., 2002). Although the vertebrate RXR phylogenetic tree is much more difficult to resolve, the fact that all three PmRXRs are equally identical to the vertebrate RXRα, RXRβ and RXRγ strongly suggests they would not group with individual RXRs but rather would form a separate group (L.A. Manzon, 2006). Preliminary developmental expression analyses for TRs indicate that PmTR2 is constitutively expressed throughout metamorphosis, but PmTR1 is up-regulated at times of tissue morphogenesis and is expressed in an organ- and tissue- specific manner throughout metamorphosis in the liver, kidney, intestine and gill (L.A. Manzon, 2006). These data are consistent with the tissue-specific upregulation of TRβ and the constitutive expression of TRα in *X. laevis* (reviewed in Shi, 2000). In the Japanese flounder (*Paralichthys olivaceus*), TRβ is expressed constitutively whereas TRα expression correlates with tissue morphogenesis (Yamano and Miwa, 1998) and in the Senegalese sole (*Solea senegalensis*) both TRα and TRβ expression correlate with metamorphosis (Isorna et al., 2009). When the relatively low TH levels during metamorphosis are viewed in light of increased nuclear $T_3$ binding and the upregulation of PmTR2 they indicate that the components necessary for gene regulation are present and, thus, TH might function to drive morphogenesis in a fashion similar to that observed in other vertebrates.

When all data are considered it is feasible that TH have a dual role in lamprey development (R.G. Manzon, In Press). During the larval period high TH levels promote feeding, larval growth and lipid accumulation, but act as a juvenilizing hormone to inhibit metamorphosis (Youson, 1997). Subsequently, following some unknown signal, metamorphosis begins and much lower TH levels are necessary to drive the morphogenetic processes associated with metamorphosis as observed in other vertebrates (R.G. Manzon, In Press).

### 2.7.4.2 Hypothalamic-pituitary hormones

The past two decades have seen an explosion of new information on the hormones of the lamprey hypothalamus and adenohypophysis (pituitary). An extensive discussion of the lamprey hypothalamic-pituitary (HP) hormones and the associated evolutionary interpretations is beyond the scope of this chapter, but can be found in other recent reviews (Kawauchi and Sower, 2006; Sower et al., 2009). Much of this work has focused on the identification and characterization of the various hormones of the HP axis and has laid the foundation for future studies on its regulation and function in lamprey metamorphosis. To date there are little data in support of a role for the lamprey HP axis in the regulation of the thyroid and most data suggest that the thyroid system is regulated by peripheral rather then central mechanisms (Eales and Brown, 1993; Youson, 2007). Likewise, most studies have failed to show a relationship between the HP axis and lamprey metamorphosis, although the adenohypophysis undergoes extensive modification during metamorphosis (Wright, 1989). One notable exception is the requirement of the adenohypophysis for normal metamorphosis of *G. australis* (Joss, 1985). Removal of the larval rostral pars distalis (RPD) resulted in complete metamorphic stasis while metamorphosis was arrested in stage 3 following the removal of the caudal pars distalis (CPD) (Joss, 1985). Some recent evidence suggests that hypothalamic GnRH may be involved in metamorphosis of *P. marinus* (Youson and Sower, 1991), *L. richardsoni* (Youson et al., 1995a), and *L. appendix* (Youson et al., 2006), and perhaps may be connected to thyroidal regulation (Youson and Sower, 2001). This latter point is consistent with the idea that there is overlap between the thyroid and reproductive axes and that reproductive maturation is one of the ancestral functions of the thyroid (see Youson and Sower, 2001).

The hypothalamus of most gnathostomes contains one or two GnRHs and two GnRH-receptors (GnRH-Rs) (Sower et al., 2009). GnRHs regulate the synthesis and secretion of two pituitary gonadotropins (GTH): follicle stimulating hormone (FSH) and lutenizing hormone (LH), each of which act via one of two gonadal glycoprotein hormone (GpH) receptors. The heterodimeric GpH family members consist of an α and β subunit, and to date two α and five β subunits have been identified in vertebrates. Other members of the GpH family include thyrotropin (TSH), thyrostimulin (TSM), and chorionic gonadotropin (CG) (see Sower et al., 2009). Lampreys have three hypothalamic GnRHs (GnRH-I, GnRH-II and GnRH-III), one GnRH-R (partial/full length clones of two additional putative receptors have been identified but not characterized), one GTH, a second GpH with homology to TSM, one gonadal GpH receptor and one thyroidal GpH receptor (Sower et al., 2009).

There is a clear trend in *P. marinus*, *L. appendix*, and *L. richardsoni* that GnRH-I and GnRH-III increase in the brain throughout metamorphosis with the most notable increase occurring towards the end of metamorphosis (Youson and Sower, 1991; Youson et al., 1995a, 2006). Although developmental data are currently lacking on the more recently identified lamprey GnRH-II, like GnRH-I and -III its transcript has been localized to the hypothalamus and it has been shown to stimulate the pituitary-gonadal axis both *in vitro* and *in vivo* (Kavanaugh et al., 2008). The concentration of GnRH-I and GnRH-III in *P. marinus* brains, as determined by radioimmunoassay following extraction, is low in larvae of year classes 2, 3, and 4, and remains low until stage 3 of metamorphosis (Youson and Sower, 1991; Youson et al., 1995a). A moderate increase in GnRH-I occurs at stage 6 followed by a dramatic increase at stage 7 and levels remain elevated in feeding parasites (Fig. 4). GnRH-III also increases at stage 6 and remains elevated through parasitic feeding although peak levels of GnRH-I are approximately 3 times greater than those of GnRH-III (Youson et al., 1995a). These data are further corroborated by both immunohistochemical (Wright et al., 1994; Tobet et al., 1995,1996,1997) and *in situ* hybridization (Root et al., 2005) studies in *P. marinus*. This rise in GnRH during metamorphosis is also observed in the non-parasitic species *L. richardsonii* (Youson et al., 1995a) and *L. appendix* (Youson et al., 2006). As with *P. marinus*, peak immunostaining for GnRH in *L. appendix* is observed at stage 7 of metamorphosis. One curious difference between the two species (which have two different life history strategies) is the observation that GnRH levels begin to increase earlier in *L. appendix* (during stage 1 and 2) and gradually increase throughout metamorphosis (Youson et al., 2006). In fact, there was an earlier activity of GnRH-I in *L. appendix* compared to *P. marinus* and this feature is most likely related to the fact that the sexual maturation of non-parasitic species commences shortly after metamorphosis and, therefore, their gonadal maturation is advanced as compared to parasitic species at the same stage of development.

There is no doubt that a rise in brain GnRH is a characteristic of metamorphosis in both parasitic and non-parasitic life history strategies. What requires further clarification is the precise function of GnRH at this time in development. Is the role of GnRH related strictly to gonadal development, does it function in the regulation of the thyroid or other endocrine axes or does it also play a more direct role in the metamorphic process? The finding that there is a rise in GnRH during spontaneous metamorphosis prompted an investigation into GnRH levels following the manipulation of the thyroid axis during $KClO_4$-induced metamorphosis in *P. marinus* (Youson and Sower, 2001). As with spontaneous metamorphosis, a rise in brain GnRH-I and III levels occurs during $KClO_4$-induced metamorphosis, however, as with other aspects of induced metamorphosis, the changes in GnRH levels do not closely parallel those that occur in spontaneous metamorphosis (Youson and Sower, 2001). A significant increase in GnRH-III is observed between stages 3 and 4 of $KClO_4$-induced metamorphosis, however during spontaneous metamorphosis this increase does not occur until stage 6. Likewise, GnRH-I in the brain increases markedly at stage 7 during spontaneous metamorphosis; not only is this increase absent in $KClO_4$-induced metamorphosis, but a significant decline occurs at stage 5 of induced metamorphosis (Youson and Sower, 2001). These differences in GnRH levels between spontaneous and induced metamorphosis may be related to the asynchronous development associated with induced metamorphosis. It is difficult to make direct comparisons between spontaneous and induced metamorphosis as staging in the latter case is at best an approximation (Holmes and Youson, 1993; R.G. Manzon and Youson, 1997; R.G. Manzon et al., 1998) (Fig. 1). Stage 5 is the most advanced stage of induced metamorphosis observed by Youson and Sower (1991), despite a 4 month $KClO_4$ exposure which is equivalent to the time required to complete spontaneous metamorphosis. A better understanding of cause and effect relationships associated with asynchrony in induced metamorphosis could prove invaluable to our understanding of the regulation of lamprey metamorphosis. For instance, the failure to complete metamorphosis in some individuals might in part be related to the need for a peak in GnRH-I and GnRH-III signaling from the hypothalamus (Youson and Sower, 2001). An alternative explanation is that peak concentrations in GnRH are not realized during induced metamorphosis because the appropriate positive or negative feedback signals are absent. Is the lack of complete metamorphosis due to the absence of a peak in GnRH or visa versa? In either case, these data support a role for GnRH in metamorphosis, they show that manipulation of the thyroid axis results in some changes in the reproductive axis, and they indicate that there is some overlap between these two axes.

That there is the potential for cross regulation between the lamprey gonadal and thyroid axes was first supported by observations that a single injection of either salmon GTH or a GnRH analog produces significant elevations in serum $T_4$ in adult *P. marinus* (Sower et al., 1985). To date, one lamprey GTH-β (Sower et al., 2006) and one GpH-α subunit, with homology to GpH-A2 (TSM-α), have been identified (Sower et al., 2009). Preliminary data also exist for a second lamprey GpH-β (lGpHB5) with homology to GpHB5 (TSM-β) (Sower et al., 2009). Sower and co-workers (Sower et al., 2009) propose that an ancestral GTH-β gave rise to one lamprey GTH-β and it was gnathostome-specific duplications that gave rise to FSH-β,

LH-β and TSH-β. When this proposition is considered in conjunction with the observations that lampreys have one functional gonadal GpH receptor (GpH-RI; Freamat et al., 2006) and a second functional thyroidal GpH receptor (GpH-RII; Freamat and Sower, 2008), it seems very likely that there is overlap between the reproductive and thyroid components of the lamprey HP axis and it is likely to be a close approximation of the ancestral axis (Sower et al., 2009). Therefore, in lampreys, it is feasible that a single GTH might function to regulate both the reproductive and thyroid axes via gonad- and thyroid-specific GpH-Rs. Many of these studies have been carried out using adult lampreys, thus, studies on these hormones and their regulatory effects in larvae and throughout metamorphosis will surely provide a better understanding of their role in metamorphosis in this ancient vertebrate and the evolution of the HP axis.

### 2.7.4.3 Lipogenic and lipolytic hormones

The accumulation of lipids prior to the onset of metamorphosis is a key indicator of immediately pre-metamorphic lampreys. As discussed above, lamprey metamorphosis is characterized by a two-phase, lipid metabolism cycle. During the lipogenic phase (larval to metamorphic stage 3/4), elevated levels of acetyl CoA carboxylase (ACC) and diacylglycerol acetyl transferase (DAGT) activity result in the accumulation of lipids in the kidney and liver, primarily in the form of triacylglycerol (TG) (Kao et al., 1997a,1997b). The lipolytic phase coincides with decreases in the activities of the aforementioned enzymes and an increase in triacylglycerol lipase activity as lipid stores are depleted between stages 4 and 7 of metamorphosis (Fig. 4; Table1; Kao et al., 1997a,1997b).

Coincident with these changes in lipid metabolism are alterations in key metabolic hormones such as insulin and somatostatin. Serum insulin concentrations are elevated in stages 6 and 7 of metamorphosis relative to earlier stages and larval levels (Youson et al., 1994). Likewise the concentration of somatostatin in pancreatic-intestinal tissue homogenates begins to increase at stage 4 of metamorphosis and peak at stage 7 of metamorphosis (Elliott and Youson, 1991). Although these changes are surely in part related to the development and expansion of the endocrine pancreatic tissues, the fact that this elevation coincides with the shift from the lipogenic phase to lipolytic phase cannot be overlooked. A role for these hormones in lipid metabolism during lamprey metamorphosis is substantiated by a series of experiments (Kao et al., 1998, 1999a). Intraperitoneal injections of somatostatin-14 (SST-14) elevated plasma fatty acid levels in larval and stage 6 metamorphosing *P. marinus* (Kao et al., 1998) and injections of insulin and alloxan (a selective pancreatic β-cell toxin) resulted in decreases and increases, respectively, of plasma fatty acid concentrations (Kao et al., 1999a).

Changes in the lipogenic and lipolytic enzymes fluctuated appropriately with the increases and decreases in plasma fatty acid levels. For instance, elevations in fatty acids following SST-14 injection coincided with increases in TG lipase activity and decreases in DAGT activity in both larval kidney and larval liver (Kao et al., 1998). At stage 6 of metamorphosis a small, non-significant increase in TG lipase activity was detected but DAGT activity in muscle was significantly suppressed (Kao et al., 1998). Intraperitoneal injections of insulin suppressed TG lipase activity in most larval tissues as well as stage 6 liver, and elevated ACC and DGAT activity levels in some tissues of larvae and stage 6 animals.

In addition to directly modulating lipid metabolism during the lamprey life cycle, insulin and somatostatin also appear to modulate lipid metabolism in concert with TH. The lipogenic and lipolytic patterns associated with $KClO_4$-induced metamorphosis closely approximate those observed in spontaneous metamorphosis (Kao et al., 1999b). Moreover, these metabolic effects can be reversed following treatment with exogenous TH which upregulate a suite of lipogenic processes. One possible theory is that the high TH levels late in the larval phase are essential to drive the lipogenic processes and fat accumulation necessary for metamorphosis and that these processes play a key role in preparing the animal for metamorphosis long before any external signs of metamorphosis are realized.

The lipogenic phase, which occurs prior to and in the first stages of metamorphosis, is required to ensure sufficient energy reserves for the long and energetically demanding non-trophic period of metamorphosis. However, it is possible that this accumulation of lipids serves a secondary purpose in lamprey metamorphosis, either related to nutrition and metabolism or some other aspect of metamorphosis. In mammals it is well established that leptin is involved in the regulation of food consumption and body weight but can also influence reproduction, growth, the stress response, and thyroid function (Harvey and Ashford, 2003; Ahima and Osei, 2004, 2008). Leptin and its receptor are also expressed during embryonic development and have been shown to correlate with birth weight and fat accumulation prior to birth (Cetin, 2000; Lepercq et al., 2001). These findings prompted the search for a lamprey leptin and its potential function in lamprey metamorphosis. Using a polyclonal antibody against the C-terminal end of human leptin, four immunoreactive proteins were identified in lamprey tissue including: a 65 kDa protein in sera, 100 and 50 kDa proteins in muscle and fat column, and 50 and 16 kDa proteins in the nephric fold (Yaghoubian et al., 2001). Of particular interest is the presence of an immunoreactive protein, similar in size to the mammalian leptin (16 kDa), in the nephric fold early in metamorphosis (Yaghoubian et al., 2001). The nephric fold is the primary site of fat storage in larval and early metamorphic *P. marinus* (Youson et al., 1979) and, therefore, it is feasible this leptin-like immunoreactivity is reflecting important functions of a leptin-like protein in metamorphosis of this species. Although definitive identification of lamprey leptin is still pending, recent findings in other anamniotes provide good rationale to continue the search for lamprey leptin and the investigation into its possible function in lamprey metabolism and development, particularly in metamorphosis of species like *P. marinus*.

## 2.8 Internal Changes During Metamorphosis

Youson (1980) provided a comprehensive summary of the anatomical and physiological changes that occur during lamprey metamorphosis. This review illustrated both the magnitude of the changes and the fact that lamprey metamorphosis has been of fascination to biologists for nearly two centuries. There is barely an organ or organ system that does not show some alteration. These changes are either a complete regression or loss of a larval structure, a transformation of a larval structure to the adult state, redifferentiation of a larval structure or development of an adult structure from anlagen that resided in the larva and waited for the metamorphic stimulus. Needless to say, these anatomical changes were necessitated to permit, in adult life, a new life style, perhaps in a new environment. The changes to be described are mostly related to a parasitic and suctorial feeding, but also to potential for saltwater acclimation, in adults. However, the anatomical changes during metamorphosis and some physiological changes that appear in species with a parasitic adult life history also occur in those with a non-parasitic adult life history. Potter et al. (1982) feel these common features of development in lamprey metamorphosis are a reflection of the origins of non-parasitic species from a parasitic ancestor. As mentioned above, internal anatomical and physiological changes during lamprey metamorphosis have had a long history of study and the reader is referred to Youson (1980) for a historical review to the date of that publication. The present report, given space restrictions, will focus on highlighting and updating only a few of the events that surround the anatomical and physiological changes.

### 2.8.1 Some selected anatomical and physiological changes

#### 2.8.1.1 Skeletal system

Lampreys have an extensive cartilaginous endoskeleton most of which is confined to the cranial and gill regions. Given the differences in their mode of life, including feeding and swimming behaviours and habitat, between larval and adult lampreys, it might be expected that there would be skeletal differences. As can be seen in Fig. 9 (from Armstrong et al., 1987) the cartilages of the head and branchial regions are much more complex in adults and new cartilages must appear between these two intervals of the life cycle.

There have been numerous studies on the nature of this skeleton in the past three decades. Fine structural studies indicated that the fibrous component of the extracellular matrix (ECM) of annular, piston, cranial, and dorsal plate cartilages was not collagen but a network of randomly arranged and branched fibrils, 15–40 nm in diameter (Wright and Youson, 1983). However, there are some fundamental differences in the ultrastructure of nasal, branchial and pericardial cartilages compared to trabecular, piston, and annular cartilages with respect to the composition and extent of the ECM and the degree of cellular composition (Wright et al., 1988). Although all the lamprey cartilage matrices stained with Verhoeff's elastic stain, other biochemical features and the amino acid composition of the ECM proteins, suggested that the major protein of the ECM is similar to, but not identical to elastin and it was called lamprin (Wright et al., 1983). Subsequently, three lamprin cDNAs were isolated and their derived protein sequences described (Robson et al., 1993). The structure of the lamprin protein variants showed randomly repeated sequences that are similar to sequences in mammalian and avian elastin and some invertebrate structural proteins, such as spider dragline silk (Robson et al., 1993). Repeated specific hydrophobic domains within the lamprin protein may be responsible for the elastic staining. There are multiple genes for lamprin in the two species that have been studied (Robson et al., 2000). The primary difference between ECMs of the two groups of cartilages listed above is that nasal, branchial, and pericardial cartilages also contain elastin-like fibers and elastin-immunoreactive material in a portion of their ECM (Wright et al., 1988). The addition of elastin-like material is consistent with the need for flexibility and elasticity to cartilages that surround or support structures that undergo periodic tensile stresses such as the heart (pericardial) and the gill (branchial). Furthermore, molecular analyses have shown that the major matrix proteins of branchial and pericardial cartilages do not contain lamprin, but a related protein containing hydroxyproline (Robson et al., 1997).

A unique skeletal structure, called mucocartilage, is present in larval lampreys (Wright and Youson, 1982) with only the larval neurocranium and branchial arches composed of cartilage (Fig. 9). The larval cartilages (except mucocartilage) contain lamprin but the branchial cartilages have fewer fibrils and more extracellular proteoglycan, supposedly giving the "springiness" attributed to adult branchial cartilage (Wright et al., 1988). The mucocartilage in larva is avascular and consists of a proteoglycan aggregate, microfibrils, and a few fibroblasts. Mucocartilage does not stain for elastin or have other biochemical properties shown for lamprin (Wright and Youson, 1982). However, there seems to be differential expression of the

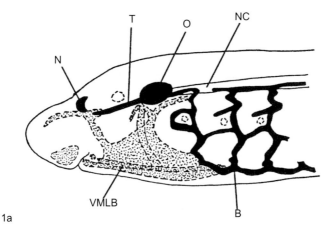

1a

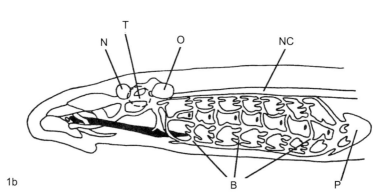

1b

**Figure 9.** Diagrammatic representations of the elements of the skeleton of the larval (upper figure) and adult lamprey. In the larva the cartilaginous structures are shaded black. The neurocranium consists of the nasal capsule (N), trabeculae (T), and otic capsule (O). The branchial cartilages (B) compose the branchial arches (basket). A special connective tissue, mucocartilage, appears stippled. A region of mucocartilage, the ventromedial longitudinal bar (VMLB), develops into the piston cartilage (see adult below). The same neurocranial cartilages are present in the adult but there are more cartilages supporting the suctorial disc. The piston cartilage (seen in black) develops from the larval (VMLB), the branchial cartilages are more elaborate, and there is a new pericardial cartilage (P). NC, notochord. (From Armstrong et al., 1987).

genes for lamprin during metamorphosis (Robson, 1998), for certain areas occupied by mucocartilage in larva become replaced by lamprin-containing cartilage (Armstrong et al., 1987). Thus, mucocartilage is not a true cartilage but a form of loose, embryonic connective tissue that, once metamorphosis is initiated, demonstrates its embryonic nature, by dedifferentiating into a mesenchyme-like tissue that redifferentiates into chondroblasts that lay down the lamprin-containing ECM. This process has been described in detail in the transformation of mucocartilage in the ventro-medial longitudinal bar (VMLB) of larva *P. marinus* into the piston cartilage of adults (Armstrong et al., 1987). Dedifferentiation of mucocartilage occurs during stages 1 and 2 of metamorphosis and is typical vertebrate mesenchyme by the end of the latter stage. The process involves migration of unspecialized peripheral cells into the presumptive site of cartilage production and degradation of mucocartilage ECM by these cells. Stage 3 is characterized by the development of a blastema of precartilage cells in the core of the VMLB with much mitosis and staining of the ECM suggesting secretion of sulfated proteoglycan. Secretion of the lamprin component into the ECM seems to coincide with the completion of differentiation of chondroblasts in stage 4 as well as both increased cellular degeneration and proliferation. As the piston cartilage matures over the final three stages of metamorphosis, there is a decline in cell proliferation and degeneration with the production of an extremely cellular cartilage surrounded by an increasingly dense ECM. The mechanism of development of the piston cartilage during lamprey metamorphosis follows a similar pattern to that seen in hyaline cartilage of higher vertebrates. Since the tools for specific identification of expression of the various isoforms of lamprin are now available (Robson et al., 1993; Robson et al., 2000), it would be of value to study this system of development at the molecular level.

### 2.8.1.2 Excretory system

Past reviews have provided a detailed description of the many changes that occur in the kidneys during metamorphosis (Youson, 1980; Youson, 1981b; Youson, 1985). In summary, the paired larval opisthonephroi that are located in the anterior region of the coelomic cavity undergo a complete regression and are replaced by more posterior adult kidneys that develop from undifferentiated nephrogenic tissue. A cord of nephrogenic tissue appeared during embryogenesis and consists of two parallel rows of cell condensations attached to the peritoneum at their proximal end (Fig. 10); the cord runs parallel to the archinephric duct towards the cloaca (Ooi and Youson, 1976,1977; Youson and Ooi, 1979). This tissue, an anlage, has waited for the stimuli of metamorphosis to begin differentation. Since the above descriptions, ultrastructural studies have provided the morphological details of the development and differentiation of the renal tubules from the rudimentary nephron units through the seven stages of metamorphosis in *P. marinus* (Youson, 1984). This model for renal tubulogenesis provides an excellent opportunity to study features of the morphogenesis of microvilli, the endocytotic apparatus, mitochondria, and a smooth tubular network, but is particularly valuable for studies of basal body replication and ciliogenesis

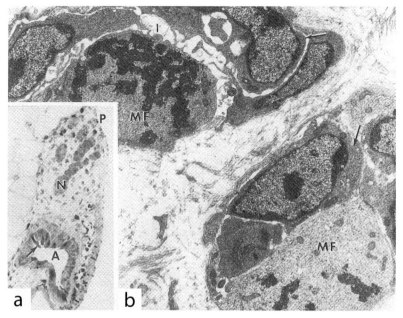

**Figure 10.** (1) Light micrograph of a portion of the nephric fold in a stage 1 metamorphosing *P. marinus* showing rod-shaped clusters of neprhogenic cells (N) extending from their proximal point of attachment with the peritoneum (P) towards the archinephric duct (A). X200. b. Electron micrograph of nephrogenic cells at stage 1 separated by wide intercellular spaces (I). They possess only a small amount of cytoplasm (arrows) and many are in mitosis (MF). X 6,000. (From Youson, 1984).

(Youson, 1982). A pair of pronephric kidneys persist throughout the life cycle of lampreys in various forms but the trend is for gradual regression in structure and function of renal activity throughout larval life (Ellis and Youson, 1990). The changes in the pronephroi during the seven stages of metamorphosis have been described and relate to the fact that they become encased within the pericardium with the development of the adult pericardial cartilage (Ellis, 1993).

## 2.8.1.3 Digestive system

Past reviews of the digestive system during metamorphosis (Youson, 1980) and throughout the life cycle (Youson, 1981a, 1981c, 1985) have emphasized the complex set of changes that occur in the alimentary canal and liver. There have been further studies of these organs since these earlier reviews and we will provide a short summary of the findings.

## 2.8.1.4 Alimentary canal

The larval oesophagus leads from the posterior end of the pharynx and terminates near the caudal tip of the liver as it unites with the anterior intestine (Fig. 11); the latter eventually extends into a posterior intestine and a hindgut that empties into a cloaca. A new oesophagus develops during metamorphosis from the epithelium covering the dorsal ridge of the pharynx, but this oesophagus is totally independent of the pharynx (Fig. 11). This arrangement is necessitated by the tidal ventilation (pumping in and out of the branchiopores) that is required for the suctorial feeding habit. The features of development of this new oesophagus have received considerable attention in early literature (for review see, Youson, 1980, 1981a). Detailed light and electron microscopic studies have taken place in *P. marinus* during the seven stages of metamorphosis of the developing adult oesophagus and the changing larval oesophagus (Elliott, 1989). Evidence is provided that the most anterior region of the larval oesophagus plays some role in the formation of the adult oesophagus (Fig. 11). A summary and description of the formation of the new oesophagus and the alteration of the junction of the oesophagus and anterior intestine is provided in Fig. 12. The remainder of the larval alimentary canal undergoes a complete transformation that increases the surface area for absorption by the development of longitudinal

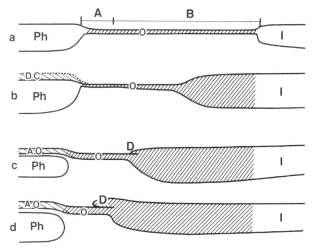

**Figure 11.** Diagrammatic representation showing the forward movement of the oesophagus-intestinal junction and the development of the adult oesophagus (AO) from a dorsal cord (DC) of tissue during metamorphosis in *P. marinus*. a. larva; b. metamorphosing stage 3; c. metamorphosing stage 4; d. juvenile. Whereas the larval oesophagus (O) leads from the caudal end of the pharynx (Ph), the adult oesophagus is independent of the pharynx. A and B denotes regions of the larval oesophagus and how it contributes to development of the adult oesophagus and intestine (I) and a small diverticulum (D). (From Elliott, 1989).

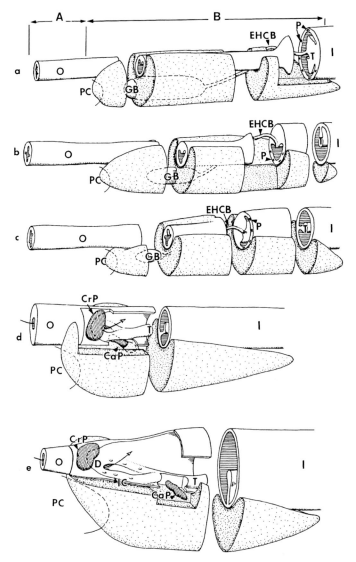

**Figure 12.** Diagrammatic representation of the larval oesophagus (O), and anterior intestine (I) during the metamorphosis of *P. marinus*. The larval oesophagus is separated into A and B regions in the larva (a) to show the contributions of this structure to the changing alimentary canal at stage 2 (b), stage 3 (c), stage 4 (d), and in the adult (e). The diagrams also show the larval liver with intrahepatic gall bladder (GB), the extrahepatic common bile duct (EHCB), and the location of the pancreatic islets (P). The cranial principal islet (CrP) appears and moves towards the developing pericardial cavity (PC) with the forward movement of the larval oesophagus, whereas the caudal principal islet (CaP) appears at the site where the EHCB entered the larval oesophagus-intestinal junction. Note the disappearance of the gall bladder and EHCB and the development of a diverticulum (D). (From Elliott, 1989).

mucosal folds (Youson and Connelly, 1978). The features of degeneration, differentiation, and proliferation in the transformation of the epithelium and submucosal tissues over seven stages of metamorphosis are provided in a light and electron microscopic study (Youson and Horbert, 1982). The intestinal epithelium of adult lampreys develops from surviving cells of the larval primary epithelium; this is in contrast to the situation in amphibians where the adult epithelium (secondary epithelium) develops from nests of undifferentiated cells in the larval epithelium.

### 2.8.1.5 Endocrine pancreas

The larval lamprey has insulin-containing cells arranged as follicles within the submucosal tissue near the junction of the oesophagus and the intestine. (Fig. 12). There is some variability of the distribution of these follicles between northern and southern hemisphere lampreys relative to their intestinal diverticula (Youson and Al Mahrouki, 1999; Youson, 2000, 2007). As noted above, the extrahepatic common bile duct in larva drains the liver of bile products and empties into the intestine at its junction with the oesophagus in northern hemisphere species but into the cephalic end of the left diverticulum in southern hemisphere species (Fig. 13). Adults of the southern hemisphere species have a cranial principal islet near the cardiac region that is believe to develop from larval follicles that migrate and proliferate during metamorphosis (Hilliard et al., 1985). In contrast, northern hemisphere species have both a cranial and caudal principal islet. Although some larval follicles may contribute to the development

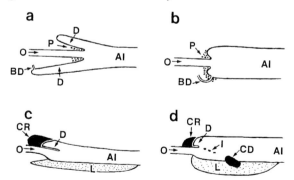

**Figure 13.** Diagrammatic representation of the distribution of pancreatic tissue in islets (P) of a larva (a) and an adult (c) of *G. australis* and a larva (b) and an adult (d) of *P. marinus*. Note the difference in position of the bile duct (BD) and the size of diverticula (D) in larva of the two species and that *G. australis* adult has only a single, large cranial pincipal islet (CR). The *P. marinus* adult has both a CR and a caudal principal islet (CD) connected by a stand of intermediate (I) pancreatic islets. The CD arose during metamorphosis from the epithelium of the BD, but no such contribution occurred during metamorphosis in *G. australis* (Modified from Youson and Elliott, 1989).

of the cranial principal islet in *P. marinus*, a major contribution is through budding and profileration of cell clusters from the epithelium of the intestinal diverticulum (Elliott and Youson, 1993a, 1993b). The caudal principal islet in *P. marinus*, and likely most other northern hemisphere species (Youson et al., 1988; Youson and Elliott, 1989), develops from a transformation/dedifferentiation and proliferation of the extrahepatic, and some intrahepatic, epithelial cells of the larval common bile duct (Fig. 14). Insulin (B) cells appear first and then eventually are accompanied by somatostatin-containing (D) cells (Elliott and Youson, 1987,1993b). The development of the caudal principal islet is highly synchronized and any deviation can alter the normal distribution of pancreatic islets that make up this principal islet (Youson and Cheung, 1990). The absence of a caudal principal islet in southern hemisphere species is due to the complete regression of the larval bile duct epithelium, seemingly because it enters the cephalic end of an intestinal diverticulum (Hilliard et al., 1985; Youson, 2000).

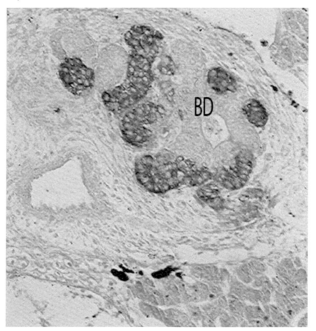

**Figure 14.** Light micrograph showing the caudal principal islet developing from the bile duct epithelium (BD) during stage 4 of metamorphosis in *P. marinus*. Immunohistochemical staining shows positive brown staining of islet tissue with an insulin antiserum. X270.

There have been only a few studies on the concentration of either insulin or somatostatin during lamprey metamorphosis (for review see Youson, 2000). Elliott and Youson (1991) were able to correlate, through a heterologous radioimmunoassy (RIA), the increased tissue levels of somatostatin in intestinal-pancreatic extracts with increasing development of the cranial and caudal principal islets in *P. marinus*. A homologous RIA was used to show significantly higher insulin concentrations appear over larval levels by stage 6 of metamorphosis in *P. marinus* (Youson et al., 1994). These latter data indicate that the developing principal islets release their newly synthesized hormone into the blood stream. Although the relevance of these hormones to metamorphosis requires further study, there is some evidence (see above discussion under *Lipogenic and Lipolytic Hormones*) that they may be involved in intermediary metabolism in an animal that is still relying on stored lipid reserves during this non-trophic phase of the life cycle (Kao et al., 1998, 1999a).

### 2.8.1.6 Liver

The metamorphic events within the liver are probably among the most fascinating and complex within the entire developmental process (Youson, 1981c, 1985). The larval liver has bile ducts and an intrahepatic gall bladder and, in holarctic lamprey species, an extrahepatic bile duct empties into the anterior intestine at its junction with the oesophagus (Fig. 12). This entire biliary tree and the bile canaliculi between the liver hepatocytes completely disappear during metamorphosis in all lamprey species. The complete process of degeneration of the bile ducts, bile canaliculi and gall bladder have been described and the consequences of these events on liver metabolism and bile product excretion reported for *P. marinus*. The reader is referred to the most recent review on this subject (Youson, 1993). It is noteworthy, and described above under endocrine pancreas, that detailed autoradiographic, fine structural and immunohistochemical observations have shown that in some species, such as *P. marinus*, the epithelium of the extrahepatic and intrahepatic common bile duct undergoes a dedifferentiation to produce cells of the caudal principal islet of the endocrine pancreas (Elliott and Youson, 1993a, 1993b). There have been recent studies to show that apoptosis is an important feature of the loss of bile ducts, lamprey biliary atresia, during metamorphosis (Boomer et al., 2010; Morii et al., 2010). This topic is showing renewed interest in recent years because the lamprey serves as a programmed model system, for its biliary atresia has features similar to the events of human biliary atresia that effects human infants (Youson, 1993).

Another change that occurs during metamorphosis of lampreys is the deposition of iron in various tissues and the liver is one of these primary sites (Fig. 15). There is a progressive deposition of iron within the cytoplasm of hepatocytes such that by the time metamorphosis is completed and into the immediately postmetamorphic juvenile, the livers are iron loaded (Macey et al., 1982a; Youson et al., 1983a,1983b). For a further summary of this topic of liver iron see Youson (1993). Further discussion of the change in iron metabolism will appear below (*Iron Metabolism and Blood*).

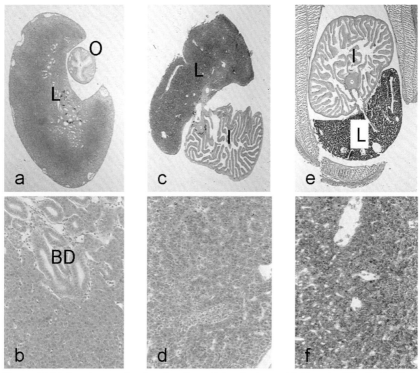

**Figure 15.** Light micrographs showing the staining for iron with Prussian Blue in the liver (L) of larva (a, b), stage 5 of metamorphosis (c, d) and in a juvenile (e, f) of *P. marinus*. Note the progressive increase of liver iron (blue stain) during metamorphosis into the iron-loaded state of the adult and the absence of this metal in the oesophagus (O) and intestine (I). There are bile ducts (BD) in the larva liver but none in the livers of metamorphosing animals or the adult. (a, c, and e. X25; b, d, and f. X250).

*Color image of this figure appears in the color plate section at the end of the book.*

## 2.8.1.7 Respiratory system

It has been referenced above that there is a different mode of ventilation in larval and adult lampreys that is related to their different modes of feeding and/or the use of adult oral disc for attachment. The pharynx of larva serves as a conduit for materials entering the oral aperture on their way to the oesophagus, whereas in adults the oral aperture is connected to an oesophagus lying above the pharynx and uniting with the intestine near the cardiac region (Fig. 11). This separate oesophagus permits for tidal ventilation in and out of the branchiopores if the suctorial disc is attached to some surface. The features of expansion of the cartilaginous branchial basket during metamorphosis has been referenced above. Transformation of the gill epithelium also takes place during metamorphosis with the appearance of adult-type, mitochondria-rich (formerly called chloride) cells that become fully mature in morphology by stage 6 or 7 (Peek and Youson, 1979a,1979b). There have numerous recent studies on the appearance of lamprey gill epithelium in larva and adults (for example, see Bartels et al., 2009) but one study with direct attention on the gill epithelium in metamorphosis is important to highlight (Reis-Santos et al., 2008). This latter study measured the expression of key ion-transporting proteins ($Na^+/K^+$-ATPase, vacuolar [V]-type $H^+$-ATPase, and carbonic anhydrase [CA]) in the examination of osmoregulatory differences between larval and metamorphosing *P. marinus*. Immunoreactivity for these proteins was also studied. As might be expected, larva did not survive even 10‰ sea water but survival in 25-35‰ increased as metamorphosis progresses into later stages. Branchial $Na^+/K^+$-ATPase was 10X higher in metamorphosing animals over larvae, irrespective of salinity, and corresponded to intense immunoreactivity to antiserum to this protein in cells in the interlamellar spaces during metamorphosis, i.e., in developing mitochondria-rich cells. Conversely, $H^+$-ATPase had a negative correlation with salinity and the immunohistochemistry indicated that this protein is not localized in the $Na^+/K^+$-ATPase-immunoreactive cells. There seems to be a correlation between the increase in salinity tolerance of metamorphosing lampreys with the expression of $Na^+/K^+$-ATPase but $H^+$-ATPase is downregulated by changes in salinity (Reis-Santos et al., 2008).

### 2.8.1.8 Blood and iron metabolism

Over the past twenty years there has been a flurry of interest in iron metabolism in lampreys due to the deposits of this metal in many tissues and organs at levels that would be considered intolerable in other vertebrates (Morgan, 1980). Although iron concentrations in the blood and tissues of lampreys vary between species and even may vary within different populations of the same species (Holmes and Youson, 1996), there seems to be a shift in the distribution of iron during metamorphosis (Fig. 15). It is beyond the scope of the present chapter to deal with all of the details of iron metabolism that have been studied to date but it is worthwhile to provide an example of changes in blood parameters that might account for the redistribution. Table 2 summarizes the results of a study where measurements were made of serum iron concentrations ($\mu$g/100 ml), haemoglobin concentrations (g/100 ml) and erythrocyte numbers ($10^6$ cells/mm$^3$) during the seven stages of metamorphosis of *P. marinus* for comparison with these parameters in larva and adults (Youson et al.,

**Table 2.** Erythrocyte numbers and haemoglobin and serum iron concentrations of lampreys, *Petromyzon marinus*, at various intervals of the life cycle. From Youson et al., 1987.

| Interval | N* | Erythrocyte number ($10^6$ cells/mm$^3$) | N | Haemoglobin (g/100ml) | N | Serum iron ($\mu$g/100ml) |
|---|---|---|---|---|---|---|
| Ammocoete | 18* | 0.830 ± 0.099† (0.360–1.210)‡ | 18 | 8.16 ± 0.54† (6.32–10.50)‡ | 37 | 5118.9 ± 1398.2† (681–15365)‡ |
| Metamorphosing stage | | | | | | |
| 1 | | | | | 9 | 5103.4 ± 2232.8 (1495–12518) |
| 2 | 9 | 0.747 ± 0.122 (0.504–1.050) | 10 | 6.79 ± 0.54 (5.33–8.13) | 7 | 3029.8 ± 1389.6 (1096–6175) |
| 3 | 7 | 1.022 ± 0.139 (0.688–1.200) | 6 | 7.36 ± 0.37 (6.88–8.13) | 17 | 8123.7 ± 1910.8 (1420–16733) |
| 4 | 10 | 0.922 ± 0.148 (0.650–1.250) | 9 | 7.12 ± 0.84 (4.80–9.00) | 12 | 4487.7 ± 1110 (1759–7465) |
| 5 | | | | | 10 | 2739.0 ± 1067.8 (1392–7078) |
| 6 | 10 | 0.625 ± 0.057 (0.510–0.725) | 10 | 5.2 ± 0.57 (4.30–7.40) | 8 | 1966.3 ± 1014 (308–4667) |
| 7 | | | | | 9 | 881 ± 624 (243–3294) |
| Juvenile adult | 12 | 0.720 ± 0.126 (0.200–0.954) | 12 | 4.58 ± 0.79 (2.66–6.99) | 32 | 3769 ± 545.8 (1095–8159) |
| Upstream migrant (F)§ | 18 | 0.888 ± 0.120 (0.54–1.350) | 19 | 10.22 ± 0.59 (8.00–12.50) | 49 | 3288.1 ± 921 (94.8–14164) |
| Upstream migrant (M)§ | 26 | 0.893 ± 0.099 (0.380–1.500) | 26 | 11.04 ± 0.90 (6.00–15.00) | 51 | 1962.3 ± 537 (298–8587) |

*Number of animals.
†Mean ± 2 SE.
‡Range.
§F = female, M = male.

1987a). The highest number of erythrocytes was seen at stages 3 and 4 but by stage 6 the number had dropped significantly to the lowest level seen in the entire life cycle. Erythrocyte numbers in larval *P. marinus* are lower than several other species (Youson et al., 1987a), but this may be correlated to their lower haematocrit (Beamish and Potter, 1972; Potter and Beamish, 1978). The increase in erythrocyte numbers in stages 3 and 4 may be related to changes in sites of haemopoiesis and sites of normal erythrophagocytosis associated with changes in organs such as alimentary canal, kidneys, and the liver. Declines in erythrocyte numbers by stage 6 correspond with the appearance of massive erythrophagocytosis in the liver by stages 5 and 6 (Youson et al., 1987b). Haemoglobin concentration also reached a significant low level by stage 6 and remained unchanged into early juvenile life. The time of decline may correlate with the time of switch from larval to adult haemoglobin (Potter and Brown, 1975). Stages 6 and 7 are suggested as the time of change from larval to adult hemoglobins in *G. australis* (Macey and Potter, 1981). Serum iron concentrations that were 5119 µg/100 ml in larvae declined abruptly to 3030 µg/100 ml by stage 2, 1966 µg/100 ml by stage 6, and only 881 µg/100 ml at stage 7. Adults show a rise in serum iron concentrations but never to the levels reached in larva. These changes in iron concentrations during the life cycle in *P. marinus* mostly parallel those described in *G. australis* but the larval values do not reach the 19,760 µg/100 ml reported for those in the latter species (Macey et al., 1982b) or the 26,773 µg/100 ml for *M. mordax* (Macey et al., 1985). These larval values, and those reported for *L. planeri* of 10,200 µg/100 ml (Macey and Potter, 1986), are no doubt a consequence of the high iron-binding capacity of ferritin, the major iron-binding protein in larval plasma (Macey et al., 1982b, 1985). The decline in serum iron concentration in *P. marinus* metamorphosis also was reported for metamorphosis in *G. australis* (Smalley et al., 1986) and it corresponded well with the overall pattern of change in the distribution of body iron (Sargent and Youson, 1986). For example, the lowest level of serum iron at stage 7 is also the time in the life cycle when liver iron concentrations are highest (Fig. 15). This is also the time in the life cycle when high iron-binding capacity of ferritin is replaced by a lower capacity transferrin-like protein (Macey et al., 1982b,1985) and when there is extensive erythrophagocytosis (Percy and Potter, 1981; Youson et al., 1987b).

### 2.8.1.9 Adenohypophysis: POM and POC

The vertebrate adenohypophysis produces a variety of bioactive peptide hormones which can be grouped into one of three families: the proopiomelanocortin (POMC) family, the glycoprotein hormone (GpH) family and the growth hormone family. Representatives of each family have been identified in lampreys, but an in-depth discussion is beyond the scope of this chapter; we will elaborate only on the POMC family of peptides and the spatial and temporal changes that occur throughout metamorphosis. For a detailed discussion on the GpH and growth hormone families the reader is referred to other recent reviews (Kawauchi and Sower, 2006; Sower et al., 2009).

In gnathostomes all members of the POMC family of peptides are derived from the preprohormone, POMC, which is encoded for by a single gene. POMC is post-translationally cleaved in a cell-specific manner to produce the bioactive hormones adrenocorticotropin (ACTH), melanotropin (MSH; γ-MSH and β-MSH), endorphin (β-END), and lipotropin (LPH) (Smith and Funder, 1988). Unlike all other vertebrates, members of the lamprey POMC family are encoded for by two distinct genes: proopiocortin (POC) and proopiomelanotropin (POM) (Heinig et al., 1995; Takahashi et al., 1995; Youson et al., 2006). POC encodes for ACTH, one MSH, β-END and nasohypophysial factor (NHF), and POM encodes for MSH-A, MSH-B and a different β-END (Takahashi et al., 1995). The POMC gene is expressed throughout the adenohypophysis as well as in the hypothalamus and other brain regions in gnathostomes, but POM and POC expression in lampreys is restricted to the pars intermedia (PI) and the pars distalis (PD), respectively (Ficele et al., 1998). Likewise, immunohistochemical data for ACTH and MSH-like peptides indicates that they are restricted to the PD and PI, respectively (Nozaki et al., 1995, 2008). Thus, the spatial distribution of POM and POC in the lamprey adenohypophysis is consistent with the localization of POMC-derived peptides in gnathostomes where MSH and ACTH peptides are restricted to the melanotropes of the PI and the corticotropes of the PD, respectively (see Sower, 1998).

POM and POC expression is regulated both temporally and spatially in the adenohypophysis of *P. marinus*. Northern blotting data indicates that the mRNA levels of both transcripts are elevated following the completion of metamorphosis, and in the parasitic juvenile and prespawning adult (Heinig et al., 1999; Youson et al., 2006). Transcript levels in larvae are lower, but tend to increase later in metamorphosis. In the non-parasitic *L. appendix*, a similar temporal expression pattern was observed with the increase in expression during metamorphosis perhaps beginning slightly earlier than in *P. marinus* (Youson et al., 2006). Youson et al. (2006) suggest that this difference in timing may be related to the earlier onset of gonadal maturation in *L. appendix*.

Detailed spatial and temporal analysis with *in situ* hybridization shows uniform POC expression in the rostral pars distalis (RPD) in larvae, throughout metamorphosis and into the spawning adult (Ficele et al., 1998). At stage 5 of metamorphosis, POC-expressing cells are also scattered throughout the caudal pars distalis (CPD) with expression becoming restricted to dorsally located cells of the CPD in prespawning adults (Fig. 16). Expression of POC in the RPD, as determined by quantitative analysis of signal density and volume, is low in larvae, increases gradually throughout metamorphosis, peaks in post-metamorphic individuals, and remains elevated into adulthood (Ficele et al., 1998). The requirement of an intact PD for metamorphosis was demonstrated for *G. australis* whereby removal

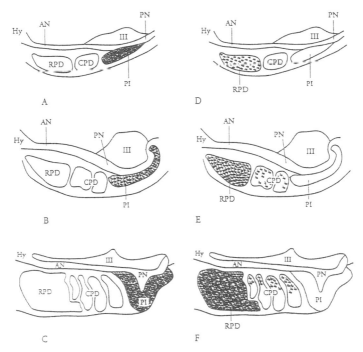

**Figure 16.** Diagrammatic representation of proopiomelanotropin (POM; A, B, C) and proopiocortin (POC; D, E, F) expression in the adenohypophysis of larval (A, D), stage 5 (B, E) and upstream migrant (C, F) *Petromyzon marinus*. Spatial and temporal expression patterns were determined using *in situ* hybridization and riboprobes specific to POM and POC mRNAs, respectively. Shading is indicative of relative expression levels. POM expression is restricted to the pars intermedia (PI) throughout development. POC is expressed in the in the rostal pars distalis (RPD) throughout development; in addition, POC expression in scattered throughout the caudal pars distalis (CPD) during metamorphosis and in the dorsal aspect of the CPD in the upstream migrant. Anterior neurohypophysis, AN; hypothalamus, Hy; posterior neurophypophysis, PN; third ventricle of the brain, III. (From Ficele et al., 1998).

of the RPD prevented metamorphosis and removal of the CPD resulted in metamorphic stasis at stage 3 (Joss, 1985). The increase in POC expression in the RPD during metamorphosis and its appearance in the CPD at stage 5 further support these findings, as does the observation that ACTH immunoreactivity increases in these cells at these stages of development (Nozaki et al., 2008). Collectively, these findings suggest that ACTH, a major POC product, may be important in lamprey metamorphosis as has been suggested for amphibians.

POM expression was detected in most cells of the PI throughout all life cycle stages (Fig. 16). POM signal density in immediately premetamorphic animals was significantly greater than in all stages examined except spawning adults. Signal density decreased during stages 1–5 of metamorphosis, then gradually increased to high levels in spawning adults (Ficele et al., 1998). Although there is a decrease in signal density during metamorphosis, the total number of POM-expressing cells as determined by volumetric measurements indicates an overall increase in POM expression through metamorphosis. The observed decrease in density is related to an increase in the size of the PI (Ficele et al., 1998). These latter findings are corroborated by northern blotting data (Heinig et al., 1999; Youson et al., 2006). The fact that POM expression is maximal in immediately pre-metamorphic larvae suggests that POM is involved in preparing larvae for metamorphosis (Ficele et al., 1998). Similarly, the elevated levels of POM late in metamorphosis coincide with, and likely function to elicit, the changes in pigmentation observed at stages 5 and 6 of metamorphosis.

## Summary and Conclusions

Lampreys are among the few fishes that have adopted the developmental strategy of metamorphosis, a late stage of ontogeny that manifests itself in a juvenile with the potential for a different life style and habitat from the pre-metamorphic larvae (Youson, 1988; Youson, 2004). There seems to be no phylogenetic relationship among the fish groups that go through this type of indirect development but in most cases the larval interval allows for wide dispersal of the species (Youson, 2004). It has been speculated that this developmental strategy in all modern-day lampreys is likely derived rather than ancient with the first lamprey-like ancestors more larva-like and perhaps paedomorphic (Youson and Sower, 2001). The speculation was based on evidence from experiments and observations on the thyroid and reproductive axes on present-day lampreys representing both of the two adult life history strategies, i.e., parasitic and non-parasitic. The non-parasitic adult life history is believed to be the more recent strategy in lamprey evolution with extant non-parasitic species having arisen from a parasitic ancestor (Hardisty, 2006).

Despite the differences in adult life history between species, they all have a protracted larval life of variable duration that has both a growth and a metamorphic phase. For the most part, the features of metamorphosis are similar in all species permitting the usage of a universal seven stages. Furthermore, the initiation of the process in all species is characterized by a sharp decline is serum concentrations of the thyroid hormones (TH), 3, 5, 3'-triiodothyronine ($T_3$) and 3, 5, 3', 5'- tetraiodothyronine (thyroxine; $T_4$). In fact, metamorphosis can be induced in the larva of both adult life history types by subjecting the animals to specific goitrogens. This chapter provides an extensive review of the earliest and recent studies of the role of the thyroid axis in lamprey metamorphosis and discusses several avenues of research that need to be explored. One of these is the question of the existence of a thyroid-stimulating hormone in lampreys and the close relationship of the hormone to gonadotropin (Sower et al., 2009). This area of study is of significance for any interpretation of the interplay between the reproductive and thyroid axes that may have influenced the evolution of metamorphosis as a developmental strategy in lampreys (Youson and Sower, 2001; Youson, 2004). There is some evidence, from both laboratory and field observations, that there is certain degree of plasticity in the initiation of the onset of metamorphosis in non-parasitic species that may reflect their pathway of evolution form parasitic species (Youson, 2004).

There are likely many factors other than THs that are critical to lamprey metamorphosis (Fig. 4). One of these is undoubtedly water temperature and, in particular, the stimulus from a rise in temperature, rather than the magnitude of the rise. Neither photoperiod nor animal density seem to be directly responsible for triggering metamorphosis, although experiments with the photosensitive pineal organ seem to suggest some involvement of this organ. Physiological preparation for the non-trophic phase of metamorphosis has been clearly demonstrated for some species where lipogenesis and fat storage are characteristic of immediately pre-metamorphic larva and early stages of metamorphosis. Whether the sites of fat storage release a factor, or hormone, as a signal of preparation that stimulates a cascade of events affecting both the brain, thyroid and reproductive systems remains to be resolved.

That lamprey metamorphosis is a true vertebrate metamorphosis (Youson, 1988) is demonstrated by the magnitude of the internal and external changes in all species. This organism has proven to be a useful model system for studying developmental change in many systems and organs. These changes include development of a new oesophagus that is independent of the pharynx to permit tidal ventilation through the pharynx when the adults are attached by the suctorial disc. The pharyngeal endostyle of larva transforms into a thyroid gland with follicles. The remainder of the alimentary canal is modified for increased rates of absorption and ingestion of a high protein diet of host flesh or blood even if the post-metamorphic animals never feed, as in non-parasitic species.

The liver loses its bile ducts and intrahepatic gall bladder in a biliary atresia that mirrors some of the events of a human disorder of the same name. Adult principal islet tissue develops during metamorphosis with new endocrine cell types. A caudal principal islet in northern hemisphere species is a consequence of transformation of bile duct cells. The larval kidneys (opisthonephroi) undergo a complete regression and are replaced by a definitive pair of opisthonephroi that develop from anlagen that have resided within a nephrogenic cord behind the larval kidneys since embryogenesis. The cartilaginous head and branchial skeleton of larva is modified and expanded to accommodate the suctorial disc, the appearance of rasping tongue (piston), changes in respiration, and added protection for the pericardial cavity. The eye of the blind larval completes its development with enlargement and the appearance of a functioning retina and a prominent pupil and iris. There are many changes in blood parameters, such as the appearance of adult erythrocytes, adult hemoglobin, adult albumin, and a new iron-binding protein. Some of these changing blood parameters are reflected in re-distribution of iron such that some organs, like the liver, become iron loaded during metamorphosis and into adult life to levels that would be toxic in many vertebrates. Changes in the epithelium of the gills, intestines and the kidney tubules are in preparation for the potential for saltwater acclimation in adults of anadromous species. Theses latter changes, and many described above that are related to adult feeding, also occur in larval of non-parasitic species during their metamorphosis. The post-metamorphic adults of these species never feed or leave their freshwater natal stream, but commence immediately into sexual maturation. The fact that these changes still occur in non-parasitic species is a reflection of both the evolutionary relationship between the two adult life history types and also the interesting evolutionary history of metamorphosis as a development strategy among lampreys.

## Acknowledgements

The authors are most grateful for the generous contribution of Dr. John R. Gosche, University of Nevada, for the photographs of metamorphosing stages of *P. marinus* following induction. Research conducted in the laboratory of R.G.M. is supported by grants from the Natural Sciences and Engineering Research Council of Canada (NSERC; Discovery, and Research Tools and Instruments Grants) and the Canada Foundation for Innovation. J.H.Y. acknowledges the long-term research support of NSERC (Discovery and Strategic Grants) and its forerunner, and the very timely funding from the Great Lakes Fishery Commission, the Medical Research Council of Canada, and the Hospital for Sick Childrens Foundation ( Sick Kids Foundation, Toronto).

## References

Ahima, R.S. and S.Y. Osei. 2004. Leptin signaling. Physiology & Behavior 81: 223–241.
Ahima, R.S. and S.Y. Osei. 2008. Adipokines in obesity. Obesity and Metabolism 36: 182–197.
Armstrong, L.A., G.M. Wright and J.H. Youson. 1987. Transformation of mucocartilage to a definitive cartilage during metamorphosis in the sea lamprey, *Petromyzon marinus*. Journal of Morphology 194: 1–21.
Barrington, E.J.W. 1972. The endostyle and thyroid gland. In: M.W. Hardisty and I.C. Potter (Eds.). The Biology of Lampreys. *Volume 2*, Academic Press, Inc., New york, pp. 105–134.

Bartels, H., A. Schmiedl, J. Rosenbruch and I.C. Potter. 2009. Exposure of the gill epithelial cells of larval lampreys to an ion-deficient environment: a stereological study. Journal of Electron Microscopy 58: 253–260.

Beamish, F.W.H. 1980. Biology of the North American anadromous sea lamprey, *Petromyzon marinus*. Canadian Journal of Fisheries and Aquatic Sciences 37: 1924–1943.

Beamish, F.W.H. and L.S. Austin. 1985. Growth of the mountain brook lamprey *Ichthyomyzon greeleyi* Hubbs and Trautman. *Copeia* 4: 881–890.

Beamish, F.W.H. and T.E. Medland. 1988. Age-determination for lampreys. *Transactions of the American Fisheries Society* 117: 63–71.

Beamish, F.W.H. and I.C. Potter. 1972. Timing of changes in blood, morphology, and behavior of *Petromyzon marinus* during metamorphosis. *Journal of the Fisheries Research Board of Canada* 29: 1277–1282.

Beamish, F.W.H. and E.J. Thomas. 1984. Metamorphosis of the southern brook lamprey, *Ichthyomyzon gagei*. *Copeia* 2: 502–515.

Beamish, R.J. 1987. Evidence that parasitic and nonparasitic life-history types are produced by one population of lamprey. Canadian Journal of Fisheries and Aquatic Sciences 44: 1779–1782.

Bergstedt, R.A. and W.D. Swink. 1995. Seasonal growth and duration of the parasitic life stage of the landlocked sea lamprey (*Petromyzon marinus*). Canadian Journal of Fisheries and Aquatic Sciences 52: 1257–1264.

Bianco, A.C., D. Salvatore, B. Gereben, M.J. Berry and P.R. Larsen. 2002. Biochemistry, cellular and molecular biology, and physiological roles of the iodothyronine selenodeiodinases. Endocrine Reviews 23: 38–89.

Bird, D.J. and I.C. Potter. 1979a. Metamorphosis in the paired species of lampreys, *Lampetra fluviatilis* (L) and *Lampetra planeri* (Bloch). 1. Description of the timing and stages. Zoological Journal of the Linnean Society 65: 127–143.

Bird, D.J. and I.C. Potter. 1979b. Metamorphosis in the paired species of lampreys, *Lampetra fluviatilis* (L) and *Lampetra planeri* (Bloch). 2. Quantitative data for body proportions, weights, lengths and sex-ratios. Zoological Journal of the Linnean Society 65: 145–160.

Bird, D.J. and I.C. Potter. 1983. Changes in the fatty-acid composition of triacylglycerols and phospholipids during the life-cycle of the lamprey *Geotria australis* Gray. Comparative Biochemistry and Physiology B-Biochemistry & Molecular Biology 75: 31–41.

Boomer, L.A., S.A. Bellister, L.L. Stephenson, S.D. Hillyard, J.D. Khoury, J.H. Youson and J.R. Gosche. 2010. Cholangiocyte apoptosis is an early event during induced metamorphosis in the sea lamprey, *Petromyzon marinus* L. Journal of Pediatric Surgery 45: 114–120.

Brown, D.D. and L.Q. Cai. 2007. Amphibian metamorphosis. Developmental Biology 306: 20–33.

Cetin, I., P.S. Morpurgo, T. Radaelli, E. Taricco, D. Cortelazzi, M. Bellotti, G. Pardi and P. Beck-Peccoz. 2000. Fetal plasma leptin concentrations: Relationship with different intrauterine growth patterns from 19 weeks to term. Pediatric Research 48: 646–651.

Clements-merlini, M. 1962a. Altered Metabolism of I-131 by the endostyle and notochord of ammocoetes larvae. 2. Effects of treatment with thiourea or potassium thiocyanate. General and Comparative Endocrinology 2: 361–368.

Clements-merlini, M. 1962b. Metabolism of I-131 by the endostyle and notochord of ammocoetes under different conditions of temperature. General and Comparative Endocrinology 2: 240–248.

Cole, W.C. and J.H. Youson. 1981. The effect of pinealectomy, continuous light, and continuous darkness on metamorphosis of anadromous sea lampreys, *Petromyzon marinus* L. Journal of Experimental Zoology 218: 397–404.

Cole, W.C. and J.H. Youson. 1982. Morphology of the pineal complex of the anadromous sea lamprey, *Petromyzon marinus* L. American Journal of Anatomy 165: 131–163.

Denver, R.J., K.A. Glennemeier and G.C. Boorse. 2002. Endocrinology of complex life cycles: amphibians. In: D. Pfaff, A. Arnold, A. Etgen, S. Fahrbach, and R. Rubin (Eds.). *Hormones, Brain and Behavior*, Elsevier Science pp. 469–513.

Docker, M.F., S.A. Sower, J.H. Youson and F.W.H. Beamish. 2003. Future sea lamprey control through regulation of metamorphosis and reproduction: a report from the SLIS II New Science and Control workgroup. Journal of Great Lakes Research 29 (Supplement 1): 801–807.

Eales, J.G. 1997. Iodine metabolism and thyroid-related functions in organisms lacking thyroid follicles: Are thyroid hormones also vitamins? Proceedings of the Society for Experimental Biology and Medicine 214: 302–317.

Eales, J.G. and S.B. Brown. 1993. Measurement and regulation of thyroidal status in teleost fish. Reviews in Fish Biology and Fisheries 3: 299–347.

Eales, J.G., J.A. Holmes, J.M. McLeese and J.H. Youson. 1997. Thyroid hormone deiodination in various tissues of larval and upstream-migrant sea lampreys, *Petromyzon marinus*. General and Comparative Endocrinology 106: 202–210.

Eales, J.G., J.M. McLeese, J.A. Holmes and J.H. Youson. 2000. Changes in intestinal and hepatic thyroid hormone deiodination during spontaneous metamorphosis of the sea lamprey, *Petromyzon marinus*. Journal of Experimental Zoology 286: 305–312.

Eddy, J.M.P. 1969. Metamorphosis and pineal complex in brook lamprey, *Lampetra planeri*. Journal of Endocrinology 44: 451–452.

Elliott, W. M. 1989. Development of the endocrine pancreas and related parts of the alimentary canal during metamorphosis in the lamprey, *Petromyzon marinus* L. Ph.D. Thesis, University of Toronto, pp. 1–357.

Elliott, W.M. and J.H. Youson. 1987. Immunohistochemical observations of the endocrine pancreas during metamorphosis of the sea lamprey, *Petromyzon marinus* L. Cell and Tissue Research 247: 351–357.

Elliott, W.M. and J.H. Youson. 1991. Somatostatin concentrations in the pancreatic intestinal tissues of the sea lamprey, *Petromyzon marinus* L, at various periods of its life-cycle. Comparative Biochemistry and Physiology A-Physiology 99: 357–360.

Elliott, W.M. and J.H. Youson. 1993a. Development of the adult endocrine pancreas during metamorphosis in the sea lamprey, *Petromyzon marinus* L. 1. Light-microscopy and autoradiography. Anatomical Record 237: 259–270.

Elliott, W.M. and J.H. Youson. 1993b. Development of the adult endocrine pancreas during metamorphosis in the sea lamprey, *Petromyzon marinus* L. 2. Electron-microscopy and immunocytochemistry. Anatomical Record 237: 271–290.

Ellis, L.C. 1993. An ultrastructural investigation of the pronephric kidney of the sea lamprey, *Petromyzon marinus* L., throughout the life cycle. Ph.D. Thesis, University of Toronto, pp. 1–217.

Ellis, L.C. and J.H. Youson. 1990. Pronephric regression during larval life in the sea lamprey, *Petromyzon marinus* L. A histochemical and ultrastructural-study. Anatomy and Embryology 182: 41–52.

Escriva, H., L. Manzon, J. Youson and V. Laudet. 2002. Analysis of lamprey and hagfish genes reveals a complex history of gene duplications during early vertebrate evolution. Molecular Biology and Evolution 19: 1440–1450.

Ficele, G., J.A. Heinig, H. Kawauchi, J.H. Youson, F.W. Keeley and G.M. Wright. 1998. Spatial and temporal distribution of proopiomelanotropin and proopiocortin mRNA during the life cycle of the sea lamprey: A qualitative and quantitative *in situ* hybridization study. General and Comparative Endocrinology 110: 212–225.

Filosa, M.F., P.A. Sargent, M.M. Fisher and J.H. Youson. 1982. An electrophoretic and immunoelectrophoretic characterization of the serum-proteins of the adult lamprey, *Petromyzon marinus* L. Comparative Biochemistry and Physiology B-Biochemistry & Molecular Biology 72: 521–530.

Filosa, M.F., P.A. Sargent and J.H. Youson. 1986. An electrophoretic and immunoelectrophoretic study of serum-proteins during the life-cycle of the lamprey, *Petromyzon marinus* L. Comparative Biochemistry and Physiology B-Biochemistry & Molecular Biology 83: 143–149.

Freamat, M. and S.A. Sower. 2008. A sea lamprey glycoprotein hormone receptor similar with gnathostome thyrotropin hormone receptor. Journal of Molecular Endocrinology 41: 219–228.

Freamat, M., H. Kawauchi, M. Nozaki and S.A. Sower. 2006. Identification and cloning of a glycoprotein hormone receptor from sea lamprey, *Petromyzon marinus*. Journal of Molecular Endocrinology 37: 135–146.

Gudernatsch, J.F. 1912. Feeding experiments on tadpoles. I. The influence of specific organs given as food on growth and differentiation. A contribution to the knowledge of organs with internal secretion. Archiv Fur Entwicklungsmechanik Der Organismen 35: 457–483.

Hardisty, M.W. 1979. *Biology of Cyclostomes*. Chapman and Hall, London pp 272.

Hardisty, M.W. 2006. *Lampreys Life Without Jaws*. Forrest Text, Credigion pp 428.

Harvey, J. and M.L.J. Ashford. 2003. Leptin in the CNS: much more than a satiety signal. Neuropharmacology 44: 845–854.

Heinig, J.A., F.W. Keeley, P. Robson, S.A. Sower and J.H. Youson. 1995. The appearance of proopiomelanocortin early in vertebrate evolution: Cloning and sequencing of POMC from a lamprey pituitary cDNA library. General and Comparative Endocrinology 99: 137–144.

Heinig, J.A., F.W. Keeley, H. Kawauchi and J.H. Youson. 1999. Expression of proopiocortin and proopiomelanotropin during the life cycle of the sea lamprey (*Petromyzon marinus*). Journal of Experimental Zoology 283: 95–101.

Henson, M.P., R.A. Bergstedt and J.V. Adams. 2003. Comparison of spring measures of length, weight, and condition factor for predicting metamorphosis in two populations of sea lamprey (*Petromyzon marinus*) larvae. Journal of Great Lakes Research 29, Supplement 1: 204–213.

Hilliard, R.W., A. Epple and I.C. Potter. 1985. The morphology and histology of the endocrine pancreas of the southern-hemisphere lamprey, *Geotria australis* Gray. Journal of Morphology 184: 253–261.

Hoheisel, G. and G. Sterba. 1963. Uber die wirkung von kaliumperchlorat (KClO4) auf ammocoeten von *Lampetra Planeri* Bloch. Zeitschrift Fur Mikroskopisch-Anatomische Forschung 70: 490–516.

Holcík, J. 1986. *Lethenteron japonicum* (Martens, 1868). In: J. Holcík (Ed.). *The Freshwater Fishes of Europe*., AULA-Verlag GmbH, Wiesbaden pp. 198–219.

Holcík, J. and C.B. Renaud. 1986. *Eudontomyzon mariae* (Berg, 1931). In: J. Holcík (Ed.). *The Freshwater Fishes of Europe*., AULA-Verlag GmbH, Wiesbaden pp. 165–185.

Hollett, A.K. 1998. Condition factor and statolith aging in assessment of metamorphosis in sea lampreys (*Petromymzon marinus*), in the Great Lakes. M.Sc. Thesis, University of Toronto. pp. 1–229.

Holmes, J.A. and J.H. Youson. 1993. Induction of metamorphosis in landlocked sea lampreys, *Petromyzon marinus*. Journal of Experimental Zoology 267: 598–604.

Holmes, J.A. and J.H. Youson. 1994. Fall condition factor and temperature influence the incidence of metamorphosis in sea lampreys, *Petromyzon marinus*. Canadian Journal of Zoology-Revue Canadienne De Zoologie 72: 1134–1140.

Holmes, J.A. and J.H. Youson. 1996. Environmental sources of trace metals in sea lamprey, *Petromyzon marinus*, larvae in New Brunswick, Canada. Environmental Biology of Fishes 47: 299–310.

Holmes, J.A. and J.H. Youson. 1997. Laboratory study of the effects of spring warming and larval density on the metamorphosis of sea lampreys. Transactions of the American Fisheries Society 126: 647–657.

Holmes, J.A. and J.H. Youson. 1998. Extreme and optimal temperatures for metamorphosis in sea lampreys. Transactions of the American Fisheries Society 127: 206–211.

Holmes, J.A., F.W.H. Beamish, J.G. Seelye, S.A. Sower and J.H. Youson. 1994. Long-term influence of water temperature, photoperiod, and food-deprivation on metamorphosis of sea lamprey, *Petromyzon marinus*. Canadian Journal of Fisheries and Aquatic Sciences 51: 2045–2051.

Holmes, J.A., H. Chu, S.A. Khanam, R.G. Manzon and J.H. Youson. 1999. Spontaneous and induced metamorphosis in the American brook lamprey, *Lampetra appendix*. Canadian Journal of Zoology-Revue Canadienne De Zoologie 77: 959–971.

Horton, F.M. 1934. On the relation of the thyroid gland to metamorphosis in the lamprey. Jounal of Experimental Biology 11: 257–261.

Isorna, E., M.J. Obregon, R.M. Calvo, R. Vazquez, C. Pendon, J. Falcon and J.A. Munoz-Cueto. 2009. Iodothyronine deiodinases and thyroid hormone receptors regulation during flatfish (*Solea senegalensis*) metamorphosis. Journal of Experimental Zoology Part B-Molecular and Developmental Evolution 312B: 231–246.

Jones, R.P. 1947. Effect of thiourea on the endostyle of ammocoetes. Nature 160: 638–639.

Joss, J.M.P. 1985. Pituitary control of metamorphosis in the southern hemisphere lamprey, *Geotria australis*. General and Comparative Endocrinology 60: 58–62.

Kao, Y-H., J.H. Youson, J.A. Holmes and M.A. Sheridan. 1997a. Changes in lipolysis and lipogenesis in selected tissues of the landlocked lamprey, *Petromyzon marinus*, during metamorphosis. Journal of Experimental Zoology 277: 301–312.

Kao, Y-H., J.H. Youson and M.A. Sheridan. 1997b. Differences in the total lipid and lipid class composition of larvae and metamorphosing sea lampreys, *Petromyzon marinus*. Fish Physiology and Biochemistry 16: 281–290.

Kao, Y-H., J.H. Youson, J.A. Holmes and M.A. Sheridan. 1998. Effects of somatostatin on lipid metabolism of larvae and metamorphosing landlocked sea lamprey, *Petromyzon marinus*. General and Comparative Endocrinology 111: 177–185.

Kao, Y-H, J.H. Youson, J.A. Holmes, A. Al Mahrouki and M.A. Sheridan. 1999a. Effects of insulin on lipid metabolism of larvae and metamorphosing landlocked sea lamprey, *Petromyzon marinus*. General and Comparative Endocrinology 114: 405–414.

Kao, Y-H., R.G. Manzon, M.A. Sheridan and J.H. Youson. 1999b. Study of the relationship between thyroid hormones and lipid metabolism during KClO4-induced metamorphosis of landlocked lamprey *Petromyzon marinus*. Comparative Biochemistry and Physiology C-Toxicology & Pharmacology 122: 363–373.

Kavanaugh, S.I., M. Nozaki and S.A. Sower. 2008. Origins of gonadotropin-releasing hormone (GnRH) in vertebrates: Identification of a novel GnRH in a basal vertebrate, the sea lamprey. Endocrinology 149: 3860–3869.

Kawauchi, H. and S.A. Sower. 2006. The dawn and evolution of hormones in the adenohypophysis. General and Comparative Endocrinology 148: 3–14.

Klenner, J.J. and A.L. Schipper. 1954. The response of the endostyle of *Lampetra lamottenii* to itrumil, thiouracil, thiourea, thiocyanate and thyroxin. *Anatomical Record* 120: 790.

Knowles, F.G.W. 1941. Duration of larval life in ammocoetes. Proceedings of the Zoological Society of London 111: 101–109.

Leach, W.J. 1946. Oxygen consumption of lampreys, with special reference to metamorphosis and phylogenetic position. Physiological Zoology 19: 365–374.

Leatherland, J.F., R.W. Hilliard, D.J. Macey and I.C. Potter. 1990. Changes in serum thyroxine and triiodothyronine concentrations during metamorphosis of the southern hemisphere lamprey *Geotria australis*, and the effect of propylthiouracil, triiodothyronine and environmental temperature on serum thyroid hormone concentrations of ammocoetes. Fish Physiology and Biochemistry 8: 167–177.

Lepercq, J., J.C. Challier, M. Guerre-Millo, M. Cauzac, H. Vidal and S. Hauguel-de Mouzon. 2001. Prenatal leptin production: Evidence that fetal adipose tissue produces leptin. Journal of Clinical Endocrinology & Metabolism 86: 2409–2413.

Lintlop, S.P. and J.H. Youson. 1983a. Binding of triiodothyronine to hepatocyte nuclei from sea lampreys, *Petromyzon marinus* L, at various stages of the life cycle. General and Comparative Endocrinology 49: 428–436.

Lintlop, S.P. and J.H. Youson. 1983b. Concentration of triiodothyronine in the sera of the sea lamprey, *Petromyzon marinus*, and the brook lamprey, *Lampetra lamottenii*, at various phases of the life cycle. General and Comparative Endocrinology 49: 187–194.

Lowe, D.R., F.W.H. Beamish and I.C. Potter. 1973. Changes in proximate body composition of landlocked sea lamprey *Petromyzon marinus* (L.) during larval life and metamorphosis. Journal of Fish Biology 5: 673–682.

Lyons, J., P.A. Cochran, O.J. Polaco, E. Merino-Nambo. 1994. Distribution and abundance of the Mexican lampreys (Petromyconttidae: *Lampetra*: subgenus *Tetrapleurodon*). The Southwestern Naturalist 39: 105–113.

Macey, D.J. and I.C. Potter. 1981. Measurements of various blood-cell parameters during the life-cycle of the southern hemisphere lamprey, *Geotria australis* Gray. Comparative Biochemistry and Physiology A-Physiology 69: 815–823.

Macey, D.J. and I.C. Potter. 1986. Concentrations of nonheme iron in ammocoetes of species representing the 3 extant lamprey families. Comparative Biochemistry and Physiology A-Physiology 84: 77–79.

Macey, D.J., J. Webb and I.C. Potter. 1982a. Distribution of iron-containing granules in lampreys, with particular reference to the southern hemisphere species *Geotria australis* Gray. Acta Zoologica 63: 91–99.

Macey, D.J., J. Webb and I.C. Potter. 1982b. Iron levels and major iron-binding proteins in the plasma of ammocoetes and adults of the southern hemisphere lamprey *Geotria australis* Gray. Comparative Biochemistry and Physiology A-Physiology 72: 307–312.

Macey, D.J., S.R. Smalley, I.C. Potter and M.H. Cake. 1985. The relationship between total nonheme, ferritin and hemosiderin iron in larvae of southern hemisphere lampreys (*Geotria australis* and *Mordacia mordax*). Journal of Comparative Physiology B-Biochemical Systemic and Environmental Physiology 156: 269–276.

Mallatt, J. 1983. Laboratory growth of larval lampreys (*Lampetra (Entosphenus) tridentata* Richardson) at different food concentrations and animal densities. Journal of Fish Biology 22: 293–301.

Manion, P.J. and H.A. Purvis. 1971. Giant American brook lampreys, *Lampetra lamottei*, in Upper Great Lakes. Journal of the Fisheries Research Board of Canada 28: 616–620.

Manzon, L. A. 2006. Cloning and developmental expression of sea lamprey (*Petromyzon marinus*) thyroid hormone and retinoid X receptors. Ph.D. Thesis, University of Toronto pp 222.

Manzon, R.G. (In Press). Thyroidal regulation of life history transitions in fish. In: T. Flatt, and A. Heyland (Eds.). *Mechanisms of Life History Evolution*, Oxford University Press Inc., Oxford.

Manzon, R.G. and J.H. Youson. 1997. The effects of exogenous thyroxine (T4) or triiodothyronine (T3), in the presence and absence of potassium perchlorate, on the incidence of metamorphosis and on serum T4 and T3 concentrations in larval sea lamrpey (*Petromyzon marinus*). General and Comparative Endocrinology 106: 211–220.

Manzon, R.G. and J.H. Youson. 1999. Temperature and KClO4-induced metamorphosis in sea lamprey (*Petromyzon marinus*). Comparative Biochemistry and Physiology C-Pharmacology Toxicology & Endocrinology 124: 253–257.

Manzon, R.G. and J.H. Youson. 2002. $KClO_4$ inhibits thyroidal activity in the larval lamprey endostyle in vitro. General and Comparative Endocrinology 128: 214–223.

Manzon, R.G., J.G. Eales and J.H. Youson. 1998. Blocking of $KClO_4$-induced metamorphosis in premetamorphic sea lampreys by exogenous thyroid hormones (TH): Effects of $KClO_4$ and TH on serum TH concentrations and intestinal thyroxine outer-ring deiodination. General and Comparative Endocrinology 112: 54–62.

Manzon, R.G., J.A. Holmes and J.H. Youson. 2001. Variable effects of goitrogens in inducing precocious metamorphosis in sea lampreys (*Petromyzon marinus*). Journal of Experimental Zoology 289: 290–303.

Manzon, R.G., T.M. Neuls and L.A. Manzon. 2007. Molecular cloning, tissue distribution, and developmental expression of lamprey transthyretins. General and Comparative Endocrinology 151: 55–65.

McGree, M., T.A. Whitesel and J. Stone. 2008. Larval metamorphosis of individual Pacific lampreys reared in captivity. Transactions of the American Fisheries Society 137: 1866–1878.

Mendel, C.M., R.A. Weisiger, A.L. Jones and R.R. Cavalieri. 1987. Thyroid hormone binding proteins in plasma facilitate uniform distribution of thyroxine within tissues: A perfused rat liver study. Endocrinology 120: 1742–1749.

Morgan, E.H. 1980. The role of plasma transferrin in iron-absorption in the rat. Quarterly Journal of Experimental Physiology and Cognate Medical Sciences 65: 239–252.

Morii, M., Y. Mezaki, N. Yamaguchi, K. Yoshikawa, M. Miura, K. Imai, H. Yoshino, T. Hebiguchi and H. Senoo. (In Press). Onset of apoptosis in the cystic duct during metamorphosis of a Japanese lamprey, *Lethenteron reissneri*. *The Anatomical Record:* Advances in Integrative Anatomy and Evolutionary Biology 293: 1155–1166.

Morman, R.H. 1987. Relationship of density to growth and metamorphosis of caged larval sea lampreys, *Petromyzon marinus* Linnaeus, in Michigan streams. Journal of Fish Biology 30: 173–181.

Morris, R. 1980. Blood composition and osmoregulation in ammocoete larvae. Canadian Journal of Fisheries and Aquatic Sciences 37: 1665–1679.

Murdoch, S.P., M.F. Docker and F.W.H. Beamish. 1992. Effect of density and individual variation on growth of sea lamprey (*Petromyzon marinus*) larvae in the laboratory. Canadian Journal of Zoology-Revue Canadienne De Zoologie 70: 184–188.

Nozaki, M., A. Takahashi, Y. Amemiya, H. Kawauchi and S.A. Sower. 1995. Distribution of lamprey adrenocorticotropin and melanotropins in the pituitary of the adult sea lamprey, *Petromyzon marinus*. General and Comparative Endocrinology 98: 147–156.

Nozaki, M., K. Ominato, T. Shimotani, H. Kawauchi, J.H. Youson and S.A. Sower. 2008. Identity and distribution of immunoreactive adenohypophysial cells in the pituitary during the life cycle of sea lampreys, *Petromyzon marinus*. General and Comparative Endocrinology 155: 403–412.

O'Boyle, R.N. and F.W.H. Beamish. 1977. Growth and intermediary metabolism of larval and metamorphosing stages of the landlocked sea lamprey, *Petromyzon marinus* L. Environmental Biology of Fishes 2: 103–120.

Ooi, E.C. and J.H. Youson. 1976. Growth of the opisthonephric kidney during larval life in anadromous sea lamprey, *Petromyzon marinus* L. Canadian Journal of Zoology-Revue Canadienne De Zoologie 54: 1449–1458.

Ooi, E.C. and J.H. Youson. 1977. Morphogenesis and growth of the definitive opisthonephros during metamorphosis of anadromous sea lamprey, *Petromyzon marinus* L. Journal of Embryology and Experimental Morphology 42: 219–235.

Peek, W.D. and J.H. Youson. 1979a. Transformation of the interlamellar epithelium of the gills of the anadromous sea lamprey, *Petromyzon marinus* L, during metamorphosis. Canadian Journal of Zoology-Revue Canadienne De Zoologie 57: 1318–1332.

Peek, W.D. and J.H. Youson. 1979b. Ultrastructure of chloride cells in young-adults of the anadromous sea lamprey, *Petromyzon marinus* L, in fresh-water and during adaptation to sea-water. Journal of Morphology 160: 143–163.

Percy, R. and I.C. Potter. 1981. Further observations on the development and destruction of lamprey blood-cells. Journal of Zoology 193: 239–251.

Potter, I.C. 1970. Life cycles and ecology of Australian lampreys of genus *Mordacia*. Journal of Zoology 161: 487–511.

Potter, I.C. 1980. Ecology of larval and metamorphosing lampreys. Canadian Journal of Fisheries and Aquatic Sciences 37: 1641–1657.

Potter, I.C. and F.W.H. Beamish. 1977. Freshwater biology of adult anadromous sea lampreys *Petromyzon marinus*. Journal of Zoology 181: 113–130.

Potter, I.C. and F.W.H. Beamish. 1978. Changes in hematocrit and hemoglobin concentration during life-cycle of anadromous sea lamprey, *Petromyzon marinus* L. Comparative Biochemistry and Physiology A-Physiology 60: 431–434.

Potter, I.C. and I.D. Brown. 1975. Changes in hemoglobin electropherograms during the life cycle of two closely related lampreys. Comparative Biochemistry and Physiology B-Biochemistry & Molecular Biology 51: 517–519.

Potter, I.C. and H.S. Gill. 2003. Adaptive radiation of lampreys. Journal of Great Lakes Research 29: 95–112.

Potter, I.C., R. Percy and J.H. Youson. 1978a. A proposal for the adaptive significance of the development of the lamprey fat column. Acta Zoologica 59: 63–67.

Potter, I.C., G.M. Wright and J.H. Youson. 1978b. Metamorphosis in the anadromous sea lamprey, *Petromyzon-marinus* L. Canadian Journal of Zoology-Revue Canadienne De Zoologie 56: 561–570.

Potter, I.C., R.W. Hilliard and D.J. Bird. 1980. Metamorphosis in the southern hemisphere lamprey, *Geotria australis*. Journal of Zoology 190: 405–430.

Potter, I.C., R.W. Hilliard and D.J. Bird. 1982. Stages in metamorphosis. In: M.W. Hardisty, and I.C. Potter (Eds.). *Biology of Lampreys. Volume 4b*, Academic Press, London, pp. 137–164.

Purvis, H.A. 1970. Growth, age at metamorphosis, and sex ratio of northern brook lamprey in a tributary of southern Lake Superior. Copeia 1970: 326–332.

Purvis, H.A. 1979. Variations in growth age at transformation and sex ratio of sea lampreys *Petromyzon marinus* reestablished in chemically treated tributaries of the Upper Great Lakes USA. Great Lakes Fishery Commission Technical Report 35: 1–36.

Purvis, H.A. 1980. Effects of temperature on metamorphosis and the age and length at metamorphosis in sea lamprey (*Petromyzon marinus*) in the Great Lakes. Canadian Journal of Fisheries and Aquatic Sciences 37: 1827–1834.

Reis-Santos, P., S.D. McCormick and J.M. Wilson. 2008. Ionoregulatory changes during metamorphosis and salinity exposure of juvenile sea lamprey (*Petromyzon marinus* L.). Journal of Experimental Biology 211: 978–988.

Renaud, C.B. 1986. Eudontomyzon hellenicus Vladykov, Renaud, Kott, and Economidis, 1982. In: J. Holčík (Ed.). *The Freshwater Fishes of Europe. Volume 1, Part 1*, AULA-Verlag GmbH, Wiesbaden, pp. 186–195.

Renaud, C.B. and P.S. Economidis. 2010. *Eudontomyzon graecus*, a new nonparasitic lamprey species from Greece (Petromyzontiformes: Petromyzontidae). Zootaxa 2477: 37–48.

Robson, P. 1998. Biochemical and molecular studies of the cartilaginous endoskeleton of adult lampreys and hagfish. Ph.D. Thesis, University of Toronto, pp. 1–190.

Robson, P., G.M. Wright, E. Sitarz, A. Maiti, M. Rawat, J.H. Youson and F.W. Keeley. 1993. Characterization of lamprin, an unusual matrix protein from lamprey cartilage. Implications for evolution, structure, and assembly of elastin and other fibrillar proteins. Journal of Biological Chemistry 268: 1440–1447.

Robson, P., G.M. Wright, J.H. Youson and F.W. Keeley. 1997. A family of non-collagen-based cartilages in the skeleton of the sea lamprey, *Petromyzon marinus*. Comparative Biochemistry and Physiology B-Biochemistry & Molecular Biology 118: 71–78.

Robson, P., G.M. Wright, J.H. Youson and F.W. Keeley. 2000. The structure and organization of lamprin genes: Multiple-copy genes with alternative splicing and convergent evolution with insect structural proteins. Molecular Biology and Evolution 17: 1739–1752.

Rodriguez-Munoz, R., A.G. Nicieza and F. Brana. 2003. Density-dependent growth of sea lamprey larvae: Evidence for chemical interference. Functional Ecology 17: 403–408.

Root, A.R., N.V. Nucci, J.D. Sanford, B.S. Rubin, V.L. Trudeau and S.A. Sower. 2005. *In situ* characterization of gonadotropin-releasing hormone-I, -III, and glutamic acid decarboxylase expression in the brain of the sea lamprey, *Petromyzon marinus*. Brain Behavior and Evolution 65: 60–70.

Sargent, P.A. and J.H. Youson. 1986. Quantification of iron deposits in several body tissues of lampreys (*Petromyzon marinus* L.) throughout the life Cycle. Comparative Biochemistry and Physiology A-Physiology 83: 573–577.

Sheridan, M.A. and Y.H. Kao. 1998. Regulation of metamorphosis-associated changes in the lipid metabolism of selected vertebrates. American Zoologist 38: 350–368.

Shi, Y.B. 2000. *Amphibian Metamorphosis*. Wiley-Liss, Toronto.

Shu, D.G., H.L. Luo, S.C. Morris, X.L. Zhang, S.X. Hu, L. Chen, J. Han, M. Zhu, Y. Li and L.Z. Chen. 1999. Lower Cambrian vertebrates from south China. Nature 402: 42–46.

Smalley, S.R., D.J. Macey and I.C. Potter. 1986. Changes in the amount of nonheme iron in the plasma, whole-body, and selected organs during the postlarval life of the lamprey *Geotria australis*. Journal of Experimental Zoology 237: 149–157.

Smith, A.I. and J.W. Funder. 1988. Proopiomelanocortin processing in the pituitary, central nervous-system, and peripheral-tissues. Endocrine Reviews 9: 159–179.

Sower, S.A. 1998. Brain and pituitary hormones of lampreys, recent findings and their evolutionary significance. American Zoologist 38: 15–38.

Sower, S.A., E. Plisetskaya and A. Gorbman. 1985. Steroid and thyroid-hormone profiles following a single injection of partly purified salmon gonadotropin or GnRH analogs in male and female sea lamprey. Journal of Experimental Zoology 235: 403–408.

Sower, S.A., S. Moriyama, M. Kasahara, A. Takahashi, M. Nozaki, K. Uchida, J.M. Dahstrom and H. Kawauchi. 2006. Identification of sea lamprey GTH beta-like cDNA and its evolutionary implications. General and Comparative Endocrinology 148: 22–32.

Sower, S.A., M. Freamat and S.I. Kavanaugh. 2009. The origins of the vertebrate hypothalamic-pituitary-gonadal (HPG) and hypothalamic-pituitary-thyroid (HPT) endocrine systems: New insights from lampreys. General and Comparative Endocrinology 161: 20–29.

Suzuki, S. 1986. Induction of metamorphosis and thyroid function in the larval lamprey. Frontiers in Thyroidology 1: 667–670.

Suzuki, S. 1987. Induction of metamorphosis and thyroid function in the larval lamprey. Proceedings of the First Congress of the Asia Oceania Society of Comparative Endocrinology 1: 220–221.

Suzuki, S. 1989. Why goitrogens are chemical triggers of metamorphosis in the lamprey relationship between thyroid-function and metamorphosis. General and Comparative Endocrinology 74: 277.

Takahashi, A., Y. Amemiya, M. Sarashi, S.A. Sower and H. Kawauchi. 1995. Melanotropin and corticotropin are encoded on 2 distinct genes in the lamprey, the earliest evolved extant vertebrate. Biochemical and Biophysical Research Communications 213: 490–498.

Tata, J.R., B.S. Baker, I. Machuca, E.M.L. Rabelo and K. Yamauchi. 1993. Autoinduction of Nuclear Receptor Genes and Its Significance. Journal of Steroid Biochemistry and Molecular Biology 46: 105–119.

Tobet, S.A., M. Nozaki, J.H. Youson and S.A. Sower. 1995. Distribution of lamprey gonadotropin-releasing hormone-III (GnRH-III) in brains of larval lampreys (*Petromyzon marinus*). Cell and Tissue Research 279: 261–270.

Tobet, S.A., T.W. Chickering and S.A. Sower. 1996. Relationship of gonadotropin-releasing hormone (GnRH) neurons to the olfactory system in developing lamprey (*Petromyzon marinus*). Journal of Comparative Neurology 376: 97–111.

Tobet, S.A., S.A. Sower and G.A. Schwarting. 1997. Gonadotropin-releasing hormone containing neurons and olfactory fibers during development: From lamprey to mammals. Brain Research Bulletin 44: 479–486.

Treble, A.J., M.L. Jones and T.B. Steeves. 2008. Development and evaluation of a new predictive model for metamorphosis of Great Lakes larval sea lamprey (*Petromyzon marinus*) populations. Journal of Great Lakes Research 34: 404–417.

Tsuneki, K. and M. Ouji. 1984. Morphometric changes during growth of the brook lamprey *Lampetra reissneri*. Japanese Journal of Ichthyology 31: 38–46.

Twohey, M.B., P.W. Sorensen and W.M. Li. 2003. Possible applications of pheromones in an integrated sea lamprey management program. Journal of Great Lakes Research 29 (Supplement 1): 794–800.

White, B. and C.S. Nicoll. 1981. Hormonal control of amphibian metamorphosis. In: L.A. Gilbert and E. Frieden (Eds.). *Metamorphosis, A Problem in Developmental Biology*, Plenum Press, New York, pp. 263–396.

Wright, G.M. 1989. Ultrastructure of the adenohypophysis in the anadromous sea lamprey, *Petromyzon marinus*, during metamorphosis. Journal of Morphology 202: 205–223.

Wright, G.M. and J.H. Youson. 1976. Transformation of the endostyle of anadromous sea lamprey, *Petromyzon marinus* L, during metamorphosis. 1. Light-microscopy and autoradiography with I[125]. General and Comparative Endocrinology 30: 243–257.

Wright, G.M. and J.H. Youson. 1977. Serum thyroxine concentrations in larval and metamorphosing anadromous sea lamprey, *Petromyzon marinus* L. Journal of Experimental Zoology 202: 27–32.

Wright, G.M. and J.H. Youson. 1980a. Transformation of the endostyle of the anadromous sea lamprey, *Petromyzon marinus* L., during metamorphosis. 2. Electron-microscopy. Journal of Morphology 166: 231–257.

Wright, G.M. and J.H. Youson. 1980b. Variation in serum levels of thyroxine in anadromous larval lampreys, *Petromyzon marinus* L. General and Comparative Endocrinology 41: 321–324.

Wright, G.M. and J.H. Youson. 1982. Ultrastructure of mucocartilage in the larval anadromous sea lamprey, *Petromyzon marinus* L. American Journal of Anatomy 165: 39–51.

Wright, G.M. and J.H. Youson. 1983. Ultrastructure of cartilage from young adult sea lamprey, *Petromyzon marinus* L: A new type of vertebrate cartilage. American Journal of Anatomy 167: 59–70.

Wright, G.M., M.F. Filosa and J.H. Youson. 1980. Immunocytochemical localization of thyroglobulin in the transforming endostyle of anadromous sea lampreys, *Petromyzon marinus* L, during metamorphosis. General and Comparative Endocrinology 42: 187–194.

Wright, G.M., J.H. Youson and F.W. Keeley. 1983. Lamprey cartilage: A new type of vertebrate cartilage. Anatomical Record 205: A221.

Wright, G.M., L.A. Armstrong, A.M. Jacques and J.H. Youson. 1988. Trabecular, nasal, branchial, and pericardial cartilages in the sea lamprey, *Petromyzon marinus*: Fine structure and immunohistochemical detection of elastin. American Journal of Anatomy 182: 1–15.

Wright, G.M., K.M. McBurney, J.H. Youson and S.A. Sower. 1994. Distribution of lamprey gonadotropin-releasing-hormone in the brain and pituitary gland of larval, metamorphic, and adult sea lampreys, *Petromyzon marinus*. Canadian Journal of Zoology-Revue Canadienne De Zoologie 72: 48–53.

Yaghoubian, S., M.F. Filosa and J.H. Youson. 2001. Proteins immunoreactive with antibody against a human leptin fragment are found in serum and tissues of the sea lamprey, *Petromyzon marinus* L. Comparative Biochemistry and Physiology B-Biochemistry & Molecular Biology 129: 777–785.

Yamano, K. and S. Miwa. 1998. Differential gene expression of thyroid hormone receptor alpha and beta in fish development. General and Comparative Endocrinology 109: 75–85.

Young, J.Z., and C.W. Bellerby. 1935. The response of the lamprey to injection of anterior lobe pituitary extract. Journal of Experimental Biology XII: 246–253.

Youson, J.H. 1980. Morphology and physiology of lamprey metamorphosis. Canadian Journal of Fisheries and Aquatic Sciences 37: 1687–1710.

Youson, J.H. 1981a. The alimentary canal. In: M.W. Hardisty, and I.C. Potter (Eds.). The Biology of Lampreys. *Volume 3*, Academic Press, London, pp. 95–189.

Youson, J.H. 1981b. The kidneys. In: M.W. Hardisty, and I.C. Potter (Eds.). The Biology of Lampreys. Volume 3, Academic Press, London, pp. 192–261.

Youson, J.H. 1981c. The liver. In: M.W. Hardisty, and I.C. Potter (Eds.). The Biology of Lampreys. Volume 3, pp. 262–332.

Youson, J.H. 1982. Replication of basal bodies and ciliogenesis in a ciliated epithelium of the lamprey. Cell and Tissue Research 223: 255–266.

Youson, J.H. 1984. Differentiation of the segmented tubular nephron and excretory duct during lamprey metamorphosis. Anatomy and Embryology 169: 275–292.

Youson, J. 1985. Organ development and specialization in lamprey species. In: R.E. Foreman, A. Gorbman, J.M. Dodd, and R. Olsson (Eds.). The Evolutionary Biology of Primitive Fishes, Plenum, New York, pp. 141–156.

Youson, J.H. 1988. First metamorphosis. In: W.S. Hoar, and D.J. Randall (Eds.). Fish Physiology: The Physiology of Developing Fish, Academic Press Inc., Toronto, pp. 135–196.

Youson, J.H. 1993. Biliary atresia in lampreys. In: C.E. Cornelius (Ed.). *Advances in Veterinary Science and Comparative Medicine, Vol. 37.* Animal Models in Liver Research, Academic Press Inc., San Diego pp. 197-255.

Youson, J.H. 1994. Environmental and hormonal cues and endocrine glands during lamprey metamorphosis. In: K.G. Davey, R.E. Peter, and S.S. Tobe (Eds.). Perspectives in Comparative Endocrinology, National Research Council of Canada, Ottawa, pp. 400–407.

Youson, J.H. 1997. Is lamprey metamorphosis regulated by thyroid hormones? American Zoologist 37: 441–460.

Youson, J.H. 1999. The lability and endocrinology of lamprey metamorphosis: An ancient developmental strategy. In: E.W. Roubos, S.E. Wendelaar-Bonga, H. Vaudry, and De Loof (Eds.). Recent Developments in Comparative Endocrinology and Neurobiology, Shaker, Maastrict, The Neatherlands, pp. 285–288.

Youson, J.H. 2000. The agnathan enteropancreatic endocrine system: Phylogenetic and ontogenetic histories, structure, and function. American Zoologist 40: 179–199.

Youson, J.H. 2003. The biology of metamorphosis in sea lampreys: endocrine, environmental, and physiological cues and events, and their potential application to lamprey control. Journal of Great Lakes Research 29 (Supplement 1): 26–49.

Youson, J.H. 2004. The impact of environmental and hormonal cues on the evolution of fish metamorphosis. In: B.K. Hall, R.D. Pearson and G.B. Muller (Eds.). Environment, Development, and Evolution. Toward a Synthesis, Massachusetts Institute of Technology, Cambridge, pp. 239–278.

Youson, J.H. 2007. Peripheral endocrine glands. I. The gastroenteropancreatic endocrine system and the thyroid gland. In: D.J. McKenzie, A.P. Farrell and C.J. Brauner (Eds.). *Primitive Fishes*, Academic Press, New York, pp. 381–455.

Youson, J.H. and A.A. Al Mahrouki. 1999. Ontogenetic and phylogenetic development of the endocrine pancreas (islet organ) in fishes. General and Comparative Endocrinology 116: 303–335.

Youson, J.H. and R.J. Beamish. 1991. Comparison of the internal morphology of adults of a population of lampreys that contains a nonparasitic life-history type, *Lampetra richardsoni*, and a potentially parasitic form, *L. richardsoni var. marifuga*. Canadian Journal of Zoology-Revue Canadienne De Zoologie 69: 628–637.

Youson, J.H. and R. Cheung. 1990. Morphogenesis of somatostatin-secreting and insulin-secreting cells in the lamprey endocrine pancreas. Fish Physiology and Biochemistry 8: 389–397.

Youson, J.H. and K.L. Connelly. 1978. Development of longitudinal mucosal folds in the intestine of anadromous sea lamprey, *Petromyzon marinus* L, during metamorphosis. Canadian Journal of Zoology-Revue Canadienne De Zoologie 56: 2364–2371.

Youson, J.H. and W.M. Elliott. 1989. Morphogenesis and distribution of the endocrine pancreas in adult lampreys. Fish Physiology and Biochemistry 7: 125–131.

Youson, J.H. and W.R. Horbert. 1982. Transformation of the intestinal epithelium of the larval anadromous sea lamprey *Petromyzon marinus* L during metamorphosis. Journal of Morphology 171: 89–117.

Youson, J.H. and E.C. Ooi. 1979. Development of the renal corpuscle during metamorphosis in the lamprey. American Journal of Anatomy 155: 201–221.

Youson, J.H. and I.C. Potter. 1979. Description of the stages in the metamorphosis of the anadromous sea lamprey, *Petromyzon marinus* L. Canadian Journal of Zoology-Revue Canadienne De Zoologie 57: 1808–1817.

Youson, J.H. and S.A. Sower. 1991. Concentration of gonadotropin-releasing-hormone in the brain during metamorphosis in the lamprey, *Petromyzon marinus*. Journal of Experimental Zoology 259: 399–404.

Youson, J.H. and S.A. Sower. 2001. Theory on the evolutionary history of lamprey metamorphosis: role of reproductive and thyroid axes. Comparative Biochemistry and Physiology B-Biochemistry & Molecular Biology 129: 337–345.

Youson, J.H., J. Lee and I.C. Potter. 1979. Distribution of fat in larval, metamorphosing, and young adult anadromous sea lampreys, *Petromyzon marinus* L. Canadian Journal of Zoology-Revue Canadienne De Zoologie 57: 237–246.

Youson, J.H., P.A. Sargent and E.W. Sidon. 1983a. Iron loading in the liver of parasitic adult lampreys, *Petromyzon marinus* L. American Journal of Anatomy 168: 37–49.

Youson, J.H., P.A. Sargent and E.W. Sidon. 1983b. Iron loading in the livers of metamorphosing lampreys, *Petromyzon marinus* L. Cell and Tissue Research 234: 109–124.

Youson, J.H., P.A. Sargent and A. Barrett. 1987a. Serum iron concentration and other blood parameters during the life cycle of the sea lamprey, *Petromyzon marinus* L. Comparative Biochemistry and Physiology A-Physiology 88: 325–330.

Youson, J.H., P.A. Sargent, K. Yamamoto, D. Ogilvie and M.M. Fisher. 1987b. Nonparenchymal liver-cells and granulomas during lamprey biliary atresia. American Journal of Anatomy 179: 155–168.

Youson, J.H., W.M. Elliott, R.J. Beamish and D.W. Wang. 1988. A comparison of endocrine pancreatic tissue in adults of four species of lampreys in British Columbia: A morphological and immunohistochemical study. General and Comparative Endocrinology 70: 247–261.

Youson, J.H., J.A. Holmes, J.A. Guchardi, J.G. Seelye, R.E. Beaver, J.E. Gersmehl, S.A. Sower and F.W.H. Beamish. 1993. Importance of condition factor and the influence of water temperature and photoperiod on metamorphosis of sea lamprey, *Petromyzon marinus*. Canadian Journal of Fisheries and Aquatic Sciences 50: 2448–2456.

Youson, J.H., E.M. Plisetskaya and J.F. Leatherland. 1994. Concentrations of insulin and thyroid hormones in the serum of landlocked sea lampreys (*Petromyzon marinus*) of three larval year classes, in larvae exposed to two temperature regimes, and in individuals during and after metamorphosis. General and Comparative Endocrinology 94: 294–304.

Youson, J.H., M.F. Docker and S.A. Sower. 1995a. Concentration of gonadotropin-releasing hormones in brain of larval and metamorphosing lampreys of two species with different adult life histories. In: F.W. Goetz and P. Thomas (Eds.). *Proceedings of the 5th International Symposium on the Reproductive Physiology of Fish.*, University of Texas Press, Austin, pp. 83.

Youson, J.H., J.A. Holmes and J.F. Leatherland. 1995b. Serum concentrations of thyroid hormones in KClO4-treated larval sea lampreys (*Petromyzon marinus* L). Comparative Biochemistry and Physiology C-Pharmacology Toxicology & Endocrinology 111: 265–270.

Youson, J.H., R.G. Manzon, B.J. Peck and J.A. Holmes. 1997. Effects of exogenous thyroxine (T4) and triiodothyronine (T3) on spontaneous metamorphosis and serum T4 and T3 levels in immediately premetamorphic sea lampreys, *Petromyzon marinus*. Journal of Experimental Zoology 279: 145–155.

Youson, J.H., J.A. Heinig, S.F. Khanam, S.A. Sower, H. Kawauchi and F.W. Keeley. 2006. Patterns of proopiomelanotropin and prooplocortin gene expression and of immunohistochemistry for gonadotropin-releasing hormones (1GnRH-I and -III) during the life cycle of a nonparasitic lamprey: relationship to this adult life history type. General and Comparative Endocrinology 148: 54–71.

# CHAPTER 3

# Metamorphosis of Elopomorphs

*Keisuke Yamano*

## 3.1 Introduction

The elopomorphs, which includes tenpounders, tarpons, bonefishes and various eels, are representative of fishes that undergo drastic morphological changes during their early life stages from larvae to juveniles. The bodies of larval elopomorphs are laterally-compressed, leaf-like, and transparent, and the head is small in relation to the body size. These peculiar attributes are reflected in the use of the special name "leptocephalus" (meaning "small head") for this stage of development. During metamorphosis, the transformation of the larva into an adult-like form is associated with a reduction in both the length and the depth of the body in most species, and the body becomes opaque by thickening and pigmentation of the skin. The external appearance of the leptocephalus is so different from their adult morphology that leptocephali and adults had previously been regarded as independent species. Nowadays, however, the metamorphosis of elopomorphs is the best-known example of metamorphosis in bony fishes (Youson, 1988). Grassi (1896) was the first to demonstrate experimentally that leptocephali (*Leptocephalus brevirostris*) metamorphose into young elvers of *Anguilla vulgaris* (synonym of *A. anguilla*) in an aquarium. At present, the occurrence of a leptocephalus stage during the larval period is the most important taxonomic trait for characterizing a fish as an elopomorph (Nelson, 2006).

Since the discovery of metamorphosis in eels, their breeding places in the open ocean have been of great interest. In the early twentieth century, Schmidt (1922) concluded, based on the distribution of small leptocephali, that the European eel (*A. anguilla*) and the American eel (*A. rostrata*) spawn

National Research Institute of Aquaculture, Fisheries Research Agency, Minamiise, Mie 516-0193, Japan.
E-mail: yamano@fra.affrc.go.jp

around the Sargasso Sea in the southwestern North Atlantic Ocean. In recent years, intensive investigations have been carried out to locate the spawning place of the Japanese eel (*A. japonica*) in the Pacific Ocean (Tsukamoto, 1992; Tsukamoto, 2006; Chow et al., 2007; Tsukamoto et al., 2011). In the course of these studies, the processes of larval growth and metamorphosis of anguillids in the ocean were largely uncovered but the life history and metamorphic processes of many elopomorphs still remain a mystery.

The anguilliforms, especially anguillids, are an important fisheries resource. However, the glass eel populations of certain catadromous species, e.g., European eel, have been decreasing over a long period. These circumstances led to the listing of the European eel in Appendix II of the Convention on International Trade in Endangered Species of Wild Fauna and Flora in 2007. The export of glass eels of this species essentially has been prohibited since 2009. On the other hand, trials aimed at producing glass eels as seed for aquaculture have been conducted since the first successful hatching of larvae in captivity (Yamamoto and Yamauchi, 1974). Tanaka et al. (2003) have finally succeeded in producing glass eels of the Japanese eel. Recent progress in the development of techniques of rearing leptocephali of eels has enabled a more detailed analysis of metamorphosis.

In this article, the first metamorphosis, namely the transition from larvae to juveniles, of elopomorphs is reviewed. The second metamorphosis, which occurs at the time of reproductive development, is reviewed elsewhere.

## 3.2 Taxonomy of the Elopomorpha: Leptocephalus as a Taxonomic Trait

As with many fish taxa, there have been historical shifts in the classification of this fish group. Older classifications were based on adult characters, as is commonly applied to other fish taxa. Greenwood et al. (1966) first established the superorder Elopomorpha, composed of the Anguilliformes, Notacanthiformes, and Elopiformes, by joining together diverse fish groups that previously had been positioned in widely different taxa. He nominated eight major characteristics for the classification of the Elopomorpha but the only characteristic common to all groups of the Elopomorpha for which the larval stages were known, was the possession of the leptocephalus (Greenwood et al., 1966). Thus, the occurrence of this specific larval stage was regarded as the most important trait for assigning fishes to the Elopomorpha. However, this classification based on the larval stage gave rise to much controversy, which continues to date. For instance, phylogenetic analyses of 12S, 16S rRNA genes and nuclear18S rRNA (Filleul and Lavoué, 2001) and of mtDNA sequences in segments of the 12S and 16S rRNA genes (Obermiller and Pfeiler, 2003) did not support the monophyly of the Elopomorpha whereas analysis of 12S ribosomal RNA sequences supported a monophyletic Elopomorpha (Wang et al., 2003). Inoue et al. (2004) expanded the phylogenetic approach into complete mitochondrial genomic sequences of 33 purposefully selected species. The analysis placed all fish groups with a leptocephalus larval stage in single clade (Clade A in Fig. 1) as a sister group of the Clupeocephala, strongly supporting the monophyly of the Elopomorpha.

Although controversy still exists, and the view of the Elopomorpha may yet change as a result of to future studies, this article follows the classification based on the common occurrence of a leptocephalus larval stage as proposed

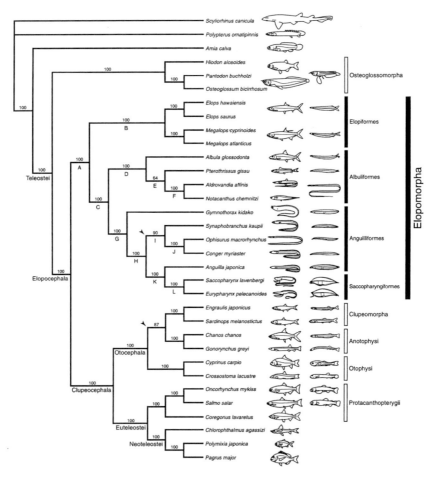

**Figure 1.** Phylogenetic relationships of 33 fishes including 30 teleosts and 3 out-group species. Bayesian phylogenetic analysis was applied to mitogenomic data (14,040 bp), and numbers above internal branches indicate Bayesian posterior probabilities. Adult and larval illustrations are shown at the right. Reproduced from Inoue et al. (2004) with permission.

by Nelson (2006). This scheme proposes that Elopomorpha is subdivided into four orders: Elopiformes (tenpounders and tarpons), Albuliformes (bonefishes, halosaurs and spiny eels), Anguilliformes (various eels), and Saccopharyngiformes (sackpharynx fishes). In total, 24 families, 156 genera, and about 856 species have been nominated in the Elopomorpha (Table 1). All but six species are marine, or primarily marine, species (Nelson, 2006).

**Table 1.** The four orders of the Elopomorpha and numbers of families, genera, and species in each order (after Nelson, 2006).

| Order | Families | Genera | Species |
|---|---|---|---|
| Elopiformens | 2 | 2 | 8 |
| Albuliformes | 3 | 8 | 30 |
| Anguilliformes | 15 | 141 | 791 |
| Saccopharyngiformes | 4 | 5 | 28 |

## 3.3 Morphological Features of Leptocephali

### 3.3.1 General features

Photographs of leptocephalus larvae of the Japanese eel are shown in Fig. 2 and drawings of various types of leptocephali are illustrated in Fig. 3. The body is laterally-compressed, leaf-like and transparent, which is considered to be adaptive to a planktonic life in the ocean. A large quantity of

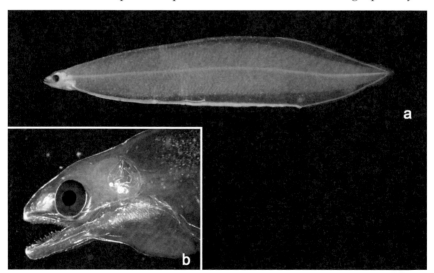

**Figure 2.** Leptocephalus of the Japanese eel (*Anguilla japonica*). (a) Late stage of wild leptocephalus (60 mm in total length) captured in the Pacific Ocean east of Okinawa Islands. The specimen was preserved in 10% formalin and was a gift from Dr. Kurogi. (b) The head of a live leptocephalus reared in an aquarium from a fertilized egg; donated by Dr. Tanaka.

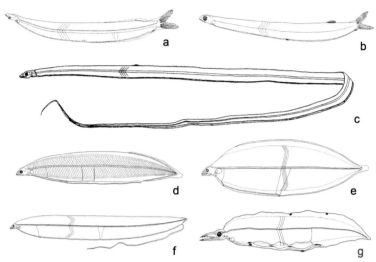

**Figure 3.** Various types of leptocephali. (a) tenpounder, *Elops hawaiensis* (33.5 mm in total length, TL); (b) bonefish, *Albula vulpes* (61.5 mm in TL); (c) halosaur, Halosauridae sp. (280 mm in TL); (d) giant mottled eel, *Anguilla marmorata* (42.0 mm in TL); (e) flat-nosed xenocongrid eel, *Chilorhinus platyrhynchus* (48.0 mm in TL); (f) congrid eel, *Ariosoma* sp. (101.0 mm in TL); (g) bobtail snipe eels, Cyematidae sp. (21.0 mm in TL). Myomeres are partially drawn except for (d). All illustrations were copied from 'An Atlas of the Early Stage Fishes in Japan' with permission. Original figures were drawn by Dr. Mochioka (a, b, c, d and f), Dr. Tabeta (e), and Dr. Kanegae (g).

transparent gelatinous matrix is accumulated under the skin. Leptocephali lack most of the ossified skeletal elements, including a vertebral column and thus the body is supported by the gelatinous matrix. The head is small in relation to the body size and the body height is relatively deep, although this latter feature varies among species. The maximum body length of the leptocephali of most species is much greater than the body length of young juveniles because of shortening that occurs during metamorphosis, although the extent of this reduction is also variable among species. At the same time, the body shape changes from leaf-like to the juvenile style. The shortening and transformation of the body are associated with the breakdown of the gelatinous tissue, which occupies most of the relatively massive trunk of leptocephali. The intestine is more or less straight and opens at an unusually posterior position on the abdomen. Prominent teeth are present on the upper and lower jaws and these are lost during metamorphosis. The eyes and olfactory organs are well-developed. V- or W-shaped myomeres are visible, and the number of myomeres is a major characteristic used in species identification.

### 3.3.2 The elopiformes

The order Elopiformes is composed of two families: the Elopidae (tenpounders) and the Metalopidae (tarpons) (Nelson, 2006). Adults of these fishes have a typical fish-like form, rather than being eel-like, but they possess a leptocephalus larval stage (Fig. 3a). The most notable characteristic of elopiform leptocephali is the well-developed, forked caudal fin. The posterior dorsal and anal fins are present and are distinct from the caudal fin. The ventral fins are formed in the later larval stages. The body height is relatively narrow. Elopiform leptocephali are the smallest (maximum length about 3 cm) of the elopomorph groups. The intestine is extremely long. Myomeres are V-shaped.

### 3.3.3 The albuliformes

The order Albuliformes is composed of three families: the Albulidae (bonefishes), the Halosauridae (halosaurs), and the Nothacanthidae (spiny eels) (Nelson, 2006). Most of the albulid fishes are coastal species whereas fishes of the other two families are deep-sea inhabitants. The external appearance of albulid leptocephali resembles that of the elopiforms but they grow to 60–180 mm, much larger than elopiform leptocephali (Fig. 3b). Halosaurid leptocephali differ greatly, both in shape and size, compared with albulid leptocephali. Halosaurid and nothacanthid leptocephali typically grow to 40–50 cm and, exceptionally, may reach a length of 2 m (Fig. 3c). The caudal fin is filamentous. The body shape is extremely slender. The base of the dorsal and ventral fins is short. The pectoral fins are small. The myomeres are V-shaped. Many halosaurid and nothacanthid leptocephali remain unidentified at the species level.

### 3.3.4 The anguilliformes

The Anguilliformes forms the largest group in the Elopomorpha, being composed of 15 families, 141 genera and 791 species (Table 1, Nelson, 2006), which includes true eels, moray eels, snake eels, conger eels and so on. Associated with vast numbers of species, the morphology of anguilliform leptocephali is diverse (Fig. 3d–f). The size is generally 5–10 cm long, but may reach 50 cm. The body shape is typically leaf-like. The caudal fin is short and round, and is merged with the dorsal and anal fins. The ventral fin is absent. The myomeres are W-shaped. The intestines are usually straight and open at the anterior edge of the anal fin. Some has a pouch-like region in the intestines or external intestines (Fig. 3f).

### 3.3.5 *The saccopharyngiformes*

The order Saccopharyngiformes is composed of four families: the Cyematidae (bobtail snipe eels), the Saccopharyngidae (swallowers), the Eurypharyngidae (gulpers or pelican eels) and the Monognathidae (one jaw gulpers) (Nelson, 2006). It is evident from their common names that all the fishes in this order are highly peculiar in their adult body shape. They are deep-sea inhabitants and until the classification of Greenwood et al. (1966), they had been placed in widely separated taxonomic positions. Although some researchers have doubted that they are true bony fishes (Tchernavin, 1946), their larvae do appear to be leptocephali (Fig. 3g). The shape of the fins of saccopharyngiform leptocephali is similar to that of the anguilliforms. The body is deep. The snout is long and straight. The myomeres are somewhat V-shaped.

## 3.4 Morphological Changes During Metamorphosis

### 3.4.1 *Elopiform metamorphosis*

The elopiforms consist of only two families and two genera, *Elops* (tenpounder) and *Megalops* (tarpon). The developmental patterns during metamorphosis of both genera have been described in detail.

Larvae of *E. saurus* at various stages, including leptocephali, metamorphosing larvae, juveniles and adults, were captured on the South Atlantic coast of the United States. Sequential changes during metamorphosis were compiled from the specimens as shown in Fig. 4 (Gehringer, 1959). The metamorphosis of *E. saurus* was divided into three phases: early, mid, and late.

At the end of the leptocephalus stage, the larvae reach 40–45 mm in total length (Fig. 4b). During the early metamorphic period (Fig. 4b–d), the body shortens to about 25 mm but the head size remains constant. During mid metamorphosis (Fig. 4e and f), a marked change in body form from a leaf-like to a typical fish-like shape occurs, and the body further shortens to about 20 mm. Subsequently, it increases in length to 25 mm. During late metamorphosis, as the fish becomes a miniature adult (Fig. 4g), the length increases from 25 to 60 mm.

At the beginning of metamorphosis, the anal and dorsal fins are rudimentary or small (Fig. 4a–c) and, subsequently during metamorphosis, these fins develop a distinctive shape and shift forward (Fig. 4d–g). The pelvic fin buds appear at the early metamorphic stage (Fig. 4d) and by the late metamorphic stage the full complement of pelvic fins has formed (Fig. 4g). The body of the leptocephalus is transparent. However, larval melanophores are sparsely scattered along the digestive tract, between the myomeres along the mid-lateral line of the body, on the caudal fin,

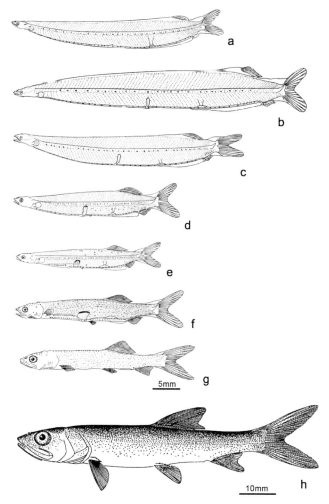

**Figure 4.** Morphological changes in a metamorphosing tenpounder (*Elops saurus*). (a) to (g) are drawn to scale. All drawings are taken from Gehringer (1959).

and between the myomeres at the anal ray bases (Fig. 4a and b). During metamorphosis the melanophores increase in size and number and also appear on the dorsal surface of the air bladder and on the head (Fig. 4c–f). At the end of metamorphosis, the pigmentation becomes denser and the body becomes opaque (Fig. 4g). The juveniles are a dense greenish-black in color, grading to silver below (Fig. 4h).

In a study by Sato and Yasuda (1980), larvae of *E. hawaiensis* captured in Ishigaki Inland, Japan were maintained in an aquarium at the salinity and temperature conditions of the collection site and were fed with rotifers,

*Branchionus plicatilis*, and water fleas, *Daphnia* sp. The process of *E. hawaiensis* metamorphosis that occurred in the reared fish was essentially similar to that of *E. saurus* except for the body length and the number of myomeres. However, the metamorphosis of *E. hawaiensis* was divided into two phases. The first and second metamorphic phases generally corresponded to the early and mid metamorphosis proposed by Gehringer (1959). The following period when the length again increased was classified as the juvenile phase. Under these captive conditions, the larvae passed through the first and second metamorphic phases in seven days, and six to eight days, respectively.

Fully-grown leptocephali of the Pacific tarpon, *M. cyprinoides*, were captured at Ohara harbor in Boso Peninsula, Japan and were maintained in the laboratory (Tsukamoto and Okiyama, 1997). Metamorphosis was examined in these cultured fish. The maximum size of the leptocephalus of this species was about 3 cm and was the smallest among the Elopomorpha. The morphological changes during metamorphosis were principally similar to those of *Elops*. The prominent characteristic of metamorphosis of this species was a relatively long delay between the metamorphic shortening and the subsequent growth in length. Thus, as shown in Fig. 5, metamorphosis was divided into three phases; leptocephalus negative growth phase (32–16 mm in length), sluggish growth phase (16–20 mm), and juvenile growth phase (20–40 mm). Although the body length was nearly constant during the sluggish phase, marked changes in body proportions took place similar to those in the mid metamorphosis of *E. saurus*. Ossification and intense pigmentation also commenced during this phase. The sluggish phase lasted about one month.

A unique feature of elopiform metamorphosis is that at the end of metamorphic shrinkage the larva is still quite immature and is still far from achieving juvenile morphology. Tsukamoto and Okiyama (1997) suggested that the most shrunken larvae of the elopiforms are similar to whitebait, larvae of the clupeiforms (herrings), gonorynchiforms (milkfishes), osmeriforms (freshwater smelts) and aulopiforms (lizardfishes), and that the subsequent growth pattern of the these larvae resembles that of whitebait-type larvae. On the basis of this style of metamorphosis, they regarded the Elopiformes as the most primitive among the Elopomorpha.

### 3.4.2 Albuliforme metamorphosis

This group of fishes possesses two different morphological types of leptocephalus with either a forked-tail (bonefishes) or a filamentous tail (halosaurs and spiny eels), as shown in Fig. 3. Only the metamorphosis of bonefishes is introduced in this section because no information is available on the metamorphosis of the filamentous-tailed leptocephali.

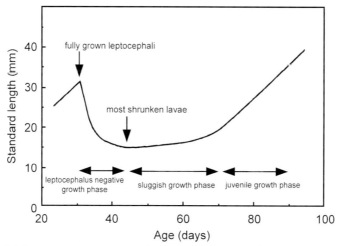

**Figure 5.** Metamorphic processes in the Pacific tarpon (*Megalops cyprinoides*). Reproduced from Tsukamoto and Okiyama (1997) with permission.

The report by Rasquin (1955) is the first detailed study of the metamorphosis of the bonefish (*Albula vulpes*). This author classified metamorphosing fish from Stage I (the largest leptocephalus, 55 mm) to Stage VII (completion of metamorphosis) based on observations of the sequence of morphological changes in the external appearance of reared fish, although the criteria for each stage were obscure. A series of studies by Pfeiler (reviewed by Pfeiler, 1986 and 2008) further examined physiological and biochemical changes during metamorphosis of the bonefish (*Albula* sp. A, undescribed species). In this study, the metamorphic period was simply divided into two stages termed Phases IIa and IIb (Fig. 6). Stages I to IV and Stages IV to VII of Rasquin (1955) appear to correspond to Phase IIa and IIb, respectively. The maximum size of the leptocephalus of *Albula* sp. A (>70 mm) was larger than that of *Albula vulpes*.

Histochemical examination revealed that the gelatinous matrix consisted of acid mucopolysaccharides and neutral fats (Rasquin, 1955). This tissue gradually disappeared as metamorphosis progressed and was considered to be utilized as a nutritive reserve. The rate of shrinkage of body length and the loss of wet mass were more rapid during Phase IIa than Phase IIb (Pfeiler, 2008). Despite the rapid reduction of body mass, the leaf-like body shape was mostly maintained during Phase IIa and then rapidly changed into the juvenile style during Phase IIb. By the end of metamorphosis, the body length had been reduced to about 40%, its dry mass had decreased to about half, and wet mass to 24% (Pfeiler, 1984a). Metamorphosis was complete within about 10 days (Rasquin, 1955; Pfeiler, 1984a). Unlike elopiforms, the body shape of albuliformes is juvenile-like when their body length was minimal at the end of metamorphosis.

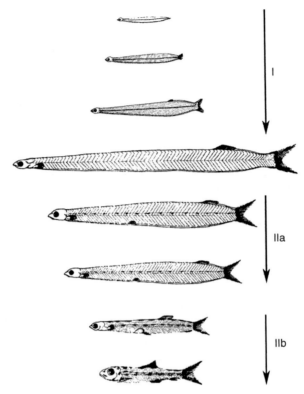

**Figure 6.** Morphological changes in larval and metamorphosing bonefish (*Albula* sp. A). All fish are drawn to scale and the largest larva is 63 mm in standard length. Reproduced from Pfeiler (2007) with permission.

Besides the change in water content, metamorphosis was accompanied by the changes in various biochemical conditions. Approximately 80–90% of the $Na^+$, $Cl^-$ and carbohydrate, and about 50% of the lipids, were lost during metamorphosis and were considered to be closely associated with the breakdown of the gelatinous matrix (Pfeiler, 1984a and b; Pfeiler and Luna, 1984). These changes in water and salt content were not affected by external salinities varying between 8 and 35‰ (Pfeiler, 1984c).

### 3.4.3 Anguilliform metamorphosis

This group comprises the largest number of species in the Elopomorpha and, consequently, they vary greatly in shape, size, habitat, and life history. Leptocephali of the anguilliforms are also diverse in their morphology. Although there are numerous diagrams of a variety of anguilliform leptocephali, the sequence of metamorphic change is poorly understood because of the inadequacy of collections of metamorphosing larvae and because of the difficulty of obtaining live leptocephali to examine metamorphosis in an aquarium. Information about metamorphic changes is therefore limited to a small number of species such as true eels and conger eels.

Numerous late larval stage leptocephali of the Japanese conger eel, *Conger myriaster*, appear in spring in coastal waters around Japan. They readily complete metamorphosis into elvers in an aquarium without feeding. For this reason, the conger eel is one of the best examined species in this order. As shown in Fig. 7, late stage leptocephali reached a maximum length of about 120 mm. During metamorphosis, body length was reduced to about 60% of its maximum, the leaf-like and transparent body of the leptocephalus became rod-shaped and opaque, and the anus was translocated to a position relatively much further forward. In association with the transformation of their body shape their habitat changed from pelagic to benthic. These metamorphic events were completed within 3 weeks (Yamano et al., 1991).

Takai (1959) divided metamorphosis into early, middle, and late stages based on statistical analysis of measurements of various parts of the body. During early metamorphosis, the body started to reduce in length while marked elongation of the dorsal and anal fins occurred. During the middle stage of metamorphosis, the ratios of head length, head height, eye diameter, interorbital width, snout length (distance from the tip of snout to the anterior edge of the eye), dorsal fin base length (distance from the origin of the dorsal fin to the end of the fin), and upper jaw length, relative to body length, all increased. During late metamorphosis, the ratios of head length, interorbital

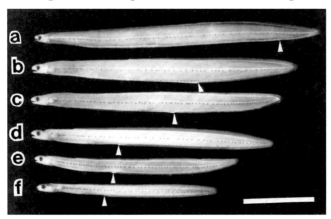

**Figure 7.** Morphological changes in metamorphosing conger eel (*Conger myriaster*). Fish were fixed in 10% formalin. (a) leptocephalus at the maximum size. (b)–(e) metamorphosing larvae. (f) fish at the end of metamorphosis. Arrowheads show the positions of the anus. Scale indicates 3 cm. Reproduced from Yamano et al. (1991) with permission.

width, body width, and pectoral fin length, relative to body length increased, whereas that of body height to body length decreased.

In the classification of larva to adult stages described by Kubota (1961), the metamorphic period ranged from stage IV to VII. Stage IV included larvae at the early phase of metamorphosis with a leptocephalus-type body (<126 mm in body length). Stage V (95–110 mm) was characterized by a body appearance intermediate between leptocephalus and elver. At stage VI (85–105 mm), the leptocephalus body shape was completely transformed. The development of internal organs and erythrocytes were visible through the skin. At stage VII (< 90 mm), the body became almost juvenile-like although the pigmentation was incomplete. The gas bladder was filled with gas.

Kubota (1961) examined the development of various tissues and organs during conger eel metamorphosis. Adult-type melanophores (smaller than larval melanophores) generally appeared from stage V or VI. However, individual differences in the stage of the pigmentation were relatively large compared with changes in other organs. Erythrocytes appeared at stage VI. The epidermal layer of the skin became thick and mucous cells in the skin increased in number at stages V and VI. The kidney of leptocephalus larvae was composed of a protonephridium and a mesonephric kidney lacking glomeruli. During metamorphosis, the protonephridium degenerated from its posterior end and a number of glomeruli developed in the mesonephric kidney. By the end of metamorphosis the internal structure of the mesonephric kidney became similar to that of the adult. The digestive tract transformed markedly during metamorphosis associated with enlargement and torsion of the stomach and shortening and thickening of the esophagus and intestines. A functional stomach was thus formed. Ossification of the vertebra, of the caudal regions, and of the head (except for a few parts in the skull) took place at the beginning of metamorphosis and was essentially complete by the end of metamorphosis. The ossification in the vertebra proceeded faster in the anterior and posterior regions than in the mid region.

The metamorphic processes of *Anguilla* sp. are similar to those of the conger eel except for the rate of body reduction; metamorphosing *Anguilla* maintain their body length while body shape drastically changes. Such metamorphic processes were first elucidated from the collections of wild metamorphosing Atlantic eels. Recently, after a challenging long-term research program, glass eels of the Japanese eel, *A. japonicus*, were successfully produced from artificially fertilized eggs by feeding larvae with a slurry-type diet made from shark-egg powder and supplements (Tanaka et al., 2003; Tanaka, 2003). This review introduces observations on metamorphosis that occurred under aquarium conditions during this study.

Fully grown leptocephali reached 50–60 mm in length approximately 250 days after hatching and then commenced metamorphosis (Fig. 8). During metamorphosis, the positions of the anus, and of the anterior edges of the dorsal and anal fins, moved forward. Body depth was drastically reduced. Black pigment and the color of the blood became visible. Metamorphosis was completed within about 2 weeks. The development of digestive system was examined using artificially produced leptocephali (Kurokawa et al., 1995; Kurokawa et al., 1996; Kurokawa and Pedersen, 2003). The segmentation of the gut into the esophagus, intestine and rectum, together with the synthesis of pancreatic the digestive enzymes trypsin, amylase, and

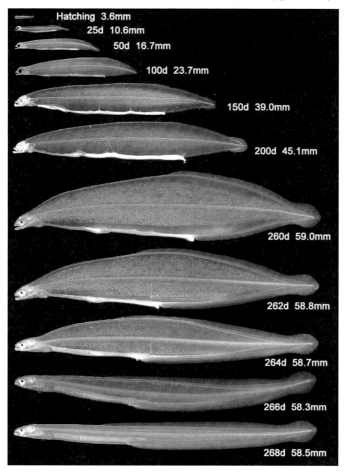

**Figure 8.** Morphological changes in larval and metamorphosing Japanese eel (*Anguilla japonica*). Live larvae were photographed. Fish at stages from hatching to 200 days are representatives of reared fish. Metamorphosing fish from 260–269 days were the same individual. Provided from Dr. Tanaka.

lipase, all occurred before the start of feeding (8 days after hatching). The stomach and gastric glands developed much later at metamorphosis. This suggests that changes in dietary habits may accompany metamorphosis.

This success in producing glass eels is of great importance, both for future commercial utilization of artificially produced glass eels for the seed in aquaculture, and for understanding the developmental biology of eels. However, improvements in the techniques for rearing leptocephali are still required, because there still remain significant problems that must be solved. The survival rate was only 4% 50 days after hatching and declined to 1% by 100 days after hatching (Tanaka, 2003). Moreover, the growth rate of reared larvae was less than half of that estimated for wild larvae (Tsukamoto et al., 1994). The proportions of body dimension, particularly the ratio of body depth to body length, and the number of pre-anal myomeres, also were different between wild and artificial larvae (Tanaka et al., 2001).

## 3.5 Endocrine Control of Eel Metamorphosis

### 3.5.1 Thyroidal regulation

The involvement of the thyroid axis in fish metamorphosis in a manner comparable to that in amphibian metamorphosis has been established in various teleosts and is reviewed in other chapters in this book. Although endocrinological studies of elopomorph metamorphosis are scarce, and are limited to anguilliforms, all of the data strongly support the regulation of eel metamorphosis by thyroid hormone.

Firstly, histological observations of leptocephali before and during metamorphosis clearly indicated that the development and activation of the thyroid gland were related to the occurrence of metamorphosis. Thyroid follicles were observed in all wild leptocephali of the Japanese eel ranging from 19.8–32.6 mm in total length but they were smaller and less abundant than in glass eels (Ozaki et al., 2000). A detailed examination using artificially produced leptocephali of the Japanese eel further revealed that a single thyroid follicle first appeared in larvae 12.2 mm in total length and that the number of follicles increased in proportion to total length from 25 to 45 mm (Yamano et al., 2007). During metamorphosis of the conger eel, the thyroid follicular cells became columnar with larger nuclei (Fig. 9b) (Yamano et al., 1991). Active uptake of colloid by the epithelial cells was also notable (Fig. 9c). These morphological features of the thyroid gland are evidence of increased function of the gland. After the completion of metamorphosis the thyroid gland decreased its histological activity (Fig. 9d). From histological observations of wild *Anguilla* leptocephali, it has also been suggested that TSH-producing cells in the pituitary appear just prior to the initiation of metamorphosis (Ozaki et al., 2000).

Secondly, whole body concentrations of thyroxine ($T_4$) and triiodothyronine ($T_3$) became elevated during metamorphosis. In the conger eel, the levels of $T_4$ increased from less than 5 ng/g to the maximum level about 30 ng/g during the early metamorphic stage, and then declined toward the end of metamorphosis (Yamano et al., 1991). On the other hand, levels of $T_3$ (less than 0.15 ng/g in premetamorphic stage) increased gradually during early metamorphosis and then steeply increased to about 2 ng/g in

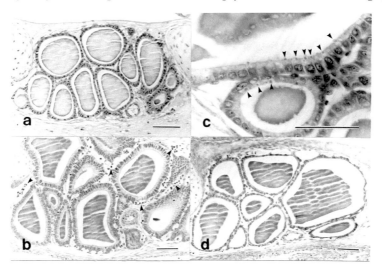

**Figure 9.** Histological sections of the thyroid glands of the conger eel (*Conger myriaster*). (a) thyroid follicles of premetamorphic larva, (b) thyroid follicles of early metamorphic larva. Note columnar cells and abundant capillaries (arrowheads) surrounding follicles. (c) active uptake of colloid (arrowheads) by follicular cells. (d) thyroid follicles of late metamorphic larva. Scale bars represent 50 μm. Reproduced from Yamano et al. (1991) with permission.

late metamorphosis (the profile is illustrated in the upper figure of Fig. 13). Thyroid hormone levels also increased during Japanese eel metamorphosis but the patterns were different (Yamano et al., 2007). $T_4$ continued increasing during metamorphosis and reached the highest level in fish at the end of metamorphosis, or in fish that had just completed metamorphosis (Fig. 10, lower figure). In contrast, $T_3$ levels peaked during the late stages of metamorphosis and declined toward the end of metamorphosis (Fig. 10, upper figure). The significance of the difference in the thyroid hormone profiles between these two species, as well as functional divergence of $T_3$ and $T_4$, are unknown.

Thirdly, the administration of thyroid hormone to premetamorphic leptocephali induced or accelerated metamorphosis. Immersion of premetamorphic leptocephali in seawater containing $T_4$ at 2 ppm and 6 ppm for one week stimulated the initiation of metamorphosis in the conger eel (Kitajima et al., 1967). Sequential metamorphic transitions of the conger eel for four weeks induced by the treatment either with $T_4$, or with the anti-thyroidal drug thiourea, are shown in Fig. 11 (Yamano unpublished

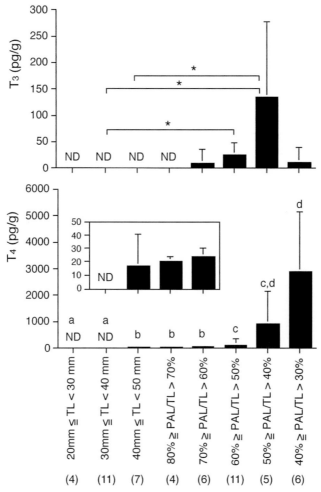

**Figure 10.** Changes in the levels of thyroxine ($T_4$) and triiodothyronine ($T_3$) during larval and metamorphic development of the Japanese eel (*Anguilla japonica*). In order to group the larvae, either TL (total length), or the ratio PAL (length from the proboscis to anus)/ TL % , were used for the fish, before and during metamorphosis, respectively. Numerals in parentheses indicate the numbers of animals in each group. Bars indicate the means and standard deviations. Asterisks in the upper graph and different letters on bars in the lower graph indicate significant differences ($P < 0.05$, Mann-Whitney $U$-test). The inset in the lower graph is shown on a different scale of Y values. ND, non-detectable. Reproduced from Yamano et al. (2007) with permission.

data). On the first day of the experiment, the value of anal length/total length was > 85% in all but two of the leptocephali (upper panel), and the blood color was indiscernible (lower panel). These leptocephali were separated into three tanks with either normal seawater (n=11), seawater with $T_4$ at 100 $\mu$M (n=12), or with thiourea at 5 mM (n=11). One week later, half of the control fish had naturally started metamorphosis, as revealed by the decrease in the value of anal length/total length, and by erythrocyte production, which was recognizable through the transparent skin. By four weeks, nine fish had completed metamorphosis and 2 fish remained as leptocephali. The treatment with $T_4$ induced metamorphosis in all fish. The induced metamorphosis progressed faster than metamorphosis of controls. Thiourea barely inhibited the change in body shape, but delayed the production of erythrocytes.

### 3.5.2 Thyroid hormone receptor (TR)

It has been well demonstrated, in higher vertebrates, that the ligand-TR complex forms a homodimer or heterodimer with the retinoid receptor and then modulates gene expression through binding to a specific regulatory nucleotide sequence of a target gene. Accordingly, eel metamorphosis must be triggered by the modification of gene expression by TR. Fish TR genes have already been cloned from more than ten fish species, including anguilliform eels (Marchand et al., 2001; Kawakami et al., 2003a; Kawakami et al., 2003b; Kawakami et al., 2007).

Phylogenetic relationships based on the amino acid sequences of TRs are illustrated in Fig. 12. Two types of TR, TRα and TRβ were widely distributed in fishes, as observed in higher vertebrates. The TRα of a given species has a higher similarity to the TRα of other species than to its own TRβ, even between fish and humans. A corresponding situation exists for TRβ. With respect to TRα, two subtypes, TRαA and TRαB, were identified in the more advanced fish groups, providing distinct groups in the phylogenetic analysis. Although the conger eel also has both TRαA and TRαB, the origin of these subtypes in the conger eel and in the more advanced fishes seems to be different. Conger eel TRαA and TRαB are positioned apart from those of more advanced fish groups and possessed a stronger similarity with each other. Thus, phylogenetic analysis strongly suggested that a TRα gene in the conger eel lineage may have been duplicated independently from the duplication in the more advanced fish groups. It remains to be resolved whether this is a feature common to elopomorph fishes or is peculiar to specific taxonomic groups of the Elopomorpha.

Unlike TRα, the single TRβ gene has been identified in all fishes except for the conger eel. Therefore, a gene duplication event, specific for the conger eel lineage, also seems to have taken place in the TRβ gene. The TRβ that

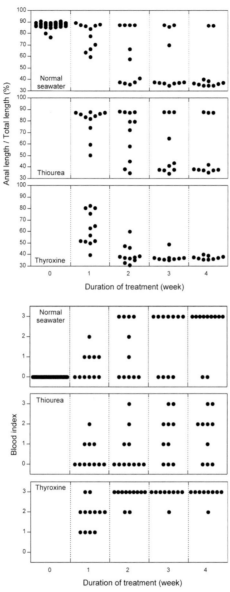

**Figure 11.** Effects of exogenous thyroxine and the goitrogen thiourea on the metamorphosis of the conger eel (*Conger miriaster*). Leptocephali almost at their largest size were kept in a tank with either normal seawater (n=11), seawater with thyroxine (100 nM, n=12) or seawater with thiourea (5 mM, n=11) for four weeks. PAL (length from the proboscis to anus)/TL (total length) % (upper figure) and the production of erythrocytes (lower figure) were measured every week. Dots indicate the values for each individual. Blood index: 0, completely invisible; 1, erythrocytes in the heart become slightly visible; 2, erythrocytes in the heart become distinct; 3, erythrocytes in the dorsal aorta become distinct.

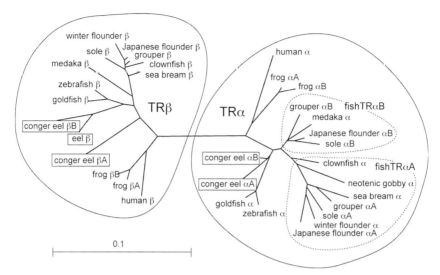

**Figure 12.** Phylogenetic tree of the thyroid hormone receptor (TR). Deduced amino acid sequences of B to D domains, which include DNA- and ligand-binding domains, were used for the comparison. The alignments were performed on the ClustalW program with a pairwise alignment algorithm. The lengths of lines indicate the genetic distances.

is specific to *A. anguilla* has the highest similarity with conger eel TRβB. Whether *A. anguilla* has a single TRβ or also possesses a second TRβ similar to TRβA of the conger eel, is open to further study.

The existence of multiple TRs within a species could imply a distinct and indispensable function of each TR type. Indeed, different expression profiles of TRs were revealed in conger eel metamorphosis (Fig. 13) (Kawakami et al., 2003a; Kawakami et al., 2003b). TRαA and TRαB transcript levels peaked at metamorphic climax and then declined toward the completion of metamorphosis. TRβA levels also increased and reached a peak during the metamorphic period but, unlike TRαs, the peak was maintained during the glass eel, elver and young eel stages. TRαA, TRαB and TRβA transcripts were ubiquitously detected in various tissues in the adult conger eel whereas TRβB transcripts were predominantly expressed in the brain and pituitary. The expression of TRβB increased in the whole head during metamorphosis and reached a peak in elvers. The specific expression of TRβB in the brain and pituitary was also seen in the Japanese eel (Kawakami et al., 2007).

These differential patterns in the levels and distribution of each TR strongly suggest diverse function for each of the TRs. Metamorphosis is, as described earlier, accompanied by various morphological and biochemical changes in the body, which can be driven by thyroid hormone. Thyroid hormone is capable of modulating its function at the receptor level by using

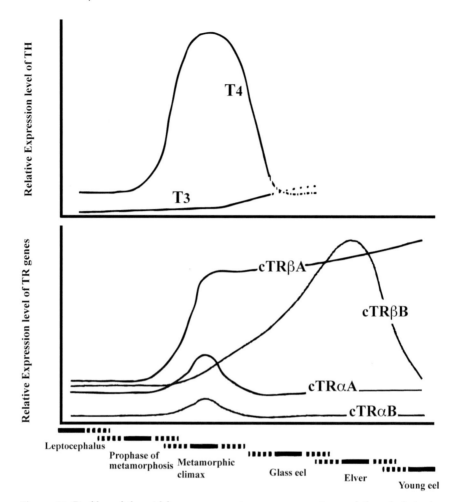

**Figure 13.** Profiles of thyroid hormone receptor gene expression and the whole body concentrations of thyroxine (T$_4$) and triiodothyronine (T$_3$) in the conger eel (*Conger miriaster*). The figure is redrawn from Kawakami et al. (2003a) with permission but the names of TRbs were changed according to the proposed re-definition of Kawakami et al. (2007).

a different TR type and this may partly explain why it is able to influence various parts of the body simultaneously. In eel metamorphosis, the hypothalamus—pituitary—thyroid axis is probably regulated principally by TRβB. Other TRs may be associated with the regulation of gene expression in various parts of the body, and thereby induce morphological and biochemical changes.

## 3.6 Timing and Duration of Metamorphosis of Catadromous Eels in the Ocean

### 3.6.1 Life cycle of catadromous eels

Almost all elopomorphs are marine or primarily marine inhabitants and only six anguillid species are catadromous (Nelson, 2006). However, the rather unusual life history of the catadromous eels has attracted our interest for many years. Fully-grown adults of catadromous eels move down from rivers or lakes to the sea for reproduction and swim thousands of kilometers toward the oceanic spawning grounds where they were born. After spawning, adults are assumed to die. Hatched larvae, termed preleptocephali, develop into leptocephali shortly after the initiation of feeding. Leptocephali grow in the ocean and drift in the ocean currents for periods of several months to more than a year. Fully-grown leptocephali reach the continental shelf near their freshwater habitat and metamorphose into glass eels in the ocean. Glass eels migrate to coastal waters and then enter the river as elvers. This life cycle was investigated in Atlantic eels (American eel, *A. rostrata* and European eels, *A. anguilla*), and then in Japanese eels (*A. japonica*). The life cycles, especially the metamorphic periods, of Atlantic eels and Japanese eels are described in this section.

### 3.6.2 Area of metamorphosis

In the early twentieth century, cruises in search of leptocephali of Atlantic eels were repeatedly carried out in order to discover the spawning places of the eels. Based upon the size distribution of *A. rostrata* and *A. anguilla* leptocephali, the spawning places of these two species were inferred to be in overlapping areas of the Sargasso Sea (Schmidt, 1922). The spawning places and the larval distribution were later re-examined by analyzing several thousands of leptocephali from these collections (Kleckner and McCleave, 1985; McCleave et al., 1987). In these studies, metamorphosing leptocephali were found near the east coast of the North American Continent for *A. rostrata*, and near the west coast of the European Continent for *A. anguilla*. These areas were distant from the spawning grounds. It can be concluded from their distribution patterns that leptocephali of *A. rostrata* are transported by the North Equatorial Current and the Gulf Stream to the continental shelf of North America, while leptocephali of *A. anguilla* are moved farther by the North Atlantic Current to the coastal regions of European countries (Fig. 14). The metamorphosis into glass eels occurs during the last stages of their oceanic life and the glass eels swim toward coastal waters.

Research into the oceanic life of *A. japonica* began late in comparison with that of Atlantic eels (reviewed by Tsukamoto et al., 2003). The first leptocephalus was captured in 1967 to the south of Taiwan (Matsui et al.,

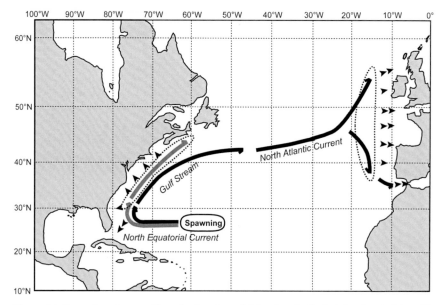

**Figure 14.** Spawning place and larval migration of Atlantic eels. Both Atlantic eels (*Anguilla anguilla* and *A. rostrata*) spawn in the Sargasso Sea. Bold grey and black lines indicate the migration pathways of *A. rostrata* and *A. anguilla*, respectively. Dotted areas are the places where metamorphosis into glass eel occurs from where they migrate toward estuaries.

1968) and a large number of leptocephali (n=958) were collected in 1991 (Tsukamoto, 1992). After repeated cruises thereafter, the spawning ground was proposed to be located near a seamount to the west of the Mariana Islands (Fig. 15). The capture of ripe males and females of *A. japonica* in the neighboring area further confirmed this postulation (Chow et al., 2009; Tsukamoto et al., 2011). Larvae are considered to be transported by the North Equatorial Current and then by Kuroshio Current. Metamorphosing larvae were found in the Kuroshio Current, typically to the east of Taiwan and to the west of Ryukyu Islands (Tsukamoto, 2006; Otake et al., 2006). Glass eels move toward the northern Philippines, Taiwan, China, North and South Korea, and Japan.

### 3.6.3 Age at metamorphosis

Estimates of the duration of the period of growth up to the glass eel stage vary widely among different researchers (Table 2). Schmidt (1922) remarked that *A. rostrata* can complete its full development from egg to

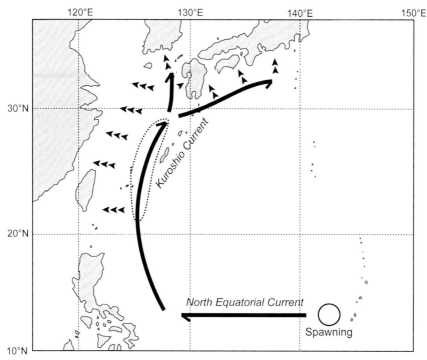

**Figure 15.** Spawning place and larval migration of the Japanese eel. Japanese eels (*Anguilla japonica*) spawn west of the Mariana Islands. The black line indicates the migration pathway of *A. japonica*. Dotted areas are the places where metamorphosis into glass eel occurs. From there, glass eels migrate toward estuaries of eastern Asian countries.

glass eel in about one year whereas *A. anguilla* takes longer than two years. Subsequently, based on the patterns in the appearance of larvae and glass eels, the timing of the metamorphosis after hatching has been estimated to be 8–12 months (Kleckner and McCleave, 1985) for *A. rostrata* and 12–15 months (BoeÈtius and Harding, 1985) for *A. anguilla*.

Analysis of otolith microstructure is another method for estimating the timing of life history stages during larval development. The daily otolith increment is marked as a ring pattern and the increment width reflects the growth rate (Fig. 16). In addition, acute physiological events such as hatching and metamorphosis are recorded as a discontinuous zone (Fig. 16). Sr : Ca ratios in the otolith, which change in association with environmental and physiological conditions, decrease sharply at the onset of eel metamorphosis (Otake et al., 1994). Thus, the larval life history can be reconstructed from the otoliths of glass eels.

From otolith analysis, the age at metamorphosis was estimated to be about 200 days (Lecomte-Finiger, 1992; Arai et al., 2000) and 350 days (Wang

**Table 2.** Timing and duration of metamorphosis and the time of arrival at the estuary for three *Anguilla* species.

| Species | Onset of metamorphosis | Duration of metamorphosis | Estuary arrival | Method of estimation | Reference |
|---|---|---|---|---|---|
| *Aniguilla anguilla* | 2.5 y | | 3 y | larval distribution | Schmidt (1922) |
| | 12–15 m | | | larval distribution | Boeĕtius and Harsing (1985) |
| | 176–196 d | 33–76 d | 216–276 d | otolith | Lecomte-Finiger (1992) |
| | 319–397(350) d | | 420–467(448) d | otolith | Wang and Tzeng (2000) |
| | 163–235(198) d | 18–52(32) d | 220–281(249) d | otolith | Arai et al. (2000) |
| *Anguilla rostrata* | 1 y | | 1.5 y | larval distribution | Schmidt (1922) |
| | 8–12 m | | | larval distribution | Kleckner and McCleave (1985) |
| | 189–214(200) d | | 220–284(255) d | otolith | Wang and Tzeng (2000) |
| | 132–191(156) d | 18–41(31) d | 171–252(206) d | otolith | Arai et al. (2000) |
| *Anguilla japonica* | 80–160 d | 20–40 d | 150–177(170) d | otolith | Arai et al. (1997) |
| | 125–142(135) d | 10–15(15) d | 155–182 d | otolith | Kawakami et al. (1998) |
| | 116–138 d | | | otolith | Cheng and Tzeng (1996) |

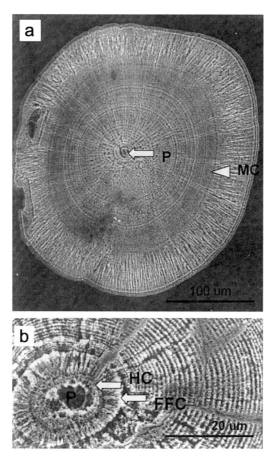

**Figure 16.** SEM micrographs illustrating the daily growth increments in otoliths of elvers of *Anguilla rostrata*. c: partial magnification of a. P: primordium; HC: hatching check; FFC: first feeding check; MC: metamorphosis check. Reproduced from Wang and Tzeng (2000) with permission.

and Tzeng, 2000) for *A. anguilla*, 160–200 days for *A. rostrata* (Wang and Tzeng, 2000; Arai et al., 2000), and 80–160 days for *A. japonica* (Cheng and Tzeng, 1996; Arai et al., 1997; Kawakami et al., 1998) (Table 2). Although these studies showed considerable variation in the age at metamorphosis, even within the same species, the timings appear to be correlated with the distance from the spawning place to the area where metamorphosis occurs. Accordingly, *A. anguilla* has a longer larval period than *A. rostrata*, and *A. japonica* has the shortest larval period among the three species. Intraspecific geographic differences in the age at metamorphosis have been found for *A. japonica*. Fish migrating to Taiwan metamorphose earlier than fish migrating to Korea or Japan (Cheng and Tzeng, 1996).

### 3.6.4 Duration of metamorphosis

The duration of metamorphosis, estimated from otolith examinations, ranged from 18–76 days for *A. anguilla*, 18–41 days for *A. rostrata* and 10–40 days for *A. japonica* (Table 2). Assuming that estimations by otolith examination are reliable, such flexibility in the duration of metamorphosis appears to be unique to *Anguilla* species. Hitherto, such a high variability had not previously been observed in any other elopomorph. Despite the fact that each *Anguilla* species spawns in a single specific area of the ocean, their habitats during freshwater life, which are remote from their spawning grounds, are relatively widespread. Therefore, variation in metamorphic duration, as well as in the day of metamorphic onset, may be adaptive to their distribution over wide areas.

The daily otolith increments in the zone from the first feeding check (FFC; Fig. 16) to the metamorphosis check (MC) are wide. However, the increments become very narrow and undistinguishable at MC. The increment width then becomes wider outside of the MC zone. This implies that the growth rate of otoliths is slower in metamorphosing larvae than in premetamorphic larvae and glass eels. The energy for metabolism during metamorphosis is, perhaps, utilized for remodeling rather than growth.

## Conclusion

The elopomorph fishes are characterized by the occurrence of a distinctive leptocephalus larval stage despite great differences in the morphology of adult elopomorphs. Recent phylogenetic analyses appear to support the monophyly of the Elopomorpha. Because leptocephalus-like larvae are not seen in any other teleosts, the elopomorphs have developed a strategy specific for this taxonomic group. The fact that the elopomorphs are distributed all over the ocean from the deep sea to coastal waters implies that the strategy has been profitable in the evolution and divergence of the elopomorphs. The morphology of leptocephali is characterized by a particular leaf-like body shape. Compared with larva of other teleosts, leptocephali grow to a very large size (approximately 3 cm in total length even in the smallest case) and the larval period is exceptionally long, lasting several months in cases where the life cycle has been elucidated. These features strongly suggest adaptation to their distribution by ocean currents, from their spawning grounds to habitats spread over a wide area.

When larvae cease planktonic life to settle in the adult habitat, they metamorphose into juveniles. The metamorphosis is accompanied by the degradation of larval tissues such as the gelatinous matrix, remodeling of tissues from a larval to juvenile type, and development of adult-type tissues. Thus, they acquire mobility and adaptability appropriate to their new habitats. The metamorphosis of eels is apparently triggered by thyroid hormone and similar hormonal regulation is probably present in other elopomorphs. Although other teleosts demonstrate different types of morphological changes during metamorphosis, thyroid hormone in teleosts appears to have the common function to induce a larva-juvenile transition. Although the elopomorphs are a taxonomically low group of teleosts, they possess the pituitary-thyroid axis and its specific receptor system is also present. Because the lamprey has a different mechanism of regulation of metamorphosis, it would be of great interest to examine the emergence of thyroidal function in this vertebrate lineage.

## Acknowledgements

A part of my own research was supported by a grant, The production of eel juvenles, from the Ministry of Agriculture, Forestry and Fisheries. I thank Dr. Tanaka for providing photographs of eel leptocephali and Dr. Kurogi for giving a specimen of wild eel leptocephali.

## References

Arai, T., T. Otake and K. Tsukamoto. 1997. Drastic changes in otolith microstructure and microchemistry accompanying the onset of metamorphosis in the Japanese eel *Anguilla japonica*. Marine Ecology Progress Series 161: 17–22.

Arai, T., T. Otake and K. Tsukamoto. 2000. Timing of metamorphosis and larval segregation of the Atlantic eels *Anguilla rostrata* and *A. anguilla*, as revealed by otolith microstructure and microchemistry. Marine Biology 137: 39–45.

BoeÈtius, J. and E.F. Harding. 1985. A re-examination of Johannes Schmidt's Atlantic eel investigations. *Dana* 4: 129–162.

Cheng, P.W. and W.N. Tzeng. 1996. Timing of metamorphosis and estuarine arrival across the dispersal range of the Japanese eel *Anguilla japonica*. Marine Ecology Progress Series 131: 87–96.

Chow, S., H. Kurogi, N. Mochioka, S. Kaji, M. Okazaki and K. Tsukamoto. 2009. Discovery of mature freshwater eels in the open ocean. Fisheries Science 75: 257–259.

Filleul, A. and S. Lavoué. 2001. Basal teleosts and the question of elopomorph monophyly. Morphological and molecular approaches. Comptes Rendus de l'Académie des Sciences, Series III Sciences de la Vie 324: 393–399.

Gehringer, J.W. 1959. Early development and metamorphosis of the ten-pounder. Fishery Bulletin 155: 615–647.

Grassi, G.B. 1896. The reproduction and metamorphosis of the common eel (*Anguilla vulgaris*). Journal of Cell Science 155: 371–385.

Greenwood, P.H., D.E. Rosen, S.H. Weitzman and G.S. Myers. 1966. Phyletic studies of teleostean fishes, with a provisional classification of living forms. Bulletin of the American Museum of Natural History 131: 339–456.

Inoue, J.G., M. Miya. K. Tsukamoto and M. Nishida. 2004. Mitogenomic evidence for the monophyly of elopomorph fishes (teleostei) and the evolutionary origin of the leptocephalus larva. Molecular Phylogenetics and Evolution 32: 274–286.

Kawakami, T., N. Mochioka and A. Nakazono. 1998. Immigration period and age of *Anguilla japonica* glass eels entering rivers in northern Kyushu, Japan during 1994. Fisheries Science 64: 235–239.

Kawakami, Y., M. Tanda, S. Adachi and K. Yamauchi. 2003a. cDNA cloning of thyroid hormone receptor βs from the conger eel, *Conger myriaster*. General and Comparative Endocrinology 131: 232–240.

Kawakami, Y., M. Tanda, S. Adachi and K. Yamauchi. 2003b. Characterization of thyroid hormone receptor α and β in the metamorphosing Japanese conger eel, *Conger myriaster*. General and Comparative Endocrinology 132: 321–332.

Kawakami, Y., S. Adachi, K. Yamauchi and H. Ohta. 2007. Thyroid hormone receptor β is widely expressed in the brain and pituitary of the Japanese eel, *Anguilla japonica*. General and Comparative Endocrinology 150: 386–394.

Kitajima, C., T. Sato and M. Kawanishi. 1967. On the effect of thyroxine to promote the metamorphosis of a conger eel. Bulletin of the Japanese Society of Scientific Fisheries 33: 919–922.

Kurokawa, T., H. Kagawa, H. Ohta, H. Tanaka, K. Okuzawa and K. Hirose. 1995. Development of digestive organs and feeding ability in larvae of Japnese eel (*Anguilla japonica*). Canadian Journal of Fisheries and Aquatic Sciences 52: 1030–1036.

Kurokawa, T., H. Tanaka, H. Kagawa and H. Ohta. 1996. Absorption of protein molecules by the rectal cells in eel larvae *Anguilla japonica*. Fisheries Science 62: 832–833.

Kurokawa, T. and B.H. Pedersen. 2003. The digestive system of eel larvae. In: *Eel Biology*, K. Aida, K. Tsukamoto and K. Yamauchi (Eds.). Springer-Verlag, Tokyo, pp. 435–444.

Kleckner, R.C. and J.D. McCleave. 1985. Spatial and temporal distribution of American eel larvae in relation to North Atlantic ocean current systems. Dana 4: 67–92.

Kubota, S. 1961. Studies on the ecology, growth and metamorphosis in conger eel, *Conger myriaster* (Brevoort). Journal of Faculty of Fisheries, Prefecural University of Mie 5: 190–370.

Lecomte-Finiger, R. 1992. Growth history and age at recruitment of European glass eels (*Anguilla anguilla*) as revealed by otolith microstructure. Marine Biology 114: 205–210.

MacCleave, J.D., R.C. Kleckner and M. Castonguay. 1987. Reproductive sympatry of American and European eels and implications for migration and taxonomy. American Fisheries Society Symposium 1: 286–297.

Marchand, O., R. Safi, H. Escriva, E.V. Rompaey, P. Prunet and V. Laudet. 2001. Molecular cloning and characterization of thyroid hormone receptor in teleost fish. Journal of Molecular Endocrinology 26: 51–65.

Matsui, I., T. Takai and A. Kataoka. 1968. Anguillid leptocephalus found in the Japan Current and tis adjacent waters. The Journal of the Shimonoseki University of Fisheries 17: 17-23.

Nelson, J.S. 2006. *Fishes of the World*. John Wiley & Sons, Inc., New Jersey. Fourth Edition.

Obermiller, L.E. and E. Pfeiler. 2003. Phylogenetic relationships of elopomorph fishes inferred from mitochondrial ribosomal DNA sequences. Molecular Phylogenetics and Evolution 26: 202–214.

Otake, T., T. Ishii, M. Nakahara and R. Nakamura. 1994. Drastic changes in otolith strontium/ calcium ratios in leptocephali and glass eels of Japanese eel *Anguilla japonica*. Marine Ecology Progress Series 112: 189–193.

Otake, T., M.J. Miller, T. Inagaki, G. Minagawa, A. Shinoda, Y. Kimura, S. Sasai, M. Oya, S. Tasumi, Y. Suzuki, M. Uchida and K. Tsukamoto. 2006. Evidence for migration of metamorphosing larvae of *Anguilla japonica* in the Kuroshio. Coastal Marine Science 30: 453–458.

Ozaki, Y., H. Okumura, Y. Kazeto, T. Ikeuchi, S. Ijiri, M. Nagase, S. Adachi and K. Yamauchi. 2000. Developmental changes in pituitary-thyroid axis, and formation of gonads in leptocephali and glass eels of *Anguilla* spp. Fisheries Science 66: 1115–1122.

Pfeiler, E. 1984a. Changes in water and salt content during metamorphosis of larval bonefish (*Albula*). Bulletin of Marine Science 34: 177–184.

Pfeiler, E. 1984b. Glycosaminoglycan breakdown during metamorphosis of larval bonefish *Albula*. Marine Biology Letter 5: 241–249.

Pfeiler, E. 1984c. Effect of salinity on water and salt balance in metamorphosing bonefish (*Albula*) leptocephali. Journal of Experimental Marine Biology and Ecology 82: 183–190.

Pfeiler, E. 1986. Towards an explanation of the developmental strategy in leptocephalus larvae of marine teleost fishes. Environmental Biology of Fishes 15: 3–13.

Pfeiler, E. 2008. Physiological ecology of developing bonefish larvae. In: *Biology and Management of the World Tarpon and Bonefish Fisheries*, Ault, J.S. (Ed.). CRC Press, Boca Raton London New York, pp. 179–193.

Pfeiler, E. and A. Luna. 1984. Changes in biochemical composition and energy utilization during metamorphosis of leptocephalus larvae of the bonefish (*Albula*). Environmental Biology of Fishes 10: 243–251.

Rasquin, P. 1955. Observations on the metamorphosis of the bonefish *Albula vulpes* (Linnaeus). Journal of Morphology 97: 77–117.

Sato, M. and F. Yasuda. 1980. Metamorphosis of the leptocephali of the ten-pounder, *Elops hawaiensis*, from Ishigaki Island, Japan. Japanese Journal of Ichthyology 26: 315–324.

Schmidt, J. 1922. The breeding places of the eel. Philosophical Transactions of the Royal Society of London. Series B 211: 179–208.

Takai, T. 1959. Studies on the morphology, ecology and culture of the important apodal fishes, *Muraenesox cineres* (Forskål) and *Conger myriaster* (Brevoort). The Journal of the Shimonoseki College of Fisheries 8: 209–555.

Tanaka, H., H. Kagawa and H. Ohta. 2001. Production of leptocephali of Japanese eel (*Anguilla japonica*) in captivity. Aquaculture 201: 51–60.

Tanaka, H. 2003. Techniques for larval rearing. In: *Eel Biology*, K. Aida, K. Tsukamoto and K. Yamauchi (Eds.). Springer-Verlag, Tokyo, pp. 427–434.

Tanaka, H., H. Kagawa, H. Ohta, T. Unuma and K. Nomura. 2003. The first production of glass eel in captivity: fish reproductive physiology facilitates great progress in aquaculture. Fish Physiology and Biochemistry 28: 493–497.

Tchernavin, V. 1946. A living bony fish which differs substantially from all living and fossil osteichthyes. Nature (London)158: 667–667.

Tsukamoto, K. 1992. Discovery of the spawning area for Japanese eel. Nature (London) 356: 789.

Tsukamoto, K., T.W. Lee and M. Mochioka. 1994. Age and growth of Japanese eel leptocephali. In: *Preliminary Report of the Hakuho Maru Cruise KH-91-4*, T. Otake and K. Tsukamoto (Eds.). Ocean Research Institute, University of Tokyo, Tokyo, pp. 50–54.

Tsukamoto, K., T.W. Lee and H. Fricke. 2003. Spawning area of the Japanese eel. In: *Eel Biology*, K. Aida, K. Tsukamoto, and K. Yamauchi (Eds.). Springer-Verlag, Tokyo, pp. 121–155

Tsukamoto, K. 2006. Spawning of eels near a seamount. Nature (London) 439: 929.

Tsukamoto, Y. and M. Okiyama. 1997. Metamorphosis of the pacific tarpon, *Megalops cyprinoides* (Elopiformes, Megalopidae) with remarks on development patterns in the Elopomorpha. Bulletin of Marine Science 60: 23–26.

Tsukamoto, K., S. Chow, T. Otake, H. Kurogi, N. Mochioka, M. J. Miller, J. Aoyama, S. Kimura, S. Watanabe, T. Yoshinaga, A. Shinoda, M. Kuroki, M. Oya, T. Watanabe, K. Hata, S. Ijiri, Y. Kazeto, K. Nomura and H. Tanaka. 2011. Oceanic spawning ecology of freshwater eels in the western North Pacific. Nature Communications 2:179 doi: 10.1038/ncomms1174.

Wang, C.H. and W.N. Tzeng. 2000. The timing of metamorphosis and growth rates of American and European eel leptocephali: a mechanism of larval segregative migration. Fisheries Research 46: 191–205.

Wang, C.H., C.H. Kuo, H.K. Mok and S.C. Lee. 2003. Molecular phylogeny of elopomorph fishes inferred from mitochondrial 12S ribosomal RNA sequences. Zoologica Scripta: 32: 231–241.

Yamamoto, K. and K. Yamauchi. 1974. Sexual maturation of Japanese eel and production of eel larvae in the aquarium. Nature (London) 251: 220–222.

Yamano, K., M. Tagawa, E.G. de Jesus, T. Hirano, S. Miwa and Y. Inui. 1991. Changes in whole body concentrations of thyroid hormones and cortisol in metamorphosing conger eel. Journal of Comparative Physiology B 161: 371–375.

Yamano, K., K. Nomura and H. Tanaka. 2007. Development of thyroid gland and changes in thyroid hormone levels in leptocephali of Japanese eel (*Anguilla japonica*). Aquaculture 270: 499–504.

Youson, J.H. 1988. First metamorphosis. In: *Fish Physiology XI(B)*, W.S. Hoar and D.J. Randall (Eds.). Academic Press, San Diego New York Boston London Sydney Tokyo Toronto, pp. 135–196.

# CHAPTER 4

# Metamorphosis of Flatfish (Pleuronectiformes)

*Yasuo Inui*[1] and *Satoshi Miwa*[2,*]

## 4.1 General Aspects of Flatfish Metamorphosis

Flatfish (Pleuronectiformes) are a group of teleosts. They have the most extremely asymmetrical body of all vertebrates: both eyes lie either one side of the head (Kyle, 1923; Norman, 1934; Hubbs and Hubbs, 1945; Policansky, 1982a; Ahlstrom et al., 1984). In the same species, usually the eyes are always either on the right side of the head (dextral) or on the left side (sinistral). The two species of very primitive genus *Psettodes* are the only flatfish in which the eyed-side is indiscriminately determined in each individual, and hence, there are both dextral and sinistral fish within the same species. (Hubbs and Hubbs, 1945; Ahlstrom et al., 1984). In the starry flounder, *Platichthys stellatus*, there are also both dextral and sinistral fish. However, the eyed-side is not randomized as *Psettodes* but depends on the location where the fish lives. Off the US coast they are nearly half sinistral and half dextral. In the middle of the US and Japan, off Alaska, about 70 percent of them are sinistral, and in Japanese waters nearly 100 percent are sinistral (Policansky, 1982a).

The asymmetry of the flatfish appears during the development. When the flatfish hatch from eggs, they have symmetrical bodies like many other fishes, and they keep the symmetry and upright swimming position during the pelagic, larval period. Toward the end of the larval stage, however, the one eye migrates across the top of the head to the contralateral side of the

[1]Tamaki, Mie, 519-0414, Japan.
E-mail: inuiyj@yahoo.co.jp
[2]Inland Station, National Research Institute of Aquaculture, Tamaki, Mie 519-0423, Japan.
E-mail: miwasat@affrc.go.jp
*Corresponding author

head, resulting in both eyes located on one side of the head. The whole body structure is modified accordingly. Pigmentation of the body becomes also asymmetrical: the ocular side (the side on which the eyes are localized) becomes dark with intensive pigmentation, and the blind side becomes white. Concomitant with these changes, their life style changes from free swimming, pelagic plankton feeders to sedentary carnivores, lying on the bottom with both eyes facing up. Their swimming position also changes from dorso-ventral upright to lateralized (Kyle, 1923; Norman, 1934; Hubbs and Hubbs, 1945; Policansky, 1982a; Ahlstrom et al., 1984).

For this characteristic change of body form together with the change in life style in early development, flatfish are considered to undergo a typical metamorphosis comparable to that in amphibians (Youson, 1988). Youson (1988) categorized fish metamorphosis into two types: true or first metamorphosis, which occurs in the early development from larvae to juvenile and second metamorphosis which takes place after the juvenile. Thus, metamorphosis of flatfish is considered as a representative of the first metamorphosis in teleosts.

In the last two decades, researches have shown that the flatfish metamorphosis is primarily controlled by the pituitary-thyroid axis in a comparable way to amphibian metamorphosis (Inui and Miwa, 1985; Miwa and Inui, 1987a, b; Miwa et al., 1988; Inui et al., 1989; Inui et al., 1994; Schreiber and Specker, 1998). These studies have disclosed the developmental changes in various organs controlled by hormones during metamorphosis (Miwa and Inui, 1991; Miwa et al., 1992; Yamano et al., 1991, 1994b; Huang et al., 1998; Soffientino and Specker, 2001; Schreiber and Specker, 2000).

Furthermore, recently intriguing studies have been conducted on the molecular pathway controlling the formation of left/right axis during early development of flatfish (Hashimoto et al., 2004; Hashimoto et al., 2007; Suzuki et al., 2009).

This review describes the present status and perspectives of studies of flatfish metamorphosis. As recent major progress on this field has been achieved on the Japanese flounder, *Paralichthys olivaceus*, the article mainly concerns with the results of these studies. However, we have also tried to include studies of other flatfish as much as possible.

### 4.1.1 Stage specific morphology during metamorphosis

Okiyama (1967, 1974) and Minami (1982) described developmental changes in general morphology from post-larva to juvenile of the Japanese flounder, defining 9 developmental stages with stage-specific criteria (Fig. 1). The criteria have often been used in studies of the metamorphosis of the Japanese flounder. Other criteria, such as the one which classifies metamorphic stages

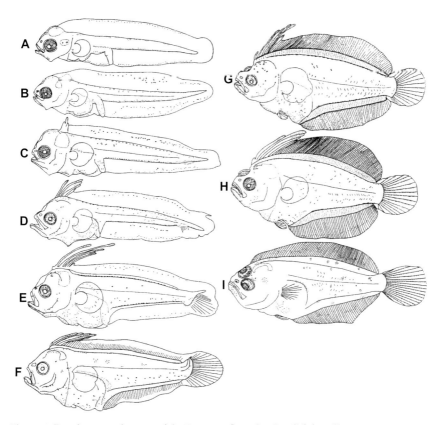

**Figure 1.** Developmental stages of the Japanese flounder, *Paralichthys olivaceus*.
**A**: BL 3.25 mm. End of the notochord is straight. The gut has a loop.
**B**: BL 5.40 mm. Anlages of elongated dorsal fin rays are visible.
**C**: BL 5.45 mm. There are 3 elongated dorsal fin rays. Anlages of caudal fin rays are visible at the ventral side of the end of notochord.
**D**: BL 5.50 mm. Eyes are still symmetrical. There are 5 elongated dorsal fin rays and 6 caudal fin rays. Body compression begins.
**E**: BL 8.30 mm. Right eye starts to move. The end of notochord is dorsally flexed by 45°. Anlages of ventral fin rays are visible. There are 11 caudal fin rays.
**F**: BL 9.20 mm. Right eye further migrates but not visible from the left side. Notochord is flexed dorsally by 90°. Additional one small fin ray appears at the anterior end of elongated dorsal fin rays. Anlages of ventral fin rays appear. Dorsal and anal fin rays are developing.
**G**: BL 10.25 mm. Part of the right eye is visible from left side. There are 6 ventral fin rays. Dorsal, anal, and caudal fin rays have developed.
**H**: BL 12.10 mm. Right eye is on the dorsal midline. Melanophores increase in number on the left side of body. Pectral fin rays start to develop. Settling behavior is observed.
**I**: BL 13.20 mm. Metamorphosis has completed. Dorsal elongated fin rays have been resorbed completely. Right eye has completely relocated to the left side. All fin rays have developed.
**A~H** larvae; **I** juvenile.
[From Minami, 1982 with permission].

into premetamorphosis, early or prometamorphosis, midmetamorphosis or climax, late or post metamorphosis and juvenile, have also been proposed (Okiyama, 1967; Miwa et al., 1988; Tanangonan et al., 1989) and used depending on the purpose of the study. Stage criteria similar to those of the Japanese flounder have been presented for spotted halibut, *Varasper variegatus* (Aritaki et al., 2001), brown sole, *Pseudopleuronectes herzenteini* (Aritaki and Seikai, 2004), barifin flounder, *Varasper moseri*, (Aritaki et al., 2000), slime flounder, *Microstomus achne* (Aritaki and Tanaka, 2003), summer flounder, *Paralichthys dentatus* (Keefe and Able, 1993; Schreiber and Specker, 1998), a plaice, *Pleuronectes platessa* (Ryland, 1966), and a turbot, *Scophthalmus maximus* (Al-Maghazachi and Gibson, 1984).

## 4.1.2 Cranium development and eye migration

The dramatic metamorphosis of the flatfish, especially eye migration, has attracted attention of many scientists. Various studies have been done to elucidate how and by what mechanisms one of the eyes of the flatfish moves to the contralateral side of the head during early development (Tranquair, 1865; Williams, 1901; Kyle, 1923; Brewster, 1987; Wagemans et al., 1998; Okada et al., 2001, 2003; Schreiber, 2006).

These studies can be grossly summarized as follows. The asymmetry of flatfish cranium is not due to any rotation of the cranium parts, but is caused by the development and relocation of anterior cranium elements on the blind side towards the ocular side (Brewster, 1987; Okada et al., 2001, 2003). During the larval stage chondrocranium is formed, and the only cranial element to ossify prior to metamorphosis is the parasphenoid bone (Brewster, 1987; Okada et al., 2001). The cranial elements such as trabecular cartilage, parasphenoid bone, lateral esmoids and supraorbital bars originally appear in the position similar to those of other teleosts, but they develop and relocate during metamorphosis, twisting toward the ocular side and ossify after the start of eye migration (Brewster, 1987; Okada et al., 2001, 2003). When the migrating eye reaches its new position, the other cranial elements ossify and these elements share the same relative position as those of other teleosts (Brewster, 1987).

Among cranial elements which become asymmetrical during eye migration, the change in supraorbital bars is especially drastic. The bars are originally located symmetrically between the eyes, but the one on the prospective ocular side grows larger, fusing with trabecular cartilage, while the blind side bar degrades and disappears by metamorphic climax (Williams, 1901; Okada et al., 2001, 2003; Schreiber, 2006), possibly allowing the passage of the migrating eye (Williams, 1901; Schreiber, 2006). Another candidate of frontal element to be directly involved in eye migration is the pseudomesial bar, a unique bone specifically found in flatfish. The bone is formed only on the blind side after the start of metamorphosis. The bone seems to originate from the ventral dermis beneath the migrating eye. It elongates toward dorsal direction during metamorphosis (Traquair, 1865; Okada et al., 2001, 2003). Several scientists strongly claim that the pseudomesial bar, especially in the process of its formation may play an important role in driving the eye migration, by pushing the migrating eye to the dorsal direction or at least preventing the eye from moving backward (Traquair, 1865; Kyle, 1923; Okada et al., 2001, 2003).

Thus, development and transformation of the cranium during eye migration has been clarified fairly well. Although the exact mechanism of eye migration needs to be elucidated by further studies, it seems clear that asymmetrical development of cranium including soft tissues moves the eye to the other side of the head.

### 4.1.3 Development of asymmetrical pigmentation

Flatfish are well known for their ability to rapidly match their body color to the background by adjusting their ocular-side pigmentation (Ramachandran et al., 1996). Such adjustment is mainly achieved by the change in distribution of melanosomes in the melanophore cytoplasm. This process is controlled through the sympathetic nervous system and pituitary hormones (Fujii and Ohshima, 1986; Sugimoto, 2002). The adaptation is reversible.

Apart from above mentioned mechanism of body color change, flatfish also change pigmentation ontogenetically, especially during metamorphosis, and the change is irreversible. The overall pattern of flatfish pigmentation is based on the number and distribution of three types of pigment cells: melanophores, xantophores and iridophores (Seikai, 1992; Seikai and Matsumoto, 1994; Bolker and Hill, 2000; Bolker et al., 2005). Melanophores produce melanin and are responsible for the brown or black body color. Xantophores appear yellow and contain carotenoids and pteridines, and iridophores appear white or silvery by light-reflecting platelets of guanine (Fujii, 1969, 1993; Bagnara, 1987). Development of flatfish pigmentation has been studied mainly in the Japanese flounder *Paralichthys olivaceus* (Bolker and Hill, 2000). It is known that all pigment cells in lower vertebrates are differentiated from the chromatoblast originated from nueral crest (Bagnara, 1987). This is also confirmed in the Japanese flounder by both an *in vivo* study using neural crest markers on pigment cells of developing flounder (Matsumoto and Seikai, 1992) and an *in vitro* study on the development of pigment cells from dissociated neural cells of flounder neurula (Seikai et al., 1993). In early development of the Japanese flounder, chromatoblasts are evenly delivered to the skin of both left and right side of the body and no difference is observed between the skin structures of both sides including pigmentation. During early larval period, some chromatoblasts give rise to large-sized larval melanophores on both sides of the body. However, from the onset of metamorphosis, remaining chromatoblasts start to proliferate and differentiate into adult type pigment cells, such as small-sized adult melanophores, xantophores and iridophores on the ocular side, while on the blind side such adult-type chromatophores do not develop and even cytolysis of chromatoblasts occurs at climax. Only iridophores, which are supposed to be differentiated from surviving chromatoblasts remain on the blind side. Thus, the dramatic asymmetry of dark ocular side and white-colored blind side is established (Seikai et al., 1987; Seikai, 1992; Matsumoto and Seikai, 1992; Seikai et al., 1993; Seikai and Matsumoto, 1994). Recently, Watanabe et al. (2008) principally confirmed the above mentioned process by applying specific markers for the chromatoblast, melanoblast and xanthoblast on the Japanese flounder and spotted halibut, *Varasper variegatus* at different developmental stages.

Principally similar developmental pattern of pigmentation has been found in summer flounder, *Paralichthys dentatus* with some species-specific pigment distributions (Bolker et al., 2005).

Mucous cells also show asymmetrical development during metamorphosis: the ratio of mucous cell density on the ocular side to that of the blind side is the same before metamorphosis, but the ratio increases steadily from the onset of metamorphosis to climax (Seikai, 1992; Seikai and Matsumoto, 1994).

The squamation occurs differently on the ocular and blind side. Scale formation starts earlier on the ocular side: scales start to develop along the posterior part of the lateral line and spread to the anterior and then toward the periphery of the trunk (Seikai, 1980; Fukuhara, 1986). All scales are cycloid at first but later they change to ctenoid following the developmental order, until all the scales of the ocular side become ctenoid. Squamation on the blind side occurs similarly to the ocular side but commences later and progresses more slowly. On the blind side, all the scales are cycloid and ctenii do not develop (Seikai, 1980; Kikuchi and Makino, 1990). Squamation occurs after the metamorphosis (Seikai, 1980; Kikuchi and Makino, 1990). However, the squamation seems to be related to the metamorphosis in a way, since the squamation is affected by the adult-type pigmentation, which occurs during early metamorphosis: severe abnormal pigmentation, which occurs during metamorphosis strongly affects the subsequent development of the scales (Seikai, 1980).

## 4.2 Hormonal Control of Metamorphosis

### *4.2.1 Dynamics of thyroid gland activity and pituitary thyroid-stimulating hormone (TSH) cells during metamorphosis*

One of the most spectacular actions of thyroid hormone is the stimulation of metamorphosis in amphibian larvae (Dodd and Dodd, 1976). It had also long been known that the thyroid gland is activated during the metamorphosis of eel and plaice, (Sklower, 1930). However, before the series of studies on the hormonal control of the metamorphosis of the Japanese flounder, *Paralichthys olivaceus* (Inui and Miwa, 1985), the role of the thyroid hormone in fish development had been ambiguous (Eales, 1979).

Miwa and Inui (1987a) found conspicuous histological activation of the thyroid organ and pituitary TSH cells toward the metamorphic climax in Japanese flounder larvae. In the larvae, immunoreactivity to anti-thyroid-hormone serum is already present in the subpharyngeal region 36 h after hatching and a few thyroid follicles have become evident in 4-day old fish (Fig. 2-3a). Thyroid follicles increase both in size and in number from premetamaorphosis to metamorphic climax, and the activity of the follicles judged by the follicular epithelial cell height becomes highest at climax (Fig. 2-3d). Thereafter, the number of the follicles decreases, and at post climax, the epithelial cells become flattened (Fig. 2-3e), indicating typical inactive state of the gland. Similar morphological activation in the thyroid follicles and strong immunoreactivity to T4 and T3 at metamorphic climax have been reported for developing Senegalese sole (*Solea senegalensis*), though the study did not refer to inactivation after metamorphosis (Ortiz Delgado et al., 2006).

On the other hand, immunoreactive (Ir) TSH cells are first found in the pituitary of 4 day old larvae of the Japanese flounder (Fig. 2-2a). The area occupied by the Ir TSH cells increases from dorsal part to ventral part of proximal pars distalis of the pituitary, and staining intensity of the cells also increases toward prometamorphosis. The area of the cells becomes largest, and the cells are stained very intensely in prometamorphosis (Fig. 2-2c), indicating that the amount of TSH stored in the cells reaches a peak at this time. The IrTSH cells is much decreased in number and in staining intensity in metamorphic climax (Fig. 2-2d), suggesting a rapid release of TSH from the cells, and show the least activity in post climax (Fig. 2-2e). These morphological changes in TSH cells suggest a strong stimulatory action of pituitary TSH on the thyroid gland in metamorphic climax.

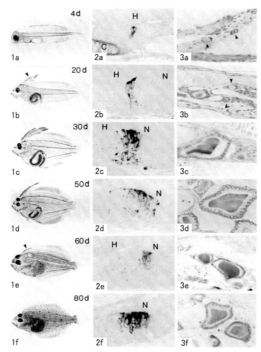

**Figure 2.** Histological changes in pituitary TSH cells and thyroid follicles during metamorphosis of the Japanese flounder.

**1a-f.** Larval development. (a) Premetamorphic larva 4 days after hatching. Absorption of yolk almost completed. (b) Symmetrical pelagic larva at premetamorphosis 20 days after hatching. Note several elongated dorsal fin rays (arrowhead). (c) Larva at prometamorphic stage 30 days after hatching. The right eye has started to migrate. (d) Larva at metamorphic climax 50 days after hatching. (e) Fish almost completed metamorphosis (post climax) 60 days after hatching. The anterior margin of the dorsal fin is blackened by melanin granules (arrowhead). (f) Jvenile 80 days after hatching. a ×16; b ×9.0; c ×5.5; d ×3.9; ex 4.1; f ×2.2.

**2a f.** Midsagittal sections of pituitary gland stained with anti-hTSHβ. Anterior direction is left. (a) Some immunoreactive TSH (Ir TSH) cells are clearly seen. Cartilage is stained with methyl green. (b) Ir TSH cells are located in the dorsal part of the proximal pars distalis (PPD). (c) Ir TSH cells are observed not only in the dorsal part but also in the ventral part of PPD. (d) Immunoreactivity is less than that in 30 day-old fish (2c). The Ir TSH cells are not observed in the ventral part of PPD. (e) Intensity of immunostaining is weak. the Ir TSH cell area is small and limited to the dorsal part of PPD. (f) IrTSH cells are stained strongly with the antiserum although the cell area is not so extended as in the fish at prometamorphosis (2c). C, cartilage; H, hypothalamus; N, neurohypophysis. A–f ×660.

**3a-f.** Sagittal section of the thyroid gland stained with HE. (a) Two thyroid follicles are seen (arrowheads). (b) The number of the follicles (arrowheads) is increasing. (c) The epithelial cells of the follicles are higher than in 20-day-old fish (3b). (d) Epithelial cells are tall. Note that each epithelial cell has enlarged nucleus and a vacuole-like structure. (e) Epithelial cells are flattened and have small nuclei, suggesting very low activity at this stage. The follicles are filled homogenously with eosinophillic colloid. **f,** Follicular cells are higher in 60-day old fish (3e), suggesting reactivation of the thyroid gland. A–f ×271. [From Miwa and Inui, 1987 with permission].

### 4.2.2 Changes in tissue thyroid hormone concentration

Changes in tissue thyroid hormone concentrations in whole fish body during metamorphosis of the Japanese flounder (*Paralichthys olivaceus*) were studied by using radioimmunoassay (Miwa et al., 1988; Tagawa et al., 1990). In the eggs of the flounder, a low level of thyroxine (T4 ) and a fairly high level of triiodothyronine (T3) are detected just after fertilization, and the concentrations of these hormones gradually lower towards hatching. After hatching, tissue T3 concentration sharply decreases and both hormone levels become very low (Tagawa et al., 1990). Because the thyroid follicles develop after hatching, these hormones in eggs are apparently of maternal origin. Tissue T4 concentration of the larval flounder is kept rather low during premetamorphosis, but it increases gradually during prometamorphosis. Then, it rises sharply at the beginning of climax, and the high level is maintained throughout climax but it decreases to about one half of the maximum level at postclimax (Fig. 3), thus, creating a clear surge of tissue T4 concentration in metamorphic climax (Miwa et al., 1988; Tagawa et al., 1990). This change in tissue T4 concentration during metamorphosis, especially the surge in metamorphic climax, well coincides with previously mentioned histological activation of the thyroid gland. Larval tissue T3 concentration is much lower (about one-tenth of that of T4) and shows similar change to that in T4 during metamorphosis (Fig. 3).

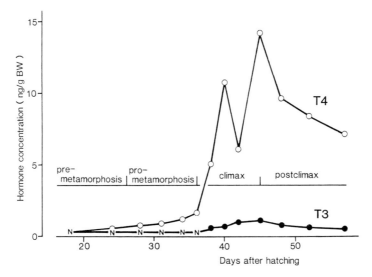

**Figure 3.** Changes in whole body tissue $T_4$ and $T_3$ concentrations during metamorphosis of the Japanese flounder. Each point represents the average of the duplicate determination of pooled samples. [From Tagawa et al., 1990, with permission].

A surge of tissue T4 concentration in metamorphic climax similar to that of the Japanese flounder has been reported for spotted halibut, *Varasper varigatus* (Hotta et al., 2001a,2001b), and summer flounder, *Paralichthys dentatus* (Schreiber and Specker, 1998).

### 4.2.3 Induction and inhibition of metamorphosis by administration of exogenous thyroid hormone and antithyroid agent

Before metamorphosis, the Japanese flounder has 5–6 characteristic elongated fin rays at the anterior end of the dorsal fin. During metamorphosis the elongated fin rays shorten and become similar length to the other fin rays.

Inui and Miwa (1985) first found that T4 administration to larval flounder induced precocious metamorphosis as evidenced by shortening

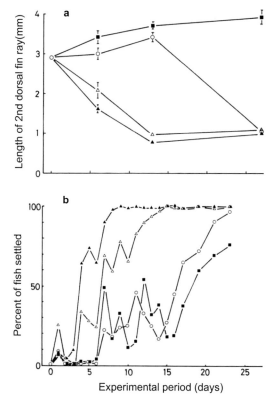

**Figure 4. a,** Effects of $T_4$ and TU on the length of the second elongated dorsal fin ray of the Japanese flounder. ○ Control; ■ 30 ppm TU; △ 0.05 ppm $T_4$; ▲ 0.1 ppm $T_4$. **b** Effects of T4 and TU on settling behavior of flounder larvae. ○ Control; ■ 30 ppm TU; △ 0.05 ppm $T_4$; ▲ 0.1 ppm $T_4$. $T_4$ treatment was terminated on day 19. [From Inui and Miwa, 1985, with permission].

of elongated fin rays (Fig. 4-a), induction of settling behavior on the bottom of the aquarium (Fig. 4-b) and migration of the right eye to the left of the head. Thus, exogenous T4 produced miniatures of naturally metamorphosed juvenile flounder (Fig. 5-b). On the other hand, administration of thiourea (TU), an inhibitor of thyroid hormone synthesis, arrested the metamorphic process and produced unnaturally large-sized larvae (Fig. 5-c). When T4 or T3 was administered to flounder larvae that were in metamorphic stasis by TU

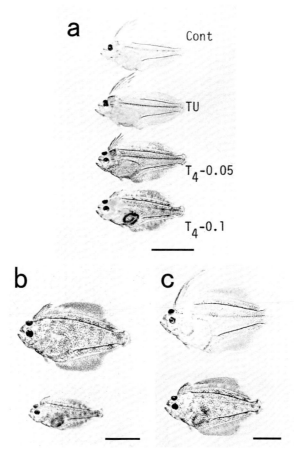

**Figure 5.** Effects of $T_4$ and TU on metamorphosis of the Japanese flounder. (a) Flounder larvae kept in normal seawater, 30ppm TU, 0.05ppm $T_4$, and 0.1ppm $T_4$ for 6 days. (b) A subnormal-sized juvenile (lower) precociously metamorphosed by $T_4$ treatment and a naturally metamorphosing larva (upper) at metamorphic climax. (c) A giant pelagic larva in metamorphic stasis induced by TU (upper) and naturally metamorphosing larva (lower), after 27 days of treatment. Each scale line represent 5mm. [From Inui and Miwa, 1985, wtih permission].

treatment, metamorphic events were induced in a dose dependent manner, with T3 being several times more potent than T4 (Miwa and Inui, 1987b).

de Jesus et al. (1990), using the *in vitro* culture of the dissected, elongated fin rays of prometamorphic larvae, also showed that both T4 and T3 administration to the culture medium accelerated shortening of the fin rays dose-dependently, with T3 being more potent than T4. The direct effect of thyroid hormones on fin ray resorption of the flounder during metamorphosis seems comparable to the case of regression of tadpole tail fin discs *in vitro* (Derby, 1968; Robinson et al., 1977).

Induction of precocious metamorphosis by T4 administration and metamorphic stasis by TU treatment have also been reported in summer flounder, *Paralichthys dentatus* (Huang et al., 1998; Schreiber and Specker, 1998) and Atlantic halibut, *Hippoglossus hipoglossus* (Solbakken et al., 1999).

In amphibian metamorphosis, each tissue has its own "readiness" and the threshold level of thyroid hormone to which the tissue responds and starts metamorphic changes (Kollros, 1961). Even in the same tissue, the sequence of the metamorphic changes sometimes requires different levels of thyroid hormone (Kollros, 1961). For the flounder, TU-treated larvae still initiated eye migration but did not complete it, while fin ray shortening was completely inhibited. On the other hand, administration of moderate concentration of T4 to TU-treated larvae completed eye migration (Miwa and Inui, 1987b). In addition, eye migration can be initiated by exogenous T4 in premetamorphic larvae. TU treatment did not completely deplete tissue T4 of the whole body (unpublished data). These results indicate that eye migration starts with a low level of T4, but a higher level of T4 is required for the completion of the migration. Thus, it seems that, also in the flounder, different tissues need different levels of T4 for the initiation of metamorphic process and that different levels of thyroid hormone are needed for the sequence of metamorphic change in each organ.

### 4.2.4 Pituitary-thyroid axis and its negative feedback regulation: effects of exogenous TSH

The activation of pituitary TSH cells in metamorphosis strongly suggests that TSH induces the surge of thyroid hormone in metamorphic climax in the Japanese flounder. To confirm this, Inui et al. (1989) injected bovine TSH to prometamorphic flounder larvae. TSH injection increased tissue T4 concentration in a dose-dependent manner, though the injection did not affect tissue T3 concentration. A single injection of TSH of a supposedly physiological dose increased tissue T4 level to the similar level of metamorphic climax (Fig. 6), and duplicate injections of TSH with

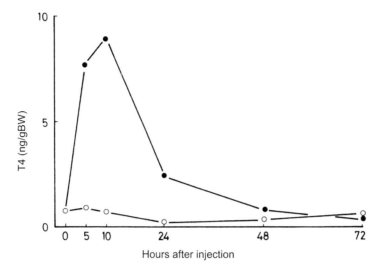

**Figure 6.** Changes in tissue T4 levels of Japanese flounder larvae after a single injection of bovine TSH at a dose of 5 mIU/g (closed circles). Open circles indicate the levels of the saline-injected control fish. [From Inui et al., 1989, with permission].

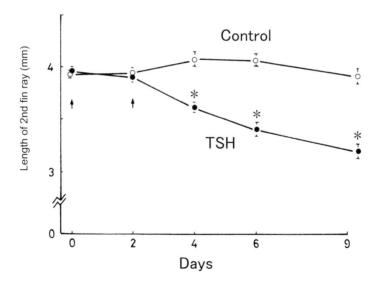

**Figure 7.** Effect of bovine TSH on the length of second fin ray of Japanese flounder larvae. Bovine TSH (5 mIU/g) was injected on day 0 and 2 as indicated by arrows. Significantly different (P<0.001) from the level of the respective saline-injected control for the given time period by t-test. [From Inui et al., 1989, with permission].

a 2-day interval accelerated metamorphosis as evidenced by shortening of elongated fin rays (Fig. 7) and right eye migration. In conclusion, it is clear that TSH stimulates thyroid organ to secrete T4 from prometamorphosis to climax and this pituitary-thyroid axis plays the major role in controlling metamorphosis of the Japanese flounder.

To examine the presence of negative feedback regulation in pituitary-thyroid axis, Miwa and Inui (1987a) treated premetamorphic larvae of the flounder with T4 and TU for 14 days. The thyroid follicles of larvae treated with TU for 2 weeks were apparently hyperactive as evidenced by tall follicle cells and well developed vacuole- like structure in the cytoplasm of the cells, but T4 was not detected immunocytochemically in the thyroid (Fig. 8-2c), indicating the inhibition of T4 synthesis by TU. On the other hand, TSH cells in the pituitary of the TU-treated larvae markedly proliferated and were stained more intensely with immunocytochemistry using anti-TSH serum compared to those of control animals (Fig. 8-1a, 1c). In contrast, T4 treatment induced typical inactivation of thyroid follicles (Fig. 8-2b) and almost completely abolished immunocytochemical reaction of pituitary to anti-TSH serum (Fig. 8-1b). These facts clearly indicate that a negative feedback regulation is functioning in the pituitary-thyroid axis during metamorphosis of the Japanese flounder.

### 4.2.5 Role of thyroid hormone in development of organs during metamorphosis

Amphibian metamorphosis is accompanied by extensive morphological and physiological remodeling of various organs (Gilbert and Frieden, 1981), and these structural and functional developments are thought to result from tissue-specific reprogramming of gene expression that is stimulated directly by thyroid hormones (Atkinson et al., 1994). Studies on flatfish metamorphosis, especially on the Japanese flounder in the last 2 decades disclosed that various internal organs of flounder larvae undergo dramatic morphological, biochemical and functional development and changes during metamorphosis, and that these development and changes are controlled by thyroid hormones.

#### 4.2.5.1 Shift of erythrocyte population

In higher vertebrates such as some mammals (Bloom and Bartelmez, 1940; Craig and Russel, 1964) and birds (Fraser, 1963), there is a shift of erythrocyte populations from primitive erythrocytes, which are specific to embryos, to definitive erythrocytes during embryonic development. In anuran amphibians, such a shift of erythrocytes is known to occur during metamorphosis from tadpoles to froglets (Broyles, 1981). A similar change in erythrocyte populations during early development has also been reported

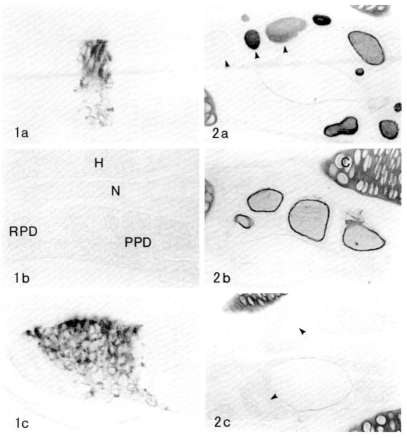

**Figure 8. 1a-c,** Midsagittal sections of the pituitary of Japanese flounder larvae treated by 0.05 ppm $T_4$ or 30 ppm TU for 2 weeks, stained with anti-hTSHβ. Anterior direction is left. (a)Control fish. Ir TSH cells show the features of prometamorphic larvae. (b) Fish treated with T4. Ir TSH cells are not evident. (c) Fish treated with TU. The number of Ir TSH cells is markedly increased. H, Hypothalamus; N, neurohypophysis; PPD, proximal pars distalis; RPD, rostral pars distalis. a–c ×886.

**2a-c,** Sagittal sections of thyroid gland of flounder larvae treated by 0.05 ppm T4 or 30 ppm TU for 2 weeks, stained with anti-T4 serum, conterstained with methyl green. (a) Control fish. The follicular cells are cuboidal and the colloid shows variable staining intensity with the antiserum (arrowheads). (b) Fish treated with T4. Follicular epithelium is flattened, and the colloid is only weakly stained with the antiserum. (c) Fish treated with TU. The follicular cells are markedly taller, and the follicles contain only a small amount of colloid which shows no reactivity to antiserum (arrowheads). C, cartilage. a–c ×738. [From Miwa and Inui, 1987a, with permission].

in rainbow trout (Iuchi, 1973a; Yamamoto and Iuchi, 1975) and angelfish (Al-Adhami and Kunz, 1976).

The Japanese flounder also shows a developmental shift of erythrocyte populations from primitive to definitive erythrocytes during metamorphosis in a way comparable to that in anuran amphibians (Miwa and Inui, 1991). Pre- and prometamorphic larvae of the Japanese flounder have only larval erythrocytes, which are large round cells with round and piknotic nuclei (Fig. 9-a). A large number of immature adult type erythrocytes, which are small and slightly elliptical cells with large nuclei, appear at metamorphic climax (Fig. 9-b). These immature cells develop into elliptical adult type (definitive) erythrocytes in fully metamorphosed juveniles (Fig. 9-c). Such a shift of erythrocyte populations can be clearly seen by measuring the

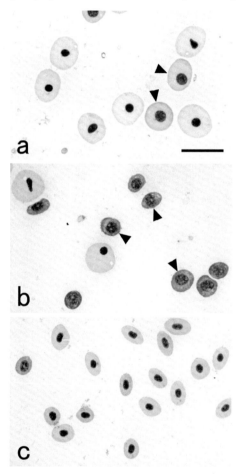

**Figure 9.** Photomicrographs of blood smears of spontaneously metamorphosing Japanese flounder larvae at different developmental stages. Wright-Giemsa. **a,** Prometamorphosis. Large larval erythrocytes with small nuclei and putative immature larval erythrocytes (arrowheads) are seen. The bar indicates 20 μm. **b,** Climax. Immature (pro-) adult erythrocytes with dark blue cytoplasm and large nuclei (arrowheads) appear in the blood at this stage. Same magnification as a. **c,** Juvenile. Small elliptic adult erythrocytes are observed. Same magnification as a. [From Miwa and Inui, 1991, with permission].

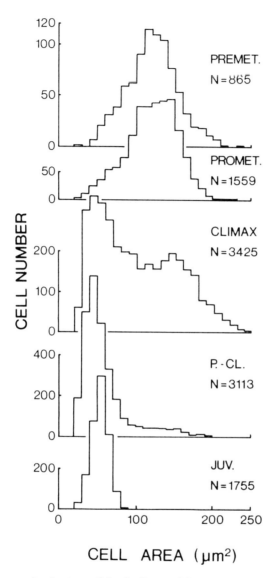

**Figure 10.** Frequency distributions of blood cell area of the spontaneously metamorphosing Japanese flounder at different developmental stages. Premet., premetamorphosis; Promet., prometamorphosis; P.-CL., post climax; Juv., juvenile; N, number of total blood cells counted. [From Miwa and Inui, 1991, with permission].

area of erythrocytes in blood-smear preparations. Figure 10 shows the distribution of erythrocyte cell area at different stages of metamorphosis. In pre- and prometamorphic larvae, the erythrocyte population is dominated by large cells, which are apparently larval type erythrocytes. At climax, in addition to the large cell population, another major population of smaller cells appears. This population is apparently composed of immature adult type erythrocytes. The relative number of large larval erythrocytes markedly decreases in the post climax, and these larval erythrocytes are totally replaced by small, adult type erythrocytes in juveniles (Fig. 10).

In rainbow trout, hemoglobin subunit components of larval type erythrocytes are different from those of adult type erythorcytes (Iuchi and Yamagami, 1969; Iuchi, 1973b, 1985) and larval type hemoglobin has higher oxygen affinity and less Bohr effect compared to adult hemoglobins (Iuchi, 1973b). Miwa and Inui (1991) also found that migrating patterns of hemoglobins of larval and adult type erythrocytes of the Japanese flounder were different in polyacrylamide gel electrophoresis. Flatfish change behavior and living habitat during metamorphosis, and hence, availability of oxygen as well as the mode of oxygen needs by the animal may change. Therefore, it is meaningful to further clarify the difference of biochemical and physiological nature of larval and adult type hemoglobins.

Thyroid hormone is involved in erythropoiesis in a variety of animals such as mammals (Popovic et al., 1977; Daniak et al., 1978), reptiles (Thapliyal and Kaur, 1976). In amphibians, the shift from tadpole erythrocytes to frog erythrocytes is also stimulated by thyroid hormone (Hollyfield, 1966; Moss and Ingram, 1965; Meints and Carver, 1973; Thomas, 1974). Miwa and Inui (1991) found that administration of T4 to prometamorphic larvae of the flounder induced precocious shift of the erythrocyte population from larval to adult type erythrocytes (Fig. 11). Thus, it is evident that the shift of erythrocytes during flounder metamorphosis is also controlled by thyroid hormone.

Sullivan et al. (1987) reported the presence of nuclear thyroid hormone receptors in adult type erythrocytes of trout. They also observed that the number of the receptors diminished during the cell development from immature erythrocytes to mature erythrocytes. These findings suggest that thyroid hormone acts directly on the erythrocyte in trout. In anuran amphibians, it is known that the number of thyroid hormone receptors in an erythrocyte nucleus increases during prometamorphosis and reaches maximum just prior to metamorphic climax, and that thyroid hormone treatment increases the number of thyroid hormone receptors in erythrocytes (Galton et al., 1994). It is plausible that a similar change in thyroid hormone receptors also occurs in flatfish metamorphosis.

In higher vertebrates, embryonic erythrocytes originate from yolk sac, while the fetal erythropoiesis is replaced by the liver. The spleen, thymus and other connective tissue organs may also produce erythroid during this time. The location of erythropoiesis is then transferred to bone marrow (Wood et al., 1979). In rainbow trout, larval erythropoiesis takes place in both the

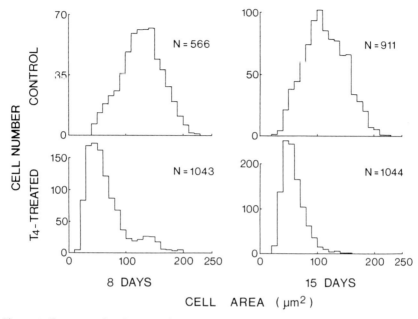

**Figure 11.** Frequency distributions of blood cell area of Japanese flounder larvae treated with either 100 ppb L-thyroxine or no hormone for 8 or 15 days. N, number of total cells counted. [From Miwa and Inui, 1991, with permission].

blood islands on the yolk sack and intermediate cell mass, which is situated between the somite and the lateral plate, and adult erythropoiesis occurs in the kidney and the spleen (Iuchi, 1985). Hence, the hematopoietic cite may also change during the shift of erythrocyte populations in the Japanese flounder, although the origin of larval erythrocytes is not known.

### 4.2.5.2 Gastric development

In most fishes, the larval alimentary canal is functional at first feeding, but is structurally and functionally less complex than that of the adults, and pinocytosis and intracellular digestion play important roles on the digestion of food during larval stages (for the review see Govoni et al., 1986). Many fish species lack the stomach in early stages of life, and the functional stomach develops during larva-juvenile transformation, i.e., metamorphosis (Govoni et al., 1986). The gastric glands rapidly develop during metamorphosis in the flounder (Miwa et al., 1992). In premetamorphic larvae, the wall of the anterior portion of the foregut, which later develops into the stomach, consists of a thin smooth muscle layer, a thin connective tissue layer and a simple cuboidal epithelium, and no glandular structure is observed (Fig. 12-a). During metamorphosis, the epithelium forms complex folds and

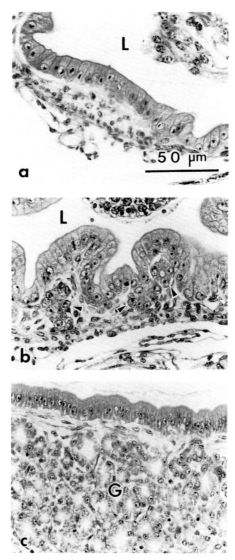

**Figure 12.** Photomicrographs of the gastric wall of spontaneously metamorphosing Japanese flounder at three different developmental stages. (a) Premetamorphosis. (b) early prometamorphosis. Note primordial crypts (arrowheads). (c) post-climax. L, gastric lumen; G, gastric glands. (a–c) Same magnification. [From Miwa et al., 1992, with permission].

crypts. The cells at the bottom of the crypt proliferate and form simple alveolar or acinar structures (Fig. 12-b). At metamorphic climax, the glandular cells rapidly proliferate, resulting in developed gastric glands at post climax stage (Fig. 12-c). The gastric wall of premetamorphic larvae shows no immunohistochemical reaction with antiserum to pepsinogen, but the tissue shows a strong immunoreaction for pepsinogen at the post climax, indicating the stomach is functional at this stage. The volume of the gastric glands continues to increase throughout the development (Miwa et al., 1992). Along with the stomach development, pyloric appendages are formed, and the larval compact pancreas start to develop into the diffuse pancreas, invading the liver along the hepatic portal vein. Thus, the digestive system of the flounder assumes the adult form in the early juvenile stage (Kurokawa and Suzuki, 1996). Morphological and functional development of the stomach during metamorphosis have also been reported in other flatfishes, such as *Solea senegalensis* (Ribeiro et al., 1999), turbot, *Scophthalmus maximus* (Segner et al., 1994), spotted halibut, *Varasper variegates* (Hotta et al., 2001a) and summer flounder, *Paralichthys dentatus* (Bisbal and Bengtson, 1995; Soffientino and Specker, 2001).

The above mentioned gastric development is also controlled by thyroid hormone, and the gastric development can be manipulated by altering the thyroid status. T4 administration to premetamorphic larvae of the Japanese flounder markedly increased the total volume of the gastric glands in the stomach, while TU treatment arrested the development of the gastric glands (Fig. 13). In addition, T4 administration induced precocious appearance of

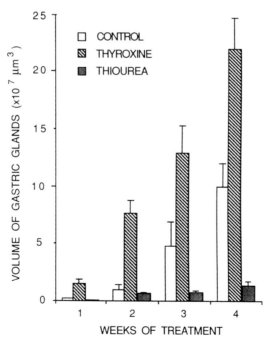

**Figure 13.** Effects of 100 ppb thyroxine ($T_4$) or 30 ppm thiourea in ambient seawater on increase in volume of gastric glands of Japanese flounder larvae. T4 treatment was terminated 2 weeks after the start of the experiment. [From Miwa et al., 1992, with permission].

immunoreactivity to anti-pepsinogen serum. In contrast, immunoreactivity of pepsinogen was maintained at a very low level in TU treated fish (Miwa et al., 1992). The results clearly show that development of functional stomach in the Japanese flounder during metamorphosis is stimulated by thyroid hormone. Precocious induction of gastric glands and pepsinogen by T4 and inhibition of the gastric development by TU in metamorphosing larvae have also been shown in the summer flounder, *Paralichthys dentatus* (Huang et al., 1998; Soffientino and Specker, 2001). Soffientino and Specker (2001), who examined the effect of exogenous T4 on the proliferation rate of stomach fundic mucosa, suggest that early increase of tissue T4 concentration stimulates proliferation of fundic mucosa and later peak T4 level promotes differentiation to gastric glands in metamorphosis of the summer flounder.

For amphibians, it is well known that the tadpole alimentary tract is reorganized in association with the transition from herbivorous to carnivorous feeding during metamorphosis (Dauca and Hourdry, 1985). Administration of T4 also induces precocious differentiation of the stomach *in vivo* (Lipson and Kaltenbach, 1965) and *in vitro* (Pouyet et al., 1983). Thus, thyroid hormone stimulates differentiation of gastric gland also in amphibians. In contrast, adrenocorticoid rather than thyroid hormone seems to be directly responsible to the gastric development of mammals. Injection of ACTH or cortisol to neonatal rats causes precocious development of pepsinogen in rat gastric mucosa (Furihata et al., 1972). Although thyroid hormone accelerates functional development of gastric glands in neonetal rats, the effect is mediated by corticosterone (Theng and Johnson, 1986; Theng et al., 1987a, b). On the other hand, it must be noted that rat gastric glands have already functionally differentiated to a certain degree at birth (Helander, 1969; Defize, 1989; Yahav, 1989). Hence, it is still possible that rat fetuses have a different hormonal regulatory system from their neonates.

### 4.2.5.3 Muscle development

Flatfish larvae are pelagic and rather slow swimmer, while juveniles, although usually sedentary on the bottom, swim quickly at feeding. The change in mobility during metamorphosis is accompanied by drastic changes in the skeletal muscle in the Japanese flounder (Yamano et al., 1991, 1994b). The muscle of premetamorphic larvae of the Japanese flounder consists of thin muscle fibers with a small number of thin myofibrils (Fig. 14-a, b). During metamorphic climax, the larval type muscle transforms into the adult type, which is characterized by thick muscle fibers with abundant myofibrils (Fig. 14-e, f) (Yamano et al., 1991). Corresponding with these morphological changes, protein components of the muscle also change. A myosin molecule of the larval type muscle is composed of two heavy

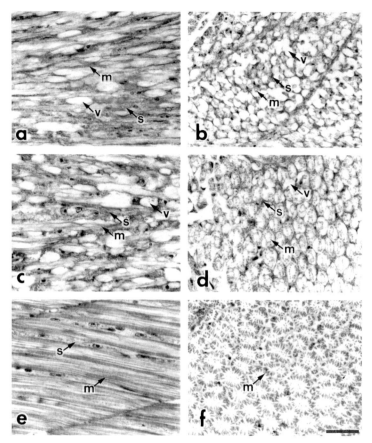

**Figure 14.** Photomicrographs of the skeletal muscle of spontaneously metamorphosing Japanese flounder. Left and right columns show longitudinal and transverse sections, respectively. (a,b) Premetamorphosis; (c,d) Prometamorphosis; (e,f) Post climax [From Yamano et al., 1991, with permission].

chains and three major light chains: two alkali light chains and one 5,5′-dithio-bis-nitrobenzoic acid (DTNB) light chain. During metamorphosis, the larval type DTNB light chain is gradually replaced by the adult type DTNB light chain, whereas the heavy chains and two alkali light chains do not change throughout metamorphosis. Troponin-T isoforms also change during metamorphosis. The skeletal muscle of premetamorphic larvae has two immunoreactive troponin-T isoforms of 41.5 and 34.0 kDa, with the former being predominant. At the metamorphic climax a new isoform appears at 33.5 kDa, and the two isoforms, 33.5 kDa and 34 kDa, become predominant.

Such morphological development and transition of molecular constituents of the muscle during ontogeny have also been reported for the plaice, *Pleuronectes platessa* (Brooks and Johnston, 1993) and turbot, *Scophthalmus maximus* (Focant et al., 2000). However, in these fishes, these muscle changes do not occur in metamorphosis but in a different growth stage after the metamorphosis from juveniles to adults (Focant et al., 2000).

Administration of T4 to premetamorphic larvae of the Japanese flounder precociously induced the morphological change in the muscle from larval to adult type: muscle fibers became thicker with abundant myofibrils, while TU administration inhibited the change (Yamano et al., 1991). T4 administration also precociously induced transition of the muscle protein components such as isoforms of myosin DTNB light chains and troponin-T (Yamano et al., 1994b). Thus, it is evident that thyroid hormone plays the primary role in controlling morphological and biochemical changes in the muscle during metamorphosis of the Japanese flounder.

As mentioned above, the transition of DTNB light chains during metamorphosis was markedly influenced by thyroid state. However, TU administration to juveniles of the flounder neither restored the larval type DTNB light chain nor depressed the expression of adult-type (Yamano et al., 1994b). In higher vertebrates, thyroid hormone induces fetal to adult transition of cardiac myosin heavy chains (MHCs) (Izumo et al., 1986). In contrast to the case of DTNB light chains in the Japanese flounder, however, the response of MHCs to thyroid hormone is reversible depending on the thyroidal states (Izumo et al., 1986). The mechanism of the thyroidal control on gene expression for the cardiac MHCs is well documented: the complex of thyroid hormone and its receptor directly binds to the thyroid-response element of the gene to regulate the expression of the gene either positively or negatively (Izumo and Mahdavi, 1988). In contrast, in flounder metamorphosis, neither the molecular mechanism of thyroid-hormone action on DTNB light chains nor the reason why the responses of DTNB light chains to thyroid hormone are irreversible is not known.

### 4.2.5.4 Chloride cell development

Sequence of chloride cell development and its control by thyroid hormone during metamorphosis is briefly presented here. A thorough review has been written on the osmoregulatory system of flatfishes during early development and metamorphosis (Schreiber, 2001).

It is well known that chloride cells (mitochondria-rich cells (MRCs) in the gills are responsible for the secretion of excess Na+ and Cl- in the body fluid in adult marine fishes (Zadunaisky, 1984). On the other hand, cutaneous chloride cells located on the epithelium covering yolk and body have been suggested as the ion-secreting site during early life stages in teleosts (Shelbourne, 1957; Roberts et al., 1973; Hwang, 1989; Hwang and

Hirano, 1985; Alderdice, 1988). Hiroi et al. (1998) studied the changes in chloride cell distribution during early development of the Japanese flounder by whole-mount immunocytochemistry using antiserum against Na+, K+-ATPase, which is supposedly located specifically on the chloride cells. In larvae immediately after hatching, the immunoreactive cells are distributed only on the yolk-sac membrane and body skin. Thereafter, premetamorphic larvae become to possess both cutaneous and newly appeared branchial chloride cells. Cutaneous chloride cells often form multicellular complexes. The complex is considered advantageous to Na+secretion (Shiraishi et al., 1997), since Na+ is transported through the paracellular pathway in the complex down its electrical gradient (McCormick, 1995). Thus, cutaneous chloride cells seem to play the major role in Na+secretion during this period. Cutaneous chloride cells decrease in size and density at the beginning of metamorphosis and disappear completely at metamorphic climax. In contrast, the number of branchial chloride cells located on gill filaments increases during metamorphosis. From these results, Hiroi et al. (1998) concluded that the site for ion secretion probably shift from cutaneous to branchial chloride cells during the metamorphosis of the Japanese flounder. Schreiber and Specker (1999a) also reported the development of chloride cells in the gills of the summer flounder during metamorphosis. At the start of metamorphosis, the gills possess one type of larval MRCs, which are characterized by large electron-lucent mitochondria, absence of well-defined apical pit, and relatively weak immunoreactivity of Na+, K+-ATPase. Towards the end of metamorphosis, Na+, K+-ATPase immunoreactivity of MRCs increases, and by the end of metamorphosis, larval type MRCs disappear, and instead, two juvenile type MRCs, which are characterized by smaller electron-dense mitochondoria, develop. One type is weakly osmiophilic with well developed apical pit, and the other is strongly osmiophillic and positioned adjacent to the former MRC. From the location of the cell and its morphology, the authors designate the latter as accessory cells of the former MRCs. At present, it is uncertain whether juvenile type MRCs develop from larval type MRCs or from a different origin. Schreiber and Specker (2000) reported that administration of T4 to premetamorphic larvae of summer flounder stimulated the development of juvenile type MRCs precociously. They also showed that, while the treatment with thiourea (TU) arrested the development, T4 administration to TU-treated fish compensated the effect of TU and stimulated the development of juvenile type MRCs. These results not only indicate that thyroid hormone stimulates the development of juvenile type gill MRCs but also strongly suggest that thyroid hormone regulates the shift of functional site of ion secretion from cutaneous chloride cells to gill chloride cells during flatfish metamorphosis.

Hiroi et al. (1997) reported that the tolerance of the Japanese flounder against low salinity was fairly high in larvae immediately after hatching, decreased toward the beginning of the metamorphosis and rapidly increased during metamorphosis, reaching the highest level thereafter. Schreiber and Specker (1999b) also reported that, in the summer flounder, tolerances to both low and high salinity decreased toward the late prometamorphosis to middle climax and subsequently increased rapidly reaching to the highest level in juveniles.

The above mentioned turning point of salinity tolerance coincides with the timing of the shift from cutaneous chloride cells to gill chloride cells. In addition, the decrease of the salinity tolerance toward the metamorphosis coincide with the decreases in size and density of cutaneous choloride cells, and the subsequent increase of the tolerance coincide with the development of juvenile type gill chloride cells, respectively. Thus, the change in salinity tolerance of flounder larvae seems to reflect the shift of chloride cells from the cutaneous to the gill and developmental changes in these chloride cells.

### 4.2.6 Thyroid hormone receptors and their dynamics during metamorphosis

The action of thyroid hormone is mediated by thyroid hormone receptors (THRs). THRs are members of nuclear receptor family (for the review see Evans, 1988), and THRs activated by thyroid hormone directly bind to a specific DNA element of a gene (thyroid hormone receptor element) to regulate expression of the gene positively or negatively (Glass et al., 1987; Chatterjee et al., 1989). Two major types of THRs ($\alpha$ and $\beta$), which are produced from distinct genes, have been identified in various classes of vertebrates such as mammals (Weinberger et al., 1986; Thompson et al., 1987), chickens (Sap et al., 1986) and amphibians ( Brooks et al., 1989; Yaoita et al., 1990). In fish, THR genes were first cloned from the Japanese flounder (Yamano et al., 1994a), and so far, four cDNAs for the flounder THRs have been cloned: two $\alpha$types ($\alpha$A and $\alpha$B) and two $\beta$types ($\beta$1 and $\beta$2) (Yamano et al., 1994a; Yamano and Inui, 1995). In mammals, two $\alpha$type THR variants ($\alpha$1 and $\alpha$2) are generated from a single gene by alternative splicing (Izumo and Mahdavi, 1988), whereas *Xenopus* possesses two distinctive genes for $\alpha$type THRs ($\alpha$A and $\alpha$B )(Yaoita et al., 1990). As is the case in *Xenopus*, the Japanese flounder appears to have distinct genes for two THR$\alpha$s. On the other hand, two $\beta$type THRs ($\beta$1 and $\beta$2) of the Japanese flounder share a constant sequence except for a 60 base additional sequence found only in THR$\beta$2, suggesting that two flounder THR$\beta$s are produced from the same gene through an alternative splicing system (Yamano et al., 1994a; Yamano and Inui, 1995).

Comparison of expression patterns of mRNAs for THRα and THRβ revealed rather ubiquitous expression of THRα and developmental stage- and tissue-specific expression of THRβ in chicken (Forrest et al., 1990, 1991) and rat (Bradley et al., 1989, 1992, 1994; Mellstrom et al., 1991). In amphibian metamorphosis, a rapid increase in THRβ mRNA has been found during spontaneous metamorphosis and also in response to exogenous thyroid hormone (Yaoita and Brown, 1990; Kawahara et al., 1991). Yamano and Miwa (1998) studied developmental changes in the expression levels and regional expression patterns of THRα and THRβ genes during metamorphosis of the Japanese flounder. Although the expression levels of flounder THR gene transcripts are very low in fertilized eggs, substantial amount of THR mRNAs are present in premetamorphic larvae, except for THRαB, which is kept low throughout larval development. THRα-gene transcripts increase in metamorphic climax and decrease rapidly in postclimax, while

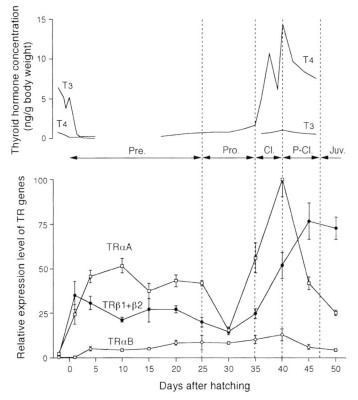

**Figure 15.** Developmental changes in the expression of TR genes during larval growth and metamorphosis of the Japanese flounder. One hundred units correspond to 36.5 amol of TR cDNA reverse transcribed from 1μg of mRNA. Changes in T4 and T3 concentrations at corresponding metamorphic stages (Tagawa et al., 1990) are shown in the uppergraph. [From Yamano and Miwa, 1998, with permission].

the expression of THRβs increases in climax, reaches its peak in post climax and remains high in metamorphosed juveniles (Fig. 15). *In situ* hybridization revealed the ubiquitous expression of THR genes and distinct tissue specificity of α and β subtypes in the fish body: THRα is strongly expressed in tissues such as the skeletal muscle (Fig. 16) and epithelial and immature glandular cells of the stomach (Fig. 17), which undergo marked change (see gastric development and muscle development of this chapter), while THRβ is strongly expressed in tissues that do not show apparent changes. Collectively, these results indicate the possibility that, unlike amphibians, αtype (αA) THR rather than βtype may play important roles in metamorphic changes in the Japanese flounder. These results also strongly suggest that the development of tissues stimulated by thyroid

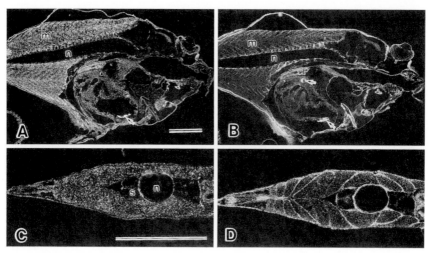

**Figure 16.** Dark-field photomicrographs of the detection of TRα (A and C) and TRβ (B and D) mRNAs in two adjacent sagittal (A and B) and transverse (C and D) sections of Japanese flounder larvae at metamorphic climax. Arrows indicate nonspecific signal from ingested food, m, skeletal muscle; n, notochord; s, spinal cord. Bar=1mm. [From Yamano and Miwa, 1998, with permission].

hormone during metamorphosis is further enhanced and/or controlled at the receptor level by stage- and region-specific expressions of THRs (Yamano and Miwa, 1998).

## 4.2.7 Modification of thyroid hormone actions by other hormones

### 4.2.7.1 Cortisol

In anuran amphibians, adrenal corticosteroids are known to enhance effects of thyroid hormones in metamorphosing tadpoles (Frieden and Naile, 1955;

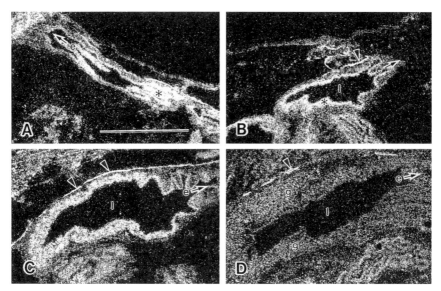

**Figure 17.** Dark-field photomicrographs showing changes in TRαgene expression in the developing mucosal epithelium of the stomach of a larva at prematamorphosis (A), a larva at prometamorphosis (B), a larva at metamorphic climax (C), and a post climax specimen (D) of the Japanese flounder. Arrowheads indicate melanophores. Asterisks indicate stomach epithelium from which the gastric glands would differentiate. These sections were treated simultaneously with the same hybridization solution. e, direction of esophagus; g, gastric gland; l, stomach lumen. Bar=0.5 mm. [From Yamano and Miwa, 1998, with permission].

Kobayashi, 1958; Kaltenbach, 1958). These hormones accelerate both T4-and T3-induced shrinkage of the cultured tail fin discs (Kikuyama et al., 1983; Kaltenbach, 1985). A similar permissive effect of cortisol, the major corticoid in teleosts, on thyroid hormone action during metamorphosis is seen in the Japanese flounder (de Jesus et al., 1990). Addition of cortisol alone to the culture medium did not affect shortening of the fin rays *in vitro*. However, when cortisol was added together with T4 or T3, it enhanced the rate of shortening compared with that of fin rays treated with T4 or T3 alone. On the other hand, the permissive effect of cortisol on thyroid hormone was not seen *in vivo*; when the prometamorphic larvae were reared in seawater containing both cortisol and T4, cortisol did not enhance the metamorphosis-promoting effect of thyroid hormone (de Jesus et al., 1990). This may be because endogenous cortisol level is sufficient so that addition of exogenous cortisol is not effective. In fact, a considerable amount of cortisol is detected in whole body extracts during flounder metamorphosis. Tissue cortisol concentration shows a change parallel to that in tissue T4 concentration; the concentration remains low during premetamorphosis, increases during prometamorphic stage, reaching a peak level at metamorphic climax and decreases to about half of the maximal level at postclimax (de Jesus et al., 1991). Taken together, it is highly possible that cortisol at least plays some role in flounder metamorphosis. Adrenal corticosteroids seem to enhance the effects of thyroid hormone by augmenting nuclear binding capacity of tadpole tail nuclei to thyroid hormone (Niki et al., 1981; Suzuki and Kikuyama, 1983). Such works are needed to know the mechanism of permissive action of cortisol in flounder metamorphosis.

### 4.2.7.2 Sex steroids

de Jesus et al. (1992) showed modulation of thyroid hormone action by sex steroids in a study using *in vitro* culture of fin rays of the Japanese flounder. Neither estradiol nor testosterone, had any direct effect on the shortening of fin rays. However, when either of the hormones was added together with thyroid hormone, fin-ray shortening was retarded compared with that treated with thyroid hormone alone. Teststerone was more potent in the inhibition than estradiol. In addition, when flounders were reared in seawater containing either estradiol or testosterone, both steroids delayed metamorphosis, again with testosterone being more potent in the inhibition of metamorphosis. On the other hand, tissue levels of these hormones are kept rather low and both steroids do not show any significant change throughout the metamorphosis (de Jesus et al., 1992). In addition, the hormone doses used in the *in vitro* study were apparently higher than the tissue hormone levels in the flounder during spontaneous metamorphosis. Thus, the physiological significance of sex steroids on modulation of thyroid hormone action in metamorphosis is not clear.

### 4.2.7.3 Prolactin and growth hormone

Prolactin is known to antagonize thyroid hormone action in amphibian metamorphosis (Yoshizato and Yasumasu, 1972a, b; Jaffe and Geschwind, 1974; Derby, 1975; Dent, 1988). When the hormone was added to *in vitro* culture of the elongated fin rays of the flounder, prolactin also exhibited an antagonistic effect on the fin-ray shortening stimulated by thyroid hormone, though prolactin alone did not exert any effect (de Jesus et al., 1994). Injection of prolactin into prometamorphic larvae also delayed the resorption of the dorsal fin rays, though other metamorphic parameters such as eye migration and settling behavior were not affected (de Jesus et al., 1994). Furthermore, the same authors found that expression level of prolactin gene in the pituitary estimated by *in situ* hybridization steadily increased during metamorphosis and became highest at post climax. Steady increase in the volume of prolactin cells in the pituitary until climax of metamorphosis of the flounder was also reported by Hiroi et al. (1997). Above mentioned antimetamorphic and/or antagonistic action of prolactin to thyroid hormone, as well as the dynamics of the hormone during flounder metamorphosis seems to be comparable to those of amphibian metamorphosis.

Expression level of growth hormone gene during metamorphosis of the Japanese flounder changed similarly to that of prolactin (de Jesus et al., 1994). However, administration of growth hormone *in vitro* did not affect shortening of fin rays stimulated by thyroid hormone, and microinjection of growth hormone to prometamorphic larvae did not affect any metamorphic parameters, including resorption of the dorsal fin rays (de Jesus et al., 1994). Thus, there is so far no evidence that growth hormone is involved in the regulation of flounder metamorphosis.

## 4.3 Effect of Temperature on Metamorphosis

In various flatfish species reared in laboratories, it is known that water temperature markedly affect metamorphosis: larvae reared at higher temperatures initiate and complete metamorphosis earlier than those reared at lower temperatures (Laurence, 1975; Policansky, 1982b; Seikai et al. 1986; Aritaki et al., 1996; Aritaki et al., 2004). The ambient water temperature also markedly affects the size of juveniles immediately after metamorphosis: higher rearing temperatures result in smaller metamorphosed juveniles in the Japanese flounder (Seikai et al., 1986) and spotted halibut, *Verasper variegates* (Aritaki et al., 2004), though Policansky (1982b) did not find such effect of temperatures on the size of metamorphosed juveniles in starry flounder, *Platichthys stellatus*. These temperature effects on flatfish metamorphosis seem to be comparable to amphibian metamorphosis. In amphibians, normal tadpoles kept in cool water grow to a larger size than they do when they are kept in warm water (Etkin, 1955). The effect of temperatures on amphibian metamorphosis is considered to be largely or entirely caused by the effect of temperatures on tissue sensitivity to thyroid hormone (Kollros, 1961). Kollros (1961) states that for any given metamorphic event at colder temperature not only much more time is required for its completion at a fixed thyroid hormone level, but also the minimal thyroid hormone levels required for the events are higher in both control and hypophysectomized *Rana pipience*.

In several laboratory-reared flatfish species, extremely low and high ambient water temperatures induced metamorphic anomalies such as ambicoloration (the blind side also becomes dark) or albinism (the dark pigmentation does not develop on the ocular side). These abnormal colorations are often accompanied by incomplete eye migration (Aritaki et al., 1996; Aritaki and Seikai, 2004). These authors suggest that the metamorphic anomalies are related to the temperature-modulated thyroid hormone action.

## 4.4 Pigmentation Abnormalities During Metamorphosis

Various morphological anomalies have been reported in a variety of wild flatfishes. These anomalies include pigmentation anomalies, eye migration anomalies, dorsal fin hooking, abnormally shaped pectoral or pelvic fins of blind side and so on (Norman, 1934; Hubbs and Hubbs, 1945; Dawson, 1962). Norman (1934) categorized pigmentation anomalies as 1) albinism 2) ambicoloration and 3) reversal. Albinism lacks pigment on a part or the whole of the ocular side of the body. Ambicoloration is further categorized into following 3 types. (1) Staining: some or greater part of the blind side looks stained with brownish or grey. (2) Spotting: the blind side bears varying number of black or brown spots. (3) True ambicoloration: whole area of the blind side is pigmented as the ocular side. In the case of true ambicoloration, various other morphological anomalies often accompany, such as the development of ctenoid scales or strongly spinulated scales on the ambicolored, blind side, characteristic dorsal fin hooking above the eye, or abnormalities in the pectoral and pelvic fins. Reversal is the condition that an individual fish is sinistral whereas the normal condition of the species is dextral, or vice versa. Classifications similar to that of Norman (1934) are also used in pigmentation anomalies of hatchery-reared fish with some modification, such as pseudoalbinism or hypomelanism for albinism.

Morphological anomalies mentioned above are found much more frequently in laboratory- or hatchery-reared flatfishes. These anomalies, especially pigmentation anomalies, are the commonest defect for the hatchery-reared flatfishes.

Among these pigmentation anomalies, albinism seems to be the most serious defect in aquaculture, since albinism decreases the commercial value of the fish, and when released for stock enhancement, it will make predators easier to find the stocked fish.

To find the causative factors for malpigmentation in hatchery produced flatfishes, various attempts have been made from genetical, environmental, and nutritional standpoints. These studies disclosed that the composition of larval diet is a critical factor for the pigmentation of flatfishes (for the review see Bolker and Hill, 2000). *Artemia* nauplii are convenient and useful diet for flatfish larval culture. However, feeding *Artemia* nauplii, especially Brazilian *Artemia*, during the early larval period often causes albinism and sometimes ambicoloration in the Japanese flounder (Seikai, 1985a), spotted halibut (Aritaki and Seikai, 2005), brown sole (Aritaki and Seikai, 2005), and Atlantic halibut (Naess et al., 1995). Feeding wild zooplankton for the substitution of *Artemia* nauplii during larval period successfully reduced albinism in the Japanese flounder (Seikai, 1985b) and Atlantic halibut (Naess et al., 1995; Naess and Lie, 1998). Two nutrients, free fatty acids and vitamin A, have been shown to be the major factors influencing pigmentation in hatchery-reared flatfishes. Concentrations of fatty acids, particularly DHA and EPA or DHA/EPA ratio, in the diet are considered to be important factors for the healthy development and normal pigmentation of flatfish larvae (Kanazawa, 1993; Estevez and Kanazawa, 1995; Naess and Lie, 1998). On the other hand, Vitamin A seems to be rather specifically related to pigmentation of cultured flatfish. Vitamin A supplementation to rotifer or *Artemia* distinctly reduced albinism in the Japanese flounder (Miki et al., 1990; Takeuchi et al., 1995) and in turbot, *Scophthalmus maximus* (Estevez and Kanazawa, 1995), though administration of too much Vitamin A induced skeletal deformity at a high rate (Estevez and Kanazawa, 1995; Takeuchi et al., 1995). Miwa and Yamano (1999) found that administration of 9-cis-retinoic acid, biologically active retinoid normally synthesized from vitamin A, in the rearing seawater at premetamorphosis or prometamorphosis induced the development of adult type chromatophores not only on the ocular side but also on the blind side, resulting in reduction of albinism and induction of ambicoloration. To enhance larval nutrition and normal pigmentation, proper enrichment of zooplankton with fatty acids and vitamin A compounds have been proposed and is now commonly used in rearing flatfishes (Takeuchi et al., 1995; Dedi et al., 1995; Bolker and Hill, 2000).

One of the most important findings shared by above mentioned studies may be that the period responsible for the modulation of pigmentation anomalies by nutrition is early developmental stage, or more specifically, the period from premetamorphosis to prometamorphosis. This period coincides with the timing of chromatoblast differentiation to adult type chromatophores (see Development of asymmetrical pigmentation in this chapter). Seikai et al. (1987) found that albinistic skins totally lack adult-type melanophores, having exclusively a small number of large-sized larval melanophores. Miwa and Yamano (1999) reported that the sensitive stage of the flounder to retinoic acid was earlier than the time when the adult chromatophores actually appear, suggesting that retinoic acid only stimulates the determination of the developmental fate of neural crest cells to adult chromatophores but does not stimulate chromatophore development itself. Seikai et al. (1987) also reported that albinism in flounder juveniles fed on Brazilian *Artemia* was caused, at least in part, by the blockage of melanoblast development into adult-type melanophores. Collectively, these studies indicate that dietary imbalance, especially that of vitamin A or its precursors induces most of albino and some ambicoloration in hatchery-reared flounder juveniles, by disturbing the differentiation of chlomatoblast to adult type chromatophores during early metamorphosis.

Administration of a high dose of thyroid hormone to the Japanese flounder, especially at prometamorphosis, induced albinism at a very high rate (Yoo et al., 2000). As thyroid hormone primarily regulate metamorphosis, it is quite conceivable that the pigmentation during metamorphosis is also influenced by the hormone. It is unknown, at present, whether anomalous pigmentation induced by thyroid hormone is caused through the same mechanism for the anomalous pigmentation induced by dietary factors. However, the similarities of the sensitive period for the two cases suggest that at least a part of the mechanism is shared for the development of albinism induced either by the manipulation of the hormone or by diet. In fact, thyroid hormone and vitamin A (retinoic acid) interact through their receptors in vertebrates. There are two biologically important isomers of retinoic acid (RA), all-trans retinoic acid (atRA) and 9-cis retinoic acid (9cRA). Biological effects of these retinoic acids are mediated by specific receptors: retinoic acid receptor (RAR) for atRA and retinoid X receptor (RXR) for 9cRA, although 9cRA can bind to both RAR and RXR (Heyman et al., 1992; Allenby et al., 1993). Both thyroid hormone receptors (TRs), and RAR are known to heterodimerize with RXR to exert pleiotropic effects of their legands. Therefore, it is quite possible that thyroid hormone and retinoic acid or its precursor, vitamin A, may affect each other in the exertion of their biological effects. Schuh et al. (1993) reported that v-erb A, an oncogene encoding TR, reduced the teratogenic effects of atRA in *Xenopus* embryogenesis and suggested a possible loss of RXR that is able to form heterodimers with RAR due to v-ervA/RXR dimmer formation. On the other hand, Forman et al. (1995) found that transcriptional activation by thyroid hormone through TR/RXR heterodimer was reduced by RXR-specific synthetic ligand LG69. The interaction of thyroid hormone and retinoids and its mechanism in flounder metamorphosis should be clarified.

Albinism and above mentioned ambicoloration are irreversible and the causes for these anomalies reside during metamorphosis. On the other hand, "staining" seems to develop after metamorphosis and at least partly reversible (Stickney and White, 1975). Staining is considered to be caused by environmental factors (for reviews see Norman, 1934; Bolker and Hill, 2000). Light irradiation to the blind side has been reported to cause staining (Cunningham, 1891, 1893, 1895; Osborn, 1940, 1941). On the other hand, some authors reported that coarse bottom surface, e.g. sand, of the rearing aquaria prevented staining (Stickney and White, 1975; Iwata and Kikuchi, 1998). Iwata and Kikuchi (1998) examined the effects of light and the surface texture of the bottom of aquaria on hypermelanosis (staining) of the blind side. They found that, when they illuminated transparent aquaria from under the bottom through glass sand, staining of the blind side did not develop. Thus, surface texture of the bottom of rearing aquaria seems more important than lighting condition as a factor affecting hypermelanosis of the blind side. These observations clearly indicate that the development of asymmetrical pigmentation of flatfish is controlled by environmental factors, such as physical stimuli on the skin. Flatfish larvae sometimes stay on the bottom on their future blind side. The period of staying on the bottom becomes longer as metamorphosis progresses. By doing so, they might suppress the development of dark pigmentation on the blind side.

## 4.5 Molecular System to Control the Asymmetry Formation

In most vertebrates, the shape or position of their internal organs such as the heart, digestive organs and some part of the brain are known to be asymmetrical. Asymmetric orientation of these internal organs is controlled by the activation of a molecular pathway (Nodal-pathway), which precedes the organogenesis (Tamura et al., 1999; Capdevila et al., 2000; Hamada et al. 2002;) including zebrafish and flounders (Concha et al., 2000; Hashimoto et al., 2004, 2007). The principal pathway of left and right (L/R)-biased signal in zebrafish and flounder is considered as follows. An embryonic organ called Kupffer's vesicle, which is teleost-specific and homologous to the "node" in higher vertebrates, is transiently formed at the tail bud in the segmental period (Cooper and D'Amico, 1996; Essner et al., 2005). The organ is considered to initiate sequential molecular signaling for L/R axis determination (Hashimoto et al., 2004, 2007; Essner et al., 2005; Kramer-Zucker et al., 2005). In the Kupffer's vesicle, two nodal pathway genes, *spaw* and *charon*, both of which are secretory signals of the transforming growth factor-B superfamily, are expressed. Leftward water flow driven by cilia in the Kupffer's vesicle transfers spaw from the Kupffer's vesicle to the posterior left lateral plate mesoderm (LPM). On the other hand, charon expressed in the Kupffer's vesicle blocks induction of spaw in the right LPM. The transferred spaw in turn leads expressions of spaw itself, lefty and pitx2, in the left LPM (Long et al., 2003; Hashimoto et al., 2004, 2007), where spaw and lefty are considered to cooperatively induce pitx2 and restrict pitx2 expression in the left side of the body (Hashimoto et al., 2004, 2007). Pitx2 is expressed first in the gut field and then at the heart field in left side LPM, and thereafter, expressed in the left side of diencephalons. Thus expressed pitx2 induces asymmetrical development of the heart, gut and diencephalons: heart jogs to the left, gut coils to rightward and dorsal diencephalons are asymmetrically formed. To determine whether the laterality of Nodal-pathway expression differs between dextral and sinistral flatfish, Hashimoto et al. (2007) examined the expression domains of charon, lefty and pitx2 in embryos of the spotted halibut, *Varasper variegates*, whose members are normally dextral. Also in this species, charon was expressed at Kupffer's vesicle, and lefty and pitx2 were expressed at left LPM and left side of the dorsal diencephalon as is the case in the Japanese flounder, which is all sinistral. Moreover, the laterality of gut looping in the spotted halibut was the same as that in the Japanese flounder. They also examined expression domains of lefty and pitx2 for another sinistral flounder, summer flounder, *Paralichthys stellatus* and dextral flathead flounder, *Hippoglossoides dubius*, with the result that left-sided expression of the Nodal-pathway genes was found in both species. Thus, the embryonic laterality determined by Nodal-pathway genes is identical in dextral and sinistral flatfish, even though external asymmetry is exactly opposite.

In contrast to asymmetrical development of these internal organs, eye migration and/or asymmetrical cranium development, as well as asymmetrical pigmentation of flatfish occur during metamorphosis: more than 2 weeks after the termination of pix2 expression in the left LPM and diencephalon at 20°C in the Japanese flounder. However, Nodal pathway also seems to be involved in triggering the development of external asymmetry. Left-sidedness in the Japanese flounder is strongly determined genetically and occurrence of reverse (dextral) is very rare (less than 1%) in the hatchery-reared fish. Hashimoto et al. (2002) isolated a clonal line "*reversed (rev)*" of the Japanese flounder by gynogenesis. The homozygous offsprings produced from *revs* exhibit reversed eye position at a high frequency of 20–30%. Hashimoto et al. (2002) examined pitx2 expression in the LPM in *rev* embryo and found that 43% of the embryo exhibited pitx2 only in the left LPM, 53% in both left and right, and 4% only in the right side LPM. After metamorphosis, 20% of the juveniles of the same siblings displayed dextral-type eye position, while the remaining fish were synistral as the normal Japanese flounder. These results strongly suggest that the molecular pathway controlling the left-sided expression of pitx2 in the LPM is genetically disturbed in *rev*, and that left-biased pitx2 expression in normal flounder is involved in the right eye migration to the left side during metamorphosis in this species.

On the other hand, 15% of the juveniles of above mentioned siblings showed reversed gut coiling (leftward coiling). However, reversed gut coiling was not always accompanied by reversal of the eye migration, suggesting that a process of Nodal-pathway that is different from that for embryonic asymmetric development of internal organs is functioning for the asymmetry formation during metamorphosis.

Finally, Suzuki et al. (2009) found that pitx2 is re-expressed at the left of the diencephalon (habenular nuclei) at premetamorphosis, earlier than the initiation of eye migration in both sinistral Japanese flounder and dextral spotted halibut, *Varasper variegatus*.

Although the eyed-side in the flatfish is usually the same within the same species, the steadiness of eye-position within a species differs by species. While left-sidedness is firmly determined in the Japanese flounder *Paralichtys olivaceus*, the spotted halibut *Varasper variegates*, which is normally dextal, often exhibits reversed eyed-side (synistral) and bilateral eye-position (each eye on both side), especially in laboratory-reared animals (Aritaki et al., 2004; Tagawa and Aritaki, 2005). Suzuki et al. (2009), utilizing this species difference, examined how pitx2 expression in embryonic stage and its re-expression in premetamorphic stage are related to the determination of eyed-side during metamorphosis. They found that pitx2 was normally expressed at left side of diencephalon and gut field in 98–99% embryo in both species. Gut looping and asymmetry-formation of diencephalon were also normal in them. However, 10% of the metamorphosed *V. varigatus* became synistral (reversed) and 24% became bilateral, while all the metamorphosed *P. olivaceus* became normally synistal. In contrast, only 40% of the *V. varigatus* larvae re-expressed pit-x2 in premetmorphosis, while 100% of *P. olivaceus* larvae re-expressed pitx2. These results strongly suggest that re-expression of pitx2 in premetamorphic stage is directly related to eye migration during metamorphosis.

Suzuki et al. (2009) concluded that although the first expression of pitx2 expression in embryonic stage forms asymmetrical formation of habenular nuclei, re-expressed pitx2 at premetamorphosis is directly responsible for lateralizing habenular nuclei to rightward in sinistral flounder and leftward in dextral flounder respectively, thus inducing asymmetric development of the brain and eyes.

Clarification of details of molecular mechanisms of pitx2 re-expression and its function in regulating metamorphosis is eagerly needed. In particular, possible involvement of thyroid hormone in the re-expression of pitx2 should be a subject of future study.

## References

Ahlstrom, E.H., K. Amaoka, D.A. Hensley, H.G. Moser and B.Y. Sumida. 1984. Pleuronectiformes: development, ontogeny and systematics of fishes. American Society of Ichthyologists and Herpetologists. Ser. 1: 640–670.

Al-Adhami, M.A. and Y.W. Kunz. 1976. Haemopoietic centres in the developing angelfish, *Pterophyllum scalare* (Cuvier and Valenciannes). Wilhelm Roux's Archives of Developmental Biology 179: 393–401.

Alderdice, D.F. 1988. Osmotic and ionic regulation in teleost eggs and larvae. 1988. In: Fish Physiology Vol. XI A. W.S. Hoar and D.J. Randall (Eds.). Academic Press, San Diego pp. 163–251.

Allenby, G., M-T. Bocquel, M. Saunders, D. Kazmer, J. Speck, M. Rosenberger, A. Lovey, P. Kastner, J.F. Grippo, P. Chambon and A.A. Levin. 1993. Retinoic acid receptors and retinoid X receptors: Interactions with endogenous retinoid acids. Proceedings of the National Academy of Sciences USA, 90: 30–34.

Al-Maghazachi, S.J. and R. Gibson. 1984. The developmental stages of larval turbot, *Scophthalmus maximus* (L.). Journal of Experimental Marine Biology and Ecology 82: 35–51.

Aritaki, M. and T. Seikai. 2004. Temperature effects on early development and  occurrence of metamorphosis-related morphological abnormalities in hatchery-reared brown sole *Peudopleuronectes herzensteini*. Aquaculture 240: 517–530.

Aritaki, M. and M. Tanaka. 2003. Morphological development and growth of laboratory-reared slime flounder *Microstomus achne*. Nippon Suisan Gakkaishi 69: 602–610.

Aritaki, M. and T. Seikai. 2005. Influence of timing of Brazilian *Artemia* nauplii feeding on occurrence of pseudoalbinism in two pleuronectid species. Nippon Suisan Gakkaishi 71: 165–171

Aritaki, M., T. Seikai and M. Kobayashi. 1996. Reduction of morphological abnormalities in brown sole by larval rearing with higher temperature and early feeding of *Artemia* nauplii. Nippon Suisan Gakkaishi 62: 857–864.

Aritaki, M., S. Suzuki and K. Watanabe. 2000. Morphological development and growth of laboratory-reared berfin flounder *Verasper moseri*. Nippon Suisan Gakkaishi 66: 446–453.

Aritaki, M., K. Ohta, Y. Hotta and M. Tanaka. 2001. Morphological development and growth of laboratory-reared spotted halibut *Verasper variegatus*. Nippon Suisan Gakkaishi 67: 58–66.

Aritaki, M., K. Ohta, Y. Hotta, M. Tagawa and M. Tanaka. 2004. Temperature effects on larval development and occurrence of metamorphosis-related morphological babnormalities in hatchery-reared spotted halibut *Verasper variegatus* juveniles. Nippon Suisan Gakkaishi 70: 8–15.

Atkinson, B.G., A.C. Helbing and Y. Chen. 1994. Reprogramming of gene expression in the liver of *Rana catesbeiana* tadpoles during spontaneous and thyroid hormone induced metamorphosis. In: Perspectives in Comparative Endocrinology. K. Davey, R. E. Peter and S.S. Tobe (Eds.). National Research Council of Canada, Ottawa, Ontario, pp. 416–423

Bagnara, J.T. 1987. The neural crest as a source of stem cells. In: Developmental and Evolutionary Aspects of the Neural Crest. P.F.A. Maderson (Ed.). John Wiley and Sons, New York, pp. 57–87.

Bisbal, G.A. and D.A. Bengtson. 1995. Development of the digestive tract in larval summer flounder. Journal of Fish Biolology 47: 277–291.

Bloom, W. and G.W. Bartelmez. 1940. Hematopoiesis in young human embryos. American Journal of Anatomy 67: 21–53.

Bolker, J.A. and C.R. Hill. 2000. Pigmentation development in hatchery-reared flatfishes. Journal of Fish Biology 56: 1029–1052.

Bolker, J.A., T.F. Hakala and J.E. Quist. 2005. Pigmentation development, defects, and patterning in summer flounder (*Paralichtys dentatus*). Zoology 108: 183–193.

Bradley, D.J., W.S. Young III and C. Weinberger. 1989. Differential expression of α and β thyroid hormone receptor genes in rat brain and pituitary. Proceedings of National Academy of Sciences USA, 86: 7250–7254.

Bradley, D.J., H.C. Towle and W.S. Young III. 1992. Spatial and temporal expression of α and β thyroid hormone receptor mRNAs, including the β2-subtype, in the developing mammalian nervous system. The Journal of Neuroscience 12: 2288–2302.

Bradley, D.J., H.C. Towle and W.S. Young III. 1994. α and β thyroid hormone receptor (TR) gene expression during auditory neurogenesis: evidence for TR isoform-specific transcriptional regulation *in vivo*. Proceedings of National Academy of Sciences USA, 91: 439–443.

Brewster, B. 1987. Eye migration and cranial development during flatfish metamorphosis: a reappraisal (Teleostei; Pleuronectiformes) Journal of Fish Biology 31: 805–833.

Brooks, S. and I.A. Johnston. 1993. Influence of development and rearing temperature on the distribution, ultrastructure and myosin sub-unit composition of myotomal muscle-fibre types in the plaice *Pleuronectes platessa*. Marine Biology 117: 501–513.

Brooks, A.R., G. Sweeney and R.W. Old. 1989. Structure and functional expression of a cloned *Xenopus* thyroid hormone receptor. Nucreic Acids Reserch 17: 9395–9405.

Broyles, R.H. 1981. Changes in the blood during amphibian metamorphosis. In Metamorphosis: A problem in Developmental Biology. L. I. Gilbert and E. Frieden (Eds.). Plenum Press, New York, pp. 461–490.

Capdevila, J., K.J. Vogan, C.J. Tabin and J.C.I. Belmonte. 2000. Mechanisms of left-right determination in vertebrates. Cell. 101: 9–21.

Chatterjee, V.K.K., J.K. Lee, A. Rentoumis and J.L. Jameson. 1989. Negative regulation of the thyroid-stimulating hormone α gene by thyroid hormone: receptor interaction adjacent to the TAT box. Proceedings of the National Academy of Sciences USA, 86: 9114–9118.

Concha, M.L., R.D. Burdine, C. Russell, A.F. Schier and S.W. Wilson. 2000. A nodal signaling pathway regulates the laterality of neuroanatomical asymmetries in the zebrafish forebrain. Neuron 28: 399–409.

Cooper, M.S. and L.A. D'Amico. 1996. A cluster of noninvoluting endocytic cells at the margin of the zebrafish blastoderm marks the site of embryonic shield formation. Developmental Biology 180: 184–198.

Craig, M.L. and E.S. Russell. 1964. A developmental change in hemoglobins correlated with an embryonic red cell population in the mouse, Developmental Biology 10: 191–201.

Cunningham, J.T. 1891. An experiment concerning the absence of color from the lower sides of flatfishes. Zoologischer Anzeiger 14: 27–32.

Cunningham, J.T. 1893. Researches on the coloration of the skins of flat-fishes. Journal of Marine Biological Association of the United kingdom 3: 111–118.

Cunningham, J.T. 1895. Additional evidence on the influence of light in producing pigments of lower sides of flatfishes. Journal of Marine Biological Association of the United kingdom 4: 53–59.

Daniak, N., R. Hoffman, L.A. Maffei and B.G. Forget. 1978. Potentiation of human erythropoiesis *in vitro* by thyroid hormone. Nature 272: 260–262.

Dauca, M. and J. Hourdry. 1985. Transformations in the intestinal epithelium during anuran metamorphosis. In: Metamorphosis. M. Balls and M. Bownes, (Eds.). Clarendon Press, Oxford, pp. 36–58.

Dawson, C.E. 1962. Note on anomalous American *heterosomata* with descriptions of five new records, Copeia 1: 138–146.

Dedi, J., T. Takeuchi, T. Seikai and T. Watanabe. 1995. Hypervitaminosis and safe levels of vitamin A for larval flounder (*Paralichthys olivaceus*) fed *Artemia* nauplii. Aquaculture 133: 135–146.

Defize, J. 1989. Development of pepsinogens. In: Human Gastrointestinal Development. E. Laventhal (Ed.). Raven Press, New York, pp. 299–324.

Dent, J.N. 1988. Hormonal interaction in amphibian metamorphosis. American Zoologist 28: 297–308.

Derby, A. 1968. An *in vitro* quantitative analysis of the response of tadpole tissue to thyroxine. Journal of Experimental Zoology 168: 147–156.

Derby, A. 1975. The effect of prolactin and thyroxine on tail resorption in *Rana pipiens*: *In vivo* and *in vitro*. Journal of Experimental Zoology 193: 15–20.

Dodd, M.H.I. and J.M. Dodd. 1976. The biology of metamorphosis. In: Physiology of the Amphibia. vol.3. B. Lofts (Ed.). Academic Press, New York, pp. 467–599.

Eales, J.G. 1979. Thyroid functions in cyclostomes and fishes. In: Hormones and Evolution Vol.1. E.J.W. Barrington (Ed.). Academic Press, London/New York, pp. 341–436.

Essner J.J., J.D. Amack, M.K. Nyholm, E.B. Harris and H.J. Yost. 2005. Kupffer's vesicle is a ciliated organ of asymmetry in the zebrafish embryo that initiates left-right development of the brain, heart and gut. Development 132: 1247–1260.

Estevez, A. and A. Kanazawa. 1995. Effect of (n-3) PUFA and vitamin A *Artemia* enrichment on pigmentation success of turbot, *Scophthalmus maximus* (L.) Aquaculture Nutrition 1: 159–168.

Etkin, W. 1955. In: Analysis of Development. B.H. Willier, P.A. Weiss and V. Hamburger (Eds.). W.G. Saunders Co., Philadelphia, pp. 631–663.

Evans, R.M. 1988. The steroid and thyroid hormone receptor superfamily. Science 240: 889–895.

Focant, B., S. Collin, P. Vandewalle and F. Huriaux. 2000. Expression of myofibrillar proteins and parvalbumin isoforms in white muscle of the developing turbot, *Scophthalmus maximus* (Pisces, Pleuronectiformes). Basic and Appled Myology 10: 269–278.

Forman, B.M., K. Umesono, J. Chen and R.M. Evans. 1995. Unique response pathways are established by allosteric interactions among nuclear hormone receptors. Cell. 81: 541–550.

Forrest, D., M. Sjoberg and B. Vennstrom. 1990. Contrasting developmental and tissue-specific expression of α and β thyroid hormone receptor genes. The EMBO Journal 9: 1519–1528.

Forrest, D., F. Hallbook, H. Persson and B. Vennstrom. 1991. Distinct functions for thyroid hormone receptors α and β in brain development indicated by differential expression of receptor genes. The EMBO Journal 10: 269–275.

Fraser, R.C. 1963. Cytochemistry of the developing chick embryo erythrocytes. Journal of Experimental Zoology 152: 297–305.

Frieden, E. and B. Naile. 1955. Biochemistry of amphibian metamorphosis. I. Enhancement of induced metamorphosis by glucocorticoids. Science 121: 37–39.

Fujii, R. 1969. Chromatophores and pigments. In: Fish Physiology Vol.III, W.S. Hoar and D. J. Randall (Eds.). Academic Press, pp. 307–353.

Fujii, 1993. Cytophysiology of fish chromaophores. International Review of Cytology 143: 191–255.

Fujii, R., and Ohshima, N. 1986. Control of chromatophore movements in teleost fishes, Zoological Science 3: 13–47.

Fukuhara, O. 1986. Morphological and functional development of Japanese flounder in early life stage. Nippon Suisan Gakkaishi 52: 81–91.

Furihata, C., T. Kawachi and T. Sugimura. 1972. Premature induction of pepsinogen in developing rat gastric mucosa by hormones. Biochemical and Biophysical Research Communications 47: 705–711.

Galton, V.A., C.D. Jennifer and M.J. Schneider. 1994. Mechanisms of thyroid hormone action in developing *Rana catesbeiana* tadpoles. In: Perspectives in Comparative Endocrinology. K. Davey, R.E. Peter and S.S. Tobe (Eds.). National Research Council of Canada, Otawa, Ontario, pp. 412–415.

Gilbert, L.I. and E. Frieden. 1981. Metamorphosis: A Problem in Developmental Biology. Plenum Press, New York, 578pp.

Glass, C.K., R. Franco, C. Weinberger, V.R. Albert, R.M. Evans and M.G. Rosenfeld. 1987. A c-erb-A binding site in rat growth hormone gene mediates trans-activation by thyroid hormone. Nature 329: 738–741.

Govoni, J.J., G.W. Boehlert and Y. Watanabe. 1986. The physiology of digestion in fish larvae. Environmental Biology of Fishes 16: 59–77.

Hamada, H., C. Meno, D. Watanabe and Y. Saijoh. 2002. Establishment of vertebrate left-right asymmetry. Nature Reviews Genetics 3: 103–113.

Hashimoto, H., A. Mizuta, N. Okada, T. Suzuki, M. Tagawa, K. Tabata, Y. Yokoyama, M. Sakaguchi, M. Tanaka and H. Toyohara. 2002. Isolation and characterization of a Japanese flounder clonal line, reversed, which exhibits reversal of metamorphic left-right asymmetry. Mechanisms of Development 111: 17–24.

Hashimoto, H., M. Rebagliati, N. Ahmad, O. Muraoka, T. Kurokawa, M. Hibi and T. Suzuki, 2004. The Cerberus/Dan-family protein charon is a negative regulator of nodal signaling during left-right patterning in zebrafish. Development 131: 1741–1753.

Hashimoto, H., M. Aritaki, K. Uozumi, S. Uji, T. Kurokawa and T. Suzuki. 2007. Embryogenesis and expression profiles of charon and nodal-pathway genes in sinistral (*Paralichthys olivaceus*) and dextral (*Verasper varigatus*) flounders. Zoological Science 24: 137–146.

Helander, H.F. 1969. Ultrastructure and function of gastric mucoid and zymogen cells in the rat during development. Gastroenterology, 56: 53–70.

Heyman, R.M., D.J. Mangelsdorf, J.A. Dyck, R.B. Stein, G. Eichele, R.M. Evans and C. Thaller. 1992. 9-cis retinoic acid is a high affinity ligand for the retinoid X receptor. Cell 68: 397–406.

Hiroi, J., Y. Sakakura, M. Tagawa, T. Seikai and M. Tanaka. 1997. Developmental changes in low-salinity tolerance and responses of prolactin, cortisol and thyroid hormones to low-salinity environment in larvae and juveniles of Japanese flounder, *Paralichthys olivaceus*. Zoological Science 14: 987–992.

Hiroi, J., T. Kaneko, T. Seikai and M. Tanaka. 1998. Developmental sequence of chloride cells in the body skin and gills of Japanese flounder (*Paralichthys olivaceus*) larvae. Zoological Science 15: 455–460.

Hollyfield, J.G. 1966. Erythrocyte replacement at metamorphosis in the frog, Rana pipiens. Journal of Morphology 119: 1–6.

Hotta, Y., M. Aritaki, K. Ohta, M. Tagawa and M. Tanaka. 2001a. Development of the digestive system and metamorphosis relating hormones in spotted halibut larvae and early juveniles. Nippon Suisan Gakkaishi 67: 40–48.

Hotta, Y., M. Aritaki, M. Tagawa and M. Tanaka. 2001b. Changes in tissue thyroid hormone levels of metamorphosing spotted halibut *Verasper variegatus* reared at different temperatures. Fisheries Science 67: 1119–1124.

Huang, L., A.M. Schreiber, B. Soffientino, D.A. Bengtson and J.L. Specker. 1998. Metamorphosis of summer flounder (*Paralichthys dentatus*): thyroid status and the timing of gastric gland formation. Journal of Experimental Zoology 280: 413–420.

Hubbs, C.L. and L.C. Hubbs. 1945. Bilateral asymmetry and bilateral variation in fishes. Papers of the Michigan Academy of Science, Arts and Letters 30: 229–311.

Hwang, P.P. 1989. Distribution of chloride cells in teleost larvae. Journal of Morphology 200: 1–8.

Hwang, P.P. and R. Hirano. 1985. Effects of environmental salinity on intracellular organization and junctional structure of chloride cells in early stages of teleost development. Journal of Experimental Zoology 236: 115–126.

Inui, Y. and S. Miwa. 1985. Thyroid hormone induces metamorphosis of flounder larvae. General and Comparative Endocrinology 60: 450–454.

Inui, Y., M. Tagawa, S. Miwa and T. Hirano. 1989. Effects of bovine TSH on the tissue thyroxine level and metamorphosis in prometamorphic flounder larvae. General and Comparative Endocrinology 74: 406–410.

Inui, Y., S. Miwa and K. Yamano. 1994. Hormonal control of flounder metamorphosis. In: Perspectives in Comparative Endocrinology. K.G. Davey, R.E. Peter and S.S. Tobe (Eds.). National Research Council Canada, pp. 408–411.

Iuchi, I. 1973a. The post-hatching transition of erythrocytes from larval to adult type in the rainbow trout, *Salmo gairdnerii irideus*. Journal of Experimental Zoology 184: 383–396.

Iuchi, I. 1973b. Chemical and physiological properties of the larval and the adult hemoglobins in rainbow trout, *Salmo gairdnerii irideus*. Comparative Biochemistry and Physiology 44B: 1087–1101.

Iuchi, I. 1985. Cellular and molecular bases of the larval-adult shift of hemoglobins in fish. Zoological Science 2: 11–23.

Iuchi, I. and K. Yamagami. 1969. Electrophoretic pattern of larval haemoglobins of the salmonid fish, Salmo gairdnerii irideus. Comparative Biochemistry and Physiology 28: 977–979.

Iwata, N. and K. Kikuchi. 1998. Effects of sandy substrate and light on hypermelanosis of the blind side in cultured Japanese flounder. *Paralichthys olivaceus*. Environmental Biology of Fishes 52: 291–297.

Izumo, S. and V. Mahdavi. 1988. Thyroid hormone receptor α isoforms generated by alternative splicing differentially activate myosin HC gene transcription. Nature 334: 539–542.

Izumo, S., B. Nadal-Ginard and V. Mahdavim. 1986. All members of the MHC multigene family respond to thyroxine hormone in a highly tissue-specific manner. Nature 231: 597–600.

Jaffe, R.C. and I.I. Geschwind. 1974. Studies on prolactin inhibition of thyroxine-induced metamorphosis in *Rana catesbeiana* tadpoles. General and Comparative Endocrinology 22: 289–295.

de Jesus, E.G., Y. Inui and T. Hirano. 1990. Cortisol enhances the stimulating action of thyroid hormones on dorsal fin-ray resorption of flounder larvae *in vitro*. General and Comparative Endocrinology 79: 167–173.

de Jesus, E.G., T. Hirnano and Y. Inui. 1991. Changes in cortisol and thyroid hormone concentrations during early development and metamorphosis in the Japanese flounder, *Paralichthys olivaceus*. General and Comparative Endocrinology 82: 369–376.

de Jesus, E.G., T. Hirnano and Y. Inui. 1992. Gonadal steroids delay spontaneous flounder metamorphosis and inhibit T$_3$-induced fin ray shortening *in vitro*. Zoological Science 9: 633–638.

de Jesus, E.G., T. Hirnano and Y. Inui. 1994. The antimetamorphic effect of prolactin in the Japanese flounder. General and Comparative Endocrinology 93: 44–50.

Kaltenbach, J.C. 1958. Direct steroid enhancement of induced metamorphosis in peripheral tissues. The Anatomical Record 131: 569–570.

Kaltenbach, J.C. 1985. Amphibian metamorphosis: influence of thyroid and steroid hormones. In: Current Trends in Comparative Endocrinology Vol. 1, B. Lofts and W.N. Holmes (Eds.). Hong Kong University Press, Hong Kong, pp. 533–534.

Kanazawa, A. 1993. Nutritional mechanisms involved in the occurrence of abnormal pigmentation in hatchery-reared flatfish. Journal of the World Aquaculture Society 24: 162–166.

Kawahara, A., B.S. Baker and J.R. Tata. 1991. Developmental and regional expression of thyroid hormone receptor genes during *Xenopus* metamorphosis. Development 112: 933–943.

Keefe, M. and K.W. Able. 1993. Patterns of metamorphosis in summer flounder, *Paralichtys dentatus*. Journal of Fish. Biolology 42: 713–728.

Kikuchi, S. N. and Makino. 1990. Characteristics of the progression of squamation and the formation of ctenii in the Japanese flounder, *Paralichthys olivaceus*. Journal of Experimental Zoology 254: 177–185.

Kikuyama, S., K. Niki, M. Mayumi, R. Shibayama, M. Nishikawa and N. Shintake. 1983. Studies on corticoid action on the toad tadpole *in vitro*. General and Comparative Endocrinology 52: 395–399.

Kobayashi, H. 1958. Effects of desoxycorticosterone acetate on metamorphosis induced by thyroxine in anuran tadpoles. Endocrinology 62: 371–377.

Kollros, J.J. 1961. Mechanisms of amphibian metamorphosis: hormones. American Zoologist 1: 107–114.

Kramer-Zucker, A.G., F. Olare, C.J. Haycraft, B.K. Yoder, A.F. Schier and L.A. Drummond. 2005. Cilia-driven fluid flow in the zebrafish pronephros, brain and Kupffer's vesicle is required for normal organogenesis. Development 132: 1907–1921.

Kurokawa, T. and T. Suzuki. 1996. Formation of the diffuse pancreas and the development of digestive enzyme synthesis in larvae of the Japanese flounder *Paralichthys olivaceus*. Aquaculture 141: 267–276.

Kyle, H.M. 1923. The asymmetry, metamorphosis and origin of flat-fishes. Philosophical Transactions of the Royal Society of London (B) 211: 75–129.

Laurence, G.C. 1975. Laboratory growth and metabolism of the winter flounder *Pseudopleuronectes americanus* from hatching through metamorphosis at three temperatures. Marine Biology (Berlin) 32: 223–229.

Lipson, M.S. and J.C. Kaltenbach. 1965. A histochemical study of the esophagus and stomach of Rana pipiens during normal and thyroxine-induced metamorphosis. American Zoologist 5: 212.

Long, S., N. Ahmad and M. Rebagliati. 2003. The zebrafish nodal-related gene southpaw is required for visceral and diencephalic left-right asymmetry. Development 130: 2303–2316.

Matsumoto, J. and T. Seikai. 1992. Asymmetric pigmentation and pigment disorders in pleuronectiformes (flounder). Pigment Cell Research Supplement 2: 275–282.

McCormick, S.D. 1995. Hormonal control of gill Na$^+$, K$^+$-ATPase and chloride cell function. In: Cellular and Molecular Approaches to Fish Ionic Regulation. C.M. Wood and T.J. Shuttleworth (Eds.). Academic Press, New York, pp. 285–315.

Meints, R.H. and F.J. Carver. 1973. Triiodothyronine and hydrocortisone effects on Rana pipiens erythropoiesis. General and Comparative Endocrinology 21: 9–15.

Mellstrom, B., J.R. Naranjo, A. Santos, A.M. Gonzalez and J. Bernal. 1991. Independent expression of the α and β c-erbA genes in developing rat brain. Molecular Endocrinology 5: 1339–1350.

Miki, N., T. Taniguchi, H. Hamakawa, Y. Yamada and N. Sakurai. 1990. Reduction of albinism in hatchery-reared flounder "Hirame" *Paralichthys olivaceus* by feeding on rotifer enriched with vitamin-A. Suisanzoshoku 38: 147–155.

Minami, T. 1982. The early life history of a flounder *Paralichthys olivaceus*. Nippon Suisan Gakkaishi 48: 1581–1588.

Miwa, S. and Y. Inui. 1987a. Histological changes in the pituitary-thyroid axis during spontaneous and artificially-induced metamorphosis of larvae of the flounder *Paralichtys olivaceus*. Cell and Tissue Research 249: 117–123.

Miwa, S. and Y. Inui. 1987b. Effects of various doses of thyroxine and triiodothyronine on the metamorphosis of flounder (*Paralichthys olivaceus*) 1987b. General and Comparative Endocrinology 67: 356–363.

Miwa, S. and Y. Inui. 1991. Thyroid hormone stimulates the shift of erythrocyte populations during metamorphosis of the flounder. Journal of Experimental Zoology 259: 222–228.

Miwa, S. and K. Yamano. 1999. Retinoic acid stimulates development of adult-type chromatophores in the flounder. Journal of Experimental Zoology 284: 317–324.

Miwa, S., M. Tagawa, Y. Inui and T. Hirano. 1988. Thyroxine surge in metamorphosing flounder larvae. General and Comparative Endocrinology 70: 158–163.

Miwa, S., K. Yamano and Y. Inui. 1992. Thyroid hormone stimulates gastric development in flounder larvae during metamorphosis. Journal of experimental Zoology 261: 424–430.

Moss, B. and V.M. Ingram. 1965. The repression and induction by thyroxine of hemoglobin synthesis during amphibian metamorphosis. Proceedings of the National Academy of Sciences USA, 54: 967–974.

Naess, T., M. Germain-Henry and K.E. Naas. 1995. First feeding of Atlantic halibut (*Hipoglossus hipoglossus*) using different combinations of *Artemia* and wild zooplankton. Aquaculture 130: 235–250.

Naess, T. and O. Lie. 1998. A sensitive period during first feeding for the determination of pigmentation pattern in Atlantic halibut, *Hippoglossus hippoglossus* L., juveniles: the role of diet. Aquaculture Research 29: 925–934.

Niki, K., K. Yoshizato and S. Kikuyama. 1981. Augmentation of nuclear binding capacity for triiodothyronine by aldosterone in tadpole tail. Proceedings of the Japan Academy Ser. B 57: 271–275.

Norman, J.R. 1934. A systematic monograph of the flatfish (*Heterosomata*). Vol.1, Psettodidae, Bothidae, Pleluronectidae. British Museum of Natural History, London, 55pp.

Okada, N., Y. Takagi, T. Seikai, M. Tanaka and M. Tagawa. 2001. Asymmetrical development of bones and soft tissues during eye migration of metamorphosing Japanese flounder, Paralichtys olivaceus. Cell and Tissue Research 304: 59–66.

Okada, N., Y. Takagi, M. Tanaka and M. Tagawa. 2003. Fine structure of soft and hard tissues involved in eye migration in metamorphosing Japanese flounder (*Paralichthys olivaceus*) The Anatomical Record 273A: 663–668.

Okiyama, M. 1967. Study on the early life history of a flounder, *Paralichtys olivaceus* (Temminck et Shregel) I. Descriptions of post larvae. Bulletin of Japan Sea Regional Fisheries Research Laboratory 17: 1–12.

Okiyama, M. 1974. Studies on the life history of a flounder, *Paralichthys olivaceus* (Temminck et Schlegel) II. Descriptions of juveniles and comparison with those of related species. Bulletin of Japan Sea Regional Fisheries Laboratory 25: 39–61.

Ortiz Delgado, J.B., M.M. Ruane, P. Pousao-Ferreira, M.T. Dinis and C. Sarasquete. 2006. Thyroid gland development in Senegalese sole (*Solea senegalensis* Kaup, 1858) during early life stages: A histochemical and immunohistochemical approach. Aquaculture 260: 346–356.

Osborn, C.M. 1940. The experimental production of melanin pigment on the lower surface of summer flounders (*Paralichthys dentatus*). Proceedings of the National Academy of Sciences USA, 26: 155–161.

Osborn, C.M. 1941. Studies on the growth of integumentary pigment in the lower Vertebrates I. The origin of artificially developed melanophores on the normally unpigmented ventral surface of the summer flounder (*Paralichthys dentatus*). The Biological Bulletin Marine Biological Laboratory, Woods Hole 81: 341–351.

Policansky, D. 1982a. The asymmetry of flounders. Scientific American 246: 96–102.

Policansky, D. 1982b. Influence of age, size, and temperature on metamorphosis in the starry flounder, *Platichthys stellatus*. Canadian Journal of Fisheries and Aquatic Sciences 39: 514–517.

Popovic,W.J., W.E. Brown and J.M. Adamson. 1977. The influence of thyroid hormones on *in vitro* erythropoiesis. The Journal of clinical Investigation 60: 907–913.

Pouyet, J.C., J. Hourdry and J. Mensnard.1983. A histological and dynamic study of the gastric region of Discoglossus pictus larvae, cultured with or without thyroxine. Journal of Experimental Zoology 225: 423–431.

Ramachandran, V.S., C.W. Tyler, R.L. Gregory, D. Rogers-Ramachandran, S. Duensing, C. Pillsbury and C. Ramachandran. 1996. Rapid adaptive camouflage in tropical flounders. Nature 379: 815–818.

Ribeiro, L., C. Sarasquete and M.T. Dinis. 1999. Histological and histochemical development of the digestive system of *Solea senegalensis* (Kaup, 1858) larvae. Aquculture 171: 293–308.

Roberts, R.J., M. Bell and H. Young. 1973. Studies on the skin of plaice (*Pleuronectes platesssa* L.) II. The development of larval plaice skin. Journal of Fish Biology 5: 103–108.

Robinson H., S. Chaffee and V.A. Galton. 1977. The sensitivity of *Xenopus* laevis tadpole tail to the action of thyroid hormones. General and Comparative Endocrinology 32: 179–186.

Ryland, J.S. 1966. Observations on the development of larvae of the plaice, *Pleuronectes platelssa* L. in aquaria. Journal du Conseil 30: 177–195.

Sap, J., A. Munoz, K. Damm, Y. Goldberg, J. Ghysdael, A. Leutz H. Beug and B. Vennstrom. 1986. The c-erb-A protein is a high-affinity receptor for thyroid hormone. Nature 324: 635–640.

Schreiber, A.M. 2001. Metamorphosis and early larval development of the flatfishes (Pleuronectiformes): an osmoregulatory perspective. Comparative Biochemistry and Physiology Part B 129: 587–595.

Schreiber, A.M. 2006. Asymmetric craniofacial remodeling and lateralized behavior in larval flatfish. The Journal of Experimental Biology 209: 610–621.

Schreiber, A.M. and J.L. Specker. 1998. Metamorphosis in the summer flounder (*Paralichthys dentatus*): stage-specific developmental response to altered thyroid status. General and Comparative Endocrinology 111: 156–166.

Shreiber, A.M. and J.L. Specker. 1999a. Metamorphosis in the summer flounder *Paralichthys dentatus*: changes in gill mitchondria-rich cells. The Journal of Experimental Biology 202: 2475–2484.

Schreiber, A.M. and J.L. Specker. 1999b. Metamorphosis in the summer flounder, *Paralichthys dentatus*: thyroidal status influences salinity tolerance. Journal of Experimental Zoology 284: 414–424.

Schreiber, A.M. and J.L. Specker. 2000. Metamorphosis in the summer flounder, *Paralichtys dentatus*: thyroid status influences gill mitchondria-rich cells. General and Comparative Endocrinology 117: 238–250.

Schuh, T.J., B.L. Hall, J.C. Kraft, M.L. Privalsky and D. Kimelman. 1993. v-ervA and citral reduce the teratogenic effects of all-trans retinoic acid and retinol, respectively, in *Xenopus* embryogenesis. Development. 119: 785–798.

Segner, H., V. Storch, M. Reinecke, W. Kloas and W. Hanke. 1994. The development of functional digestive and metabolic organs in turbot, Scophtalmus maximus. Marine Biology 119: 471–486.

Seikai, T. 1980. Early development of squamation in relation to color anomalies in hatchery-reared flounder, *Paralichthys olivaceus*. Japanese Journal of Ichthyology 27: 249–255.

Seikai, T. 1985a. Influence of feeding periods of Brazilian *Artemia* during larval development of hatchery-reared flounder Paralichtys olivaceus on the appearance of albinism. Nippon Suisan Gakkaishi 51: 521–527.

Seikai, T. 1985b. Reduction in occurrence frequency of albinism in juvenile flounder *Paralichthys olivaceus* hatchery-reared on wild zooplankton. Nippon Suisan Gakkaishi 51: 1261–1267.

Seikai, T. 1992. Process of pigment cell differentiation in the skin on the left and right sides of the Japanese flounder *Paralichthys olivaceus*, during metamorphosis. Japanese Journal of Ichthyology 39: 85–92.

Seikai, T. and J. Matsumoto. 1994. Mechanism of pseudoalbinism in flatfish: an association between pigment cell and skin differentiation. Journal of the World Aquaculture Society 25: 78–85.

Seikai, T., J.B. Tanangonan and M. Tanaka. 1986. Temperature influence on larval growth and metamorphosis of the Japanese flounder *Paralichthys olivaceus* in the laboratory. Nippon Suisan Gakkaishi 52: 977–982.

Seikai, T., J. Matsumoto, M. Shimozaki, A. Oikawa and T. Akiyama. 1987. An association of melanophores appearing at metamorphosis as vehicles of asymmetric skin color formation with pigment anomalies developed under hatchery conditions in the Japanese flounder. Pigment Cell Research 1: 143–151.

Seikai, T., E. Hirose and J. Matsumoto. 1993. Dual appearances of pigment cells from *in vitro* cultured embryonic cells of Japanese flounder: An implication for a differentiation-associated clock. Pigment Cell Research 6: 423–431.

Shelbourne, J.E. 1957. Site of chloride regulation in marine fish larvae, Nature 180: 920–922.

Shiraishi, K., T. Kaneko, S. Hasegawa and T. Hirano. 1997. Development of multicellular complexes of chloride cells in the yolk-sac membrane of tilapia (*Oreochromis mossambicus*) embryos and larvae in seawater. Cell and Tissue Research 288: 583–590.

Sklower, A. 1930. Die Bedeutung der Schilddruse fur die Metamorphose des Aales und der Plattfishe, Forschung und Fortschritte der Deutsche Wissenshaftliche. 6: 435–436.

Soffientino, B. and J.L. Specker. 2001. Metamorphosis of summer flounder, *Paralichthys dentatus*: cell proliferation and differentiation of the gastric mucosa and developmental effects of altered thyroidal status. Journal of Experimental Zoology 290: 31–40.

Solbakken, J.S., B. Norberg, K. Watanabe and K. Pittman. 1999. Thyroxine as a mediator of metamorphosis of Atlantic halibut, *Hippoglossus hippoglossus*. Environmental Biology of Fishes 56: 53–65.

Stickney, R.R. and D.B. White. 1975. Ambicoloration in tank cultured flounder, *Paralichthys dentatus*. Transactions of the American Fisheries Society 104: 158–160.

Sugimoto, M. 2002. Morphological color changes in fish: regulation of pigment cell density and morphology. Microscopy Research and Technique 58: 496–503.

Sullivan, C.V., D.S. Darling and W.W. Dickhoff. 1987. Nuclear receptors for L-triiodothyronine in trout erythrocytes. General and Comparative Endocrinology 65: 149–160.

Suzuki, M.R. and S. Kikuyama. 1983. Corticoids augment nuclear binding capacity for triiodothyronine in bullfrog tadpole tail fins. General and Comparative Endocrinology 52: 272–278.

Suzuki, T., Y. Washio, M. Aritaki, Y. Fujinami, D. Shimizu, S. Uji and H. Hashimoto. 2009. Metamorphic pitx2 expression in the left habenula correlated with lateralization of eye-sidedness in flounder. Development, Growth and Differentiation 51: 797–808.

Tagawa, M. and M. Aritaki. 2005. Production of symmetrical flatfish by controlling the timing of thyroid hormone treatment in spotted halibut *Verasper variegates*. General and Comparative Endocrinology 141: 184–189.

Tagawa, M., S. Miwa, Y. Inui, E.G. de Jesus and T. Hirano. 1990. Changes in thyroid hormone concentrations during early development and metamorphosis of the flounder, *Paralichthys olivaceus*. Zoological Science 7: 93–96.

Takeuchi, T., J. Dedi, C. Ebisawa, T. Watanabe, T. Seikai, K. Hosoya and J. Nakazoe. 1995. The effect of β-carotene and vitamin A enriched *Artemia* nauplii on the malformation and color abnormality of larval Japanese flounder. Fisheries Science 61: 141–148.

Tamura, K., S. Yonei-Tamura and J.C.I. Belmonte. 1999. Molecular basis of left-right asymmetry. Development, Growth and Differentiation 41: 645–656.

Tanangonan, J.B., M. Tagawa, M. Tanaka and T. Hirano. 1989. Changes in tissue thyroxine level of metamorphosing Japanese flounder *Paralichthys olivaceus* reared at different temperatures. Nippon Suisan Gakkaishi 55: 485–490.

Thapliyal, J.P. and R.J. Kaur. 1976. Effect of thyroidectomy, L-thyroxine, and temperature on hemopoiesis in the chequered water snake, *Natrix piscator*. General and Comparative Endocrinology 30: 182–188.

Theng, C.C. and L.R. Johnson. 1986. Role of thyroxine in functional gastric development. American Journal of Physiology 251: G111–116.

Theng, C.C., K.L. Shmidt and L.R. Johnson. 1987a. Hormonal effects on development of the secretory apparatus of chief cells. American Journal of Physiology 253: G274–283.

Theng, C.C., K.L. Shmidt and L.R. Johnson. 1987b. Hormonal effects on development of the secretory apparatus of parietal cells. American Journal of Physiology 253: G284–289.

Thomas, D. 1974. The influence of L-thyroxine in red blood cell type in the axolotl. Developmental Biology 38: 187–194.

Thompson, C.C., C. Weinberger, R. Lebo and R.M. Evans. 1987. Identification of a novel thyroid hormone receptor expressed in the mammalian central nervous system. Science 237: 1610–1614.

Traquair, R.H. 1865. On the asymmetry of the Pleuronectidae, as elucidated by an examination of the skeleton in the turbot, halibut, and plaice. Transactions of the Linnean Society of London 25: 263–296.

Wagemans, F., B. Focant and P. Vandewalle. 1998. Early development of the cephalic skeleton in the turbot. Journal of Fish Biology 52: 166–204.

Watanabe, K., Y. Washio, Y. Fujinami, M. Aritaki, S. Uji and T. Suzuki. 2008. Adult-type pigment cells, which color the ocular sides of flounders at metamorphosis, localize as precursor cells at the proximal parts of the dorsal and anal fins in early larvae. Development, Growth and Differentiation 50: 731–741.

Weinberger, C., C.C. Thompson, S. Ong, R. Lebo, D.J. Gruol and R.M. Evans. 1986. The c-erb-A gene encodes a thyroid hormone receptor. Nature 324: 641–646.

Williams, S.R. 1901. The changes in the facial cartilaginous skeleton of the flatfishes. *Pseudopleuronectes americanus* (a dextral fish) and *Bothus maculates* (sinistral) Science New Series 13: 378–379.

Wood, W.G., J. Nash, D.J. Weatherall, J.S. Robinson and F.A. Harrison. 1979. The sheep as an animal model for the switch from fetal to adult hemoglobins. In: Cellular and Molecular Regulation of Hemoglobin Switching. G. Stamatoyannopoulos and A.W. Nienhuis (Eds.). Grune and Stratton, pp. 153–167.

Yahav, J. 1989. Development of parietal cells, acid secretion, and response to secretagogues. In: Human gastrointestinal development. E. Leventhal (Ed.). Raven Press, New York, pp. 341–351.

Yamamoto, M. and I. Iuchi. 1975. Electron microscopic study of erythrocytes in developing rainbow trout, *Salmo gairdnerii irideus*, with particular reference to changes in the cell line. Journal of Experimental Zoology 191: 407–426.

Yamano, K. and Y. Inui. 1995. cDNA cloning of thyroid hormone receptor β for the Japanese flounder. General and Comparative Endocrinology 99: 197–203.

Yamano, K. and S. Miwa. 1998. Differential gene expression of thyroid hormone receptor α and β in fish development. General and Comparative Endocrinology 109: 75–85.

Yamano, K., S. Miwa, T. Obinata and Y. Inui. 1991. Thyroid hormone regulates developmental changes in muscle during flounder metamorphosis. General and Comparative Endocrinology 81: 464–472.

Yamano, K., K. Araki, K. Sekikawa and Y. Inui. 1994a. Cloning of thyroid hormone receptor genes expressed in metamorphosing flounder. Developmental Genetics 15: 378–382.

Yamano, K., H. Takano-Ohmuro, T. Obinata and Y. Inui. 1994b. Effect of thyroid hormone on developmental transition of myosin light chains during flounder metamorphosis. General and Comparative Endocrinology 93: 321–326.

Yaoita,Y. and D.D. Brown. 1990. A correlation of thyroid hormone receptor gene expression with amphibian metamorphosis. Genes and Development 4: 1917–1924.

Yaoita, Y., Y. Shi and D.D. Brown. 1990. *Xenopus* laevis α and βthyroid hormone receptors. Proceedings of the National Academy of Sciences USA, 87: 7090–7094.

Yoo, J.H., T. Takeuchi, M. Tagawa and T. Seikai. 2000. Effect of thyroid hormones on the stage-specific pigmentation of the Japanese flounder *Paralichthys olivaceus*. Zoological Science 17: 1101–1106.

Yoshizato, K. and I. Yasumasu. 1972a. Effect of prolactin on the tadpole tail fin. III. Effect of prolactin on the hydroxylation of protocollagen-proline of the tail fin. Development, Growth and Differentiation 14: 57–62

Yoshizato, K. and I. Yasumasu. 1972b. Effect of prolactin on the tadpole tail fin. IV. Effect of prolactin on the metabolic fate of hyaluronic acid, collagen and RNA with special reference to catabolic process. Development, Growth and Differentiation 14: 119–127.

Youson, J.H. 1988. First metamorphosis. In: Fish Physiology, Vol XI B. W. S. Hoar and D.J. Randall (Eds.). Academic Press, San Diego, pp. 135–196.

Zadunaisky, J.A. 1984. The chloride cell: the active transport of chloride and the paracellular pathways. In: Fish Physiology, Vol X B.W.S. Hoar and D.J. Randall (Eds.). Academic Press, Ohrando, pp. 129–176.

# Metamorphosis in an Amphidromous Goby, *Sicyopterus Lagocephalus* (Pallas 1767) (Teleostei: Gobioidei: Sicydiinae): a True Metamorphosis?

*Philippe Keith,[1,a] Laura Taillebois,[1,b],\* Clara Lord,[1,2] Céline Ellien,[1] Sylvie Dufour[3,c] and Karine Rousseau[3,d]*

## 5.1 Introduction

In the tropical Indo-Pacific and the Caribbean regions, insular river systems are mainly colonised by diadromous species, including fish, which are mainly represented by Gobiidae (Sicydiinae), molluscs and crustaceans. The Gobiidae life cycle is adapted to the conditions in these distinctive habitats, which are young oligotrophic rivers, and subject to extreme

[1]Research Unit BOREA «Biology of Aquatic Organisms and Ecosystems» Muséum national d'Histoire naturelle, CNRS 7208, IRD 207, UPMC, 57 rue Cuvier, cp26, 75231 Paris Cedex 05, France;
[a]E-mail: keith@mnhn.fr
[b]E-mail: taillebois@mnhn.fr
[2]The University of Tokyo, Atmosphere and Ocean Research Institute, Division of Marine Life Science, 5-1-5 Kashiwanoha, Kashiwa, Chiba, 277-8564, Japan.
[3]Research Unit BOREA "Biology of Aquatic Organisms and Ecosystems" Muséum National d'Histoire Naturelle, CNRS 7208, IRD 207, UPMC, 7 rue Cuvier, CP32, 75231 Paris Cedex 05, France.
[c]E-mail: dufour@mnhn.fr
[d]E-mail: rousse@mnhn.fr
\*Corresponding author

seasonal climatic and hydrological variations (Keith, 2003). These species spawn in freshwater, the free embryos drift downstream to the sea where they undergo a planktonic phase, before returning to the rivers to grow and reproduce, hence they are called amphidromous (McDowall, 1997, 2007). The practical details of their biological cycle and the parameters leading to such extreme evolution in amphidromous gobies are poorly known (Lord and Keith, 2007), despite the fact that these gobies contribute most to the diversity of fish communities in the Indo-Pacific and the Caribbean insular systems, have the highest levels of endemism (Nelson et al., 1997; Keith and Lord, 2010) and a high economic value as food resource for local populations.

*Sicyopterus lagocephalus* (Pallas, 1767), distributed from the western Indian Ocean to the eastern Pacific one, is one of them. After reproduction, the free embryos of this species are carried out to sea by the river flow (Valade et al., 2009). When reaching the sea, the embryos transform into planktonic larvae which are dispersed by oceanic current during 130 to 266 days (Lord et al., 2010). After spending several months at sea as larvae, they need to return to freshwaters to achieve their life cycle (Keith et al., 2005) (Fig. 1). Soon after entering freshwaters, the post-larvae undergo several changes in colour, body and fins shape and have a metamorphosis switching from a planktonic to a benthic feeding mode (Keith et al., 2008). This is the

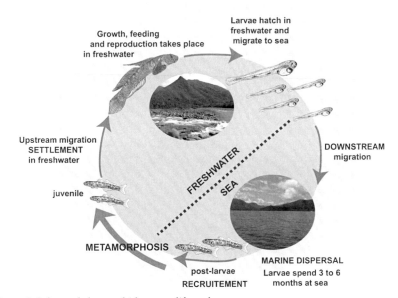

**Figure 1.** *S. lagocephalus* amphidromous life cycle.

*Color image of this figure appears in the color plate section at the end of the book.*

recruitment phase, which allows them to colonise the rivers and to grow to the reproductive adult stage.

Post-larvae migrating upstream provide an important source of food for local human populations in certain archipelagos (Bell, 1999; Keith et al., 2006), where this increasing fishing activity has a real socio-economic impact (Delacroix, 1987). However, harvesting of this food resource in estuaries is highly unsustainable, on account of the complexity of the species' life cycle and the hydrological specificities of these islands.

Keith et al. (2008) described the different post-larval and juvenile stages of *S. lagocephalus* (Gobiidae, Sicydiinae), which are fished in Reunion Island when recruiting back to rivers. The characterisation of the different stages should improve the biological and physiological knowledge needed to understand the processes involved in the metamorphosis and recruitment stages and help managers for a monitoring of stocks (Lord and Keith, 2007).

Taillebois et al. 2011 investigated the hormonal control of *S. lagocephalus* larval metamorphosis which occurs during recruitment to rivers and precisely studied the influence of thyroid hormones in the metamorphic changes.

## 5.2 Metamorphosis: Morphological Changes from a Planktonic Life to a Benthic One

*S. lagocephalus* adults are benthic and rheophile. They live on the bottom of the rivers, and are herbivorous. Their teeth and their mouth, located ventrally, are particularly adapted to grazing algae and diatoms that grow on the rocks. They have a strong pelvic sucker resulting from the fusion of the pelvic fins; it enables them to climb waterfall, to colonise rivers upstream and to resist to strong currents. When post-larvae enter freshwater from the sea, they are plankto-pelagic, with a terminal mouth and they are not adapted to the particularities of insular river habitats. They undergo a metamorphosis characterised by morphological changes, as described by Keith et al. (2008) and listed in Table 1. Keith et al. (2008) also characterised the different post-larval and juvenile stages of this species. Metamorphic changes start at sea during the transition from larvae to post-larvae. As fish recruit to the estuary, more metamorphic changes are observed as fish change through several post-larval stages, which allow them to colonise the river. As for the transformation of eel leptocephalus larva to post-larval glass eel (Chapter 3), these larval and post-larval transformation events in *S. lagocephalus* are a continuum, constituting the entire larval metamorphosis.

Several morphological characters are highly modified during the environmental shift from a plankto-pelagic larval life to a benthic one.

**Table 1.** Monitored criteria (see Keith et al., 2008).

**Standard Length (SL) and Total Length (TL)**: The total length was measured from the snout to the extremity of the caudal fin.

**Pigmentation**. The level of pigmentation was measured from a cerebral spot on the head, becoming increasingly dark in colour as the transformation progressed. Two measures were taken along the body axis, the first between the orbits, and the second on the neural chord level with the spot. The other two measures were taken from behind the orbits. The lines drawn had to be tangential to the spot shape and to stop at the spot edge.

**Corner of mouth angle**: The top of the opercula and the centre of the orbit have been used to measure this angle so that a line can be traced parallel to the body axis. The last measurement was taken at the corner of the mouth. This measure characterises the change of mouth position from a terminal to a ventral position.

**Caudal fin**: Three measurements were taken on the caudal fin in order to show that the fork disappears with time, and that the caudal fin border changes from a straight distal border to a rounded distal border. Measurements were taken from insertion of the rays to the extremity of the fin. From top to bottom, three measurements were observed: the superior length (SCFL), the length to the fork (FFL), and the inferior length (ICFL). %C=(FFL/((SCFL+ICFL)/2)).

**Pectoral fin**: The pectoral fin length was measured from the point of insertion of the inferior rays to the highest extremity of the fin.

**Scales**: The progressive appearance of scales along the body as well as their type (ctenoid/cycloid) was tracked over time.

These major changes include mainly the change in the position of the mouth with concomitant changes of the digestive system, the modification of the pectoral and caudal fins, the development of body pigmentation, and the spreading of scales to the anterior part of the body. The sucker is already formed as the fish arrive in rivers.

The most impressive changes observed during metamorphosis concern the mouth and the digestive system. The upper lip is enlarged, the mouth changes its position from a terminal to an almost ventral one, and the corner of mouth angle shows a logarithmic evolution over time (Fig. 2). During the

(a)    (b)

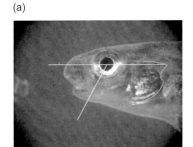 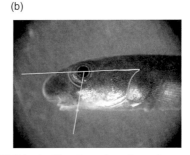

**Figure 2.** Photograph of *S. lagocephalus* (a) at day 0 (PL1) (arrival in estuary from the sea), corner of mouth angle is $63.0 \pm 0.80°$; (b) at day 37 (J1) (37 days spent in fluvarium since capture at the mouth river), corner of mouth angle is $80.4 \pm 0.68°$. Opercula is drawn on each photograph. Means are given $\pm$ SEM (n = 8 independant fish for each time point). (PL1: post-larval stage 1; J1: juvenile stage 1).

*Color image of this figure appears in the color plate section at the end of the book.*

metamorphosis, the angle changed by more than 35° (Keith et al., 2008). There is a phase of very quick modification of the angle in the first days.

In *Sicyopterus stimpsoni* from Hawaii, Schoenfuss et al. (1997) also reported observation of cranium metamorphosis. When the post-larvae of this species migrate from the ocean to rivers, they remain in the estuary while undergoing a total cranium restructuration. Forty-eight hours after entering fresh waters, both the length of the snout and the height and width of the head increase considerably. This process requires all the energy produced by the fish's metabolism to the detriment of somatic growth (Schoenfuss et al., 1997). This stop in individual growth is materialised on the otolith, which displays a metamorphosis check mark clearly visible on individuals having spent between 12–14 days in the river (Fig. 3). This check mark is therefore an indicator of the transition between the sea and the river.

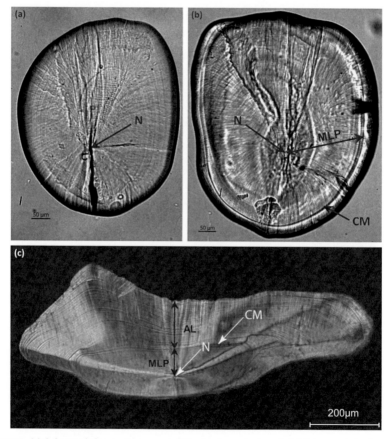

**Figure 3.** (a) *S. lagocephalus* post-larvae otolith, check mark absent. (b) *S. lagocephalus* juvenile otolith, check mark present. (c) *S. lagocephalus* adult otolith. (N: nucleus; CM: check-mark; MLP: marine larval phase; AL: adult life in freshwater).

This cranium metamorphosis is correlated to changes in diet and spatial distribution between post-larvae and juveniles (Keith, 2003). The digestive system is also modified (Schoenfuss et al., 1997). The juveniles must switch from a planktonic feeding mode (copepods) to a benthic feeding mode (periphyton), when they graze on the rocky substrate (Bielsa et al., 2003). Metamorphosis of feeding structures from the offshore larval phase is vital for the survival of the fish once it enters freshwater. The appearance of premaxillary teeth adapted to scraping diatoms off the substrate, typical of the *Sicyopterus* genus, is gradual throughout the metamorphic process. Indeed, the change in the position of the mouth and the cranium reorganisation start just before the arrival in estuaries and ends between 14 to 21 days after entrance in river. These changes are essential to allow new recruits to begin feeding on benthic algae (Schoenfuss et al., 1997).

Pectoral fin transformation was complete after only 2–3 days spent in fresh water but the caudal fin started to shift from forked to straight after *c.* 21 days; it was then straight after *c.* 28 days. The cerebral spot on the head was practically completely pigmented (> 85%) after *c.* 14 days following the arrival of individuals in freshwater. Finally, ctenoid scales spread to the anterior part of the body from *c.* 14 days after arrival.

When the transformation is complete, *Sicyopterus* is able to climb over waterfalls by using alternatively its pelvic suction cup and its lips (Blob et al., 2006; Schoenfuss and Blob, 2007). Non-metamorphosed *Sicyopterus* are unable to reach the upper part of the streams, because of high waterfalls and riffle areas that commonly block their migration (Keith, 2003).

According to these morphological observations, Keith et al. (2008) defined three post-larval stages (PL0 to PL2) and two juvenile stages (J1, J2) for *S. lagocephalus*. The true metamorphosis process takes place between stages PL0 and PL2.

## 5.2.1 Post-larval stages

The Post-Larval Stage 0 (PL0) is characterised by the beginning of the morphological transformation needed to recruit: the sucker is already formed and lateral scales start to appear. This stage takes place in salt water and individuals are then competent (i.e., physiological and morphological features required for colonisation of the adult environment are present) (Murphy and Cowan, 2007). The duration of this stage is unknown, and it probably varies according to hydrological and climatic conditions, the latter having to be favourable for migration into the river. This stage is complete as larvae enter the estuary.

The Post-Larval Stage 1 (PL1) starts as soon as larvae arrive in the estuary. It is characterised by the beginning of river colonisation and therefore by changes in osmoregulation. The individual is practically translucent, indeed the pigmentation of the spot on the head is ≤ 60%; the mouth is terminal. Metamorphosis of the head has started: the mouth corner angle is < 67°. The caudal fin is forked (%C < 63%). The length of the pectoral fin is less than 20% that of the standard length (SL). The lateral ctenoid scales generally reach, at most, half the insertion of the first dorsal fin. Cycloid scales are already present on the belly and in the predorsal zone. Fish in this stage remain in the river mouth less than 48 hours. The notches on the lips are light but visible. The rake-like teeth used by adults for grazing algae on river stones are not or just slightly visible. This stage is complete when the pectoral fin has finished transforming and when the teeth start to appear clearly.

During the Post-Larval Stage 2 (PL2), individuals show vertical coloured stripes on their sides and the pigmentation of the cerebral spot on the head now covers 60 to 90% of the observed skin surface. The mouth is sub-inferior, the head metamorphosis is ongoing: the corner of mouth angle is 67 to 80°. The caudal fin is still forked (% C < 63%). The pectoral fin has finished transforming and its length is more than 20% of that of the SL. The lateral ctenoid scales generally reach over half of the insertion place of the first dorsal fin, but do not overtake the anterior part of the insertion. PL2 individuals are found in the lower course of the river and have been in fresh water for 2 to 14 days. The notches on the lips are larger and now clearly visible. The rake-like teeth used for grazing are in place. This stage is complete when the metamorphosis of the cranium is completed.

During the two following juveniles stages (J1, J2), individuals have complete cerebral pigmentation, similar to that of non-mature adults: the spot on the head pigmentation is 95 to 100%. The caudal fin border is nearly to completely straight (%C between 63 and 100%). Lateral ctenoid scales and belly cycloid scales are in place, similar to those of adults.

## 5.3 Involvement of Thyroid Hormones in the Control of Larval Metamorphosis in *S. lagocephalus*

To study the influence of thyroid hormones in the metamorphic changes, Taillebois et al. 2011 performed an analytical study on a cohort of fish caught in estuary in Reunion Island (Indian Ocean) at PL1 stage when recruiting back to river and maintained in flume tanks (called fluvarium) which mimics as close as possible natural conditions. Biometrical parameters (total and standard length, corner of mouth angle, weight) and whole body L-tyroxine ($T_4$) and triiodothyronine ($T_3$) contents were measured on fish regularly sampled from the tank during 37 days.

   The biometrical parameters measured allowed replacing the different morphological changes of *S. lagocephalus* observed in the stages scale defined by Keith et al. (2008). Fish were collected at PL1 stage where the sucker is already formed. According to changes in the corner of mouth angle, PL1 stage was observed until day 2 when the corner of mouth angle is less than 67°. After day 2, fish entered PL2 stage, which lasted until day 10. This stage corresponds to the phase where the changes in the cranium and the changes in the corner of mouth angle are the most striking. After day 10, fish entered J1 stage when the mouth has reached its ventral position. The corner of mouth angle quickly increased during the metamorphosis, the mouth moving from a terminal position to an almost ventral one as reported in Keith et al. (2008). Standard and total lengths continuously decreased during the entire monitoring period (37 days) and the most important decrease in body length occurred during the post-larval stages. For the first time, weight and factor of condition (Fulton's $K$ factor of condition was calculated as followed: $K=100$ (standard length$^3$/body mass)) were monitored and both parameters decreased progressively during the metamorphic event (Fig. 4). The fish stops feeding when it enters freshwater; it is facing an ecological changeover due to the diet transition. No feeding behaviour was observed before day 7; after day 7 fish started grazing on pebbles that had been placed on the bottom of the fluvarium. According to this observation, fish seem

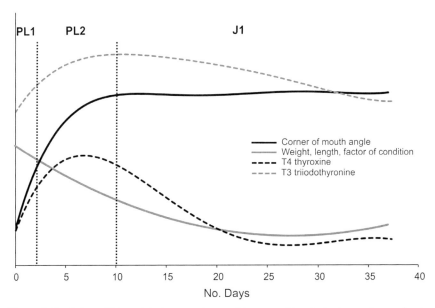

**Figure 4.** Evolution of biometrical parameters and hormonal levels during metamorphosis in fluvarium on *S. lagocephalus* post-larvae. (No. Days: number of days spent in fluvarium after capture).

not to feed during at least a week after arrival in fresh water. This, along with the energy expenditure for metamorphic changes, may contribute to the decrease in weight and factor of condition.

The modification of the position of the mouth of *S. lagocephalus* represented the best measurable metamorphic change. The higher rate of transformation occurred during the 10 first days after capture (Fig. 4). The most striking morphological changes coincided with the increase of thyroid hormones (TH) levels during post-larval stages. $T_4$ levels increased during post-larval stages and a peak was reached at the end of PL2 stage at around 10 days. The decrease of $T_4$ levels coincided with the beginning of J1 stage. $T_3$ levels slightly increased during post-larval stages, then stayed stable at the beginning of J1 stage and decreased (Fig. 4). In other teleost groups such as Elopomorphs or Pleuronectiforms, it has been shown that TH levels increased during larval metamorphosis (*Conger myriaster*: Yamano et al., 1991; *Anguilla japonica*: Yamano et al., 2007; *Paralichthys olivaceous*: Miwa and Inui, 1988; de Jesus et al., 1991; *Paralichthys dentatus*: Schreiber and Specker, 1998, *Hippoglossus hippoglossus*: Einarsdottir et al., 2006; *Verasper variegatus*: Hotta et al., 2001; *Solea senegalensis*: Fernandez-Diaz et al., 2001; Klaren et al., 2008; Manchado et al., 2008). The highest levels of whole body thyroid hormones in *S. lagocephalus* were measured when morphological changes were the deepest. This suggests an important role of the thyreotropic axis in the control of *S. lagocephalus* larval metamorphosis.

Taillebois et al. 2011 tested this hypothesis through an experimental approach, involving $T_4$ and Thiourea (TU) treatments. The hormonal treatments lasted the 10 first days of the 37 days of monitoring and where tested for each treatment on 196 fish. Variations of whole-body THs levels in *S. lagocephalus* were measured by RIA at day 10: $T_4$ and $T_3$ levels were both 4 times higher in $T_4$-treated fish than in control and 6 and 2 times, respectively, lower in TU-treated fish than in controls. The variation in $T_4$ and $T_3$ levels induced by experimental treatment in *S. lagocephalus* are thus in the same range of those induced by hormonal treatments or observed in natural physiopathological conditions in other species (*Anguilla anguilla*: Edeline et al., 2005; *Solea senegalensis*: Manchado et al., 2008; *Paralichtys dentatus*: Schreiber and Specker, 1998; *Paralichtys olivaceus*: Okada et al., 2005).

The fish which THs levels had been measured had been also morphologicaly monitored and Taillebois et al. 2011 showed that a 10-day treatment with $T_4$ accelerated and amplified the change in the corner of mouth angle in *S. lagocephalus* (Fig. 5). Inversely, treatment with TU significantly delayed the mouth change of position after 5 days of treatment (Fig. 5). The change of mouth position started again at day 18, that is 8 days after the end of the TU treatment. The corner of mouth angle then reached a value similar to the value of the controls. The fact that morphological transformations

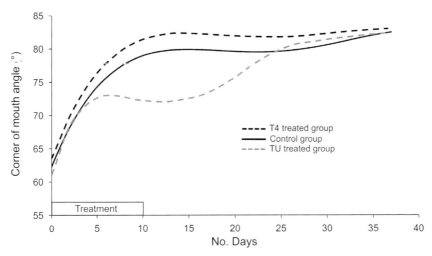

**Figure 5.** Effects of $T_4$ or TU treatments in *S. lagocephalus* post-larvae maintained 37 days in experimental aquariums as compared to controls on the the corner of mouth angle. $T_4$ or TU treatments were given in the water for 10 days.

finally occurred a few days after the treatment ended allowed to assess that the TU-treatment was reversible and non-toxic. A similar phenomenon was observed for the standard length. In other species, $T_4$ treatment accelerated quite immediately the metamorphosis stage, while TU treatment stopped the evolution of metamorphosis (*Paralichtys olivaceus*: Inui and Miwa, 1985; Miwa and Inui, 1987; *Paralichtys dentatus*: Schreiber et Specker, 1998; *Plectropomus leopardus*: Trijuno et al., 2002; *Hippoglossus hippoglossus*: Solbakken et al., 1999; *Megalops cyprinoides*: Shiao and Hwang, 2006). Moreover, it has been shown in some species that administration of TU stopped the metamorphic process and that THs were able to reverse metamorphic stasis induced by TU treatment (*Paralichtys olivaceus*: Miwa and Inui, 1987; *Verasper variegatus*: Tagawa and Aritaki, 2005; *Solea senegalensis* Manchado et al., 2008; *Megalops cyprinoides*: Shiao and Hwang, 2006). In *Paralichtys olivaceus*, abnormal non-metamorphosing fish treated with $T_4$ undergo a delayed metamorphosis (Okada et al., 2005).

Taillebois et al. 2011 findings are consistent with previous laboratory experiments conducted in some teleosts, including Elopomorphs and Pleuronectiforms, which have a true metamorphosis. They show, for the first time in a Gobioid fish, that thyroid status is involved in the control of drastic morphological changes in larval stage, which come with a switch in its habitat and feeding mode.

## Conclusion

Mechanisms of larval dispersal and recruitment to freshwater are both governed by biological, physiological, and hydrodynamical processes (Murphy and Cowan, 2007). For *S. lagocephalus*, the last steps of metamorphosis take place between the PL0 and PL2 stages. Competence is acquired at PL0 stage and post-larvae can therefore recruit at this stage. The characterisation of the different stages of *S. lagocephalus* recruits provides managers with precise knowledge of the developmental state of different cohorts observed while monitoring gobiid populations or larvae fisheries. The analysis of the fish's morphology brings information on its locomotion and feeding modes. For this type of amphidromous species the locomotion is critical as feeding and migration depend on it.

The physiological state of post-larvae has been characterised, as were the hormones responsible for triggering metamorphosis, migration and recruitment. Data show that morphological changes observed in *S. lagocephalus* during the recruitment of post-larvae and the colonisation of insular hydrosystems are under the control of the TH. The morphological changes observed in *S. lagocephalus* represent therefore a true metamorphosis as they are in Elopomorphs and Pleuronectiforms. Further studies should aim at collecting larvae in the marine environment before recruitment. This would allow investigating the initiation of metamorphosis in *S. lagocephalus*, namely the appearance of the sucker and the role of TH during this event, compulsory in transforming larvae into competent post-larvae, capable of colonising the juvenile and adult habitat.

## Acknowledgments

We thank P. Bosc, P. Valade and the staff of ARDA (Association Réunionaise pour le Développement de l'Aquaculture) for their help during sampling and for housing the experiments, and the authorities of Reunion Island.

## References

Bell, K.N.I. 1999. An overview of goby—fry fisheries. Naga ICLARM Quarterly 22: 30–36.
Bielsa, S., P. Francisco, S. Mastrorillo and J.P. Parent. 2003. Seasonal changes of periphytic nutritive quality for *Sicyopterus lagocephalus* (Pallas, 1770) (Gobiidae) in three streams of Reunion Island. Annales de Limnologie—International Journal of Limnology 39: 115–127.
Blob, R.W., R. Rai, M.L. Julius and H.L. Schoenfuss. 2006. Functional diversity in extreme environments: effects of locomotor style and substrate texture on the waterfall climbing performance of Hawaiian gobiid fishes. Journal of Zoology 268: 315–324.
Delacroix, P. 1987. Etude des "bichiques", juvéniles des *Sicyopterus lagocephalus* (Pallas), poissons Gobiidae migrateurs des rivières de la Réunion (Océan Indien): exploitation, répartition, biologie de la reproduction et de la croissance. PhD, Université de La Réunion.

Edeline, E., A. Bardonnet, V. Bolliet, S. Dufour and P. Elie. 2005. Endocrine control of *Anguilla anguilla* glass eel dispersal: effect of thyroid hormones on locomotor activity and rheotactic behavior. Hormone and Behaviour 48: 53–63.

Einarsdottir, I.E., N. Silva, D.M. Power, H. Smaradottir and B.T. Bjornsson. 2006. Thyroid and pituitary gland development from hatching through metamorphosis of a teleost flatfish, the Atlantic halibut. Anatomy and embryology 211: 47–60.

Fernandez-Diaz, C., M. Yufera, J.P. Canavate, F.J. Moyano, F.J. Alarcon and M. Diaz. 2001. Growth and physiological changes during metamorphosis of Senegal sole reared in the laboratory. Journal of Fish Biology 58: 1086–1097.

Hotta, Y., M. Aritaki, K. Ohota, M. Tagawa and M. Tanaka. 2001. Development of the digestive system and metamorphosis relating hormones in spotted halibut larvae and early juveniles. Nippon suisan gakkaishi 67: 40–48.

de Jesus, E.G., T. Hirano and Y. Inui. 1991. Changes in cortisol and thyroid hormone concentrations during early development and metamorphosis in the japanese flounder, *Paralichtys olivaceus*. General and Comparative Endocrinology 82: 369–376.

Inui, Y. and S. Miwa. 1985. Thyroid hormone induces metamorphosis of flounder larvae. General and Comparative Endocrinology 60: 450–454.

Keith, P. 2003. Review paper: Biology and ecology of amphidromous Gobiidae of the Indo-Pacific and the Caribbean regions. Journal of Fish Biology 63: 831–847.

Keith, P. and C. Lord. 2010. Tropical freshwater gobies: Amphidromy as a life cycle In: The Biology of Gobies, R.A. Patzner, J.L. Van Tassell, M. Kovacic & B.G. Kapoor (Eds.). Science Publishers Inc. (In Press).

Keith, P., T. Galewski, G. Cattaneo-Berrebi, T. Hoareau and P. Berrebi. 2005. Ubiquity of *Sicyopterus lagocephalus* (Teleostei: Gobioidei) and phylogeography of the genus Sicyopterus in the Indo-Pacific area inferred from mitochondrial cytochrome *b* gene. Molecular Phylogenetics and Evolution 37: 721–732.

Keith, P., C. Lord and E. Vigneux. 2006. *In vivo* observations on postlarval development of freshwater gobies and eleotrids from French Polynesia and New Caledonia. Ichthyological Exploration of Freshwaters 17: 187–191.

Keith, P., T. Hoareau, C. Lord, O. Ah-Yane, G. Gimmoneau, T. Robinet and P. Valade. 2008. Characterisation of post-larval to juvenile stages, metamorphosis, and recruitment of an amphidromous goby, *Sicyopterus lagocephalus* (Pallas, 1767) (Teleostei: Gobiidae: Sicydiinae). Marine and Freshwater Research 59: 876–889.

Klaren, P.H.M., Y.S. Wunderink, M. Yufera, J.M. Mancera and G. Flik. 2008. The thyroid gland and thyroid hormones in Senegalese sole (*Solea senegalensis*) during early development and metamorphosis. General and Comparative Endocrinology 155: 686–694.

Lord, C. and P. Keith. 2007. Threatened fishes of the world: *Sicyopterus sarazini* Weber & De Beaufort (Gobiidae). Environmental Biology of Fishes 83: 169–170.

Lord, C., C. Brun, M. Hautecoeur and P. Keith. 2010. Comparison of the duration of the marine larval phase estimated by otolith microstructural analysis of three amphidromous Sicyopterus species (Gobiidae: Sicydiinae) from Vanuatu and New Caledonia: insights on endemism. Ecology of freshwater fish 19: 26–38.

Manchado, M., C. Infante, E. Asensio, J.V. Planas and J.P. Canavate. 2008. Thyroid hormone down regulate thryrotropin and thyroglobulin during metamorphosis in the flatfish Senegalese sole (*Solea senegalensis* Kaup). General and Comparative Endocrinology 155: 447–455.

McDowall, R.M. 1997. Is there such a thing as amphidromy ? Micronesica 30: 3–14.

McDowall, R.M. 2007. On amphidromy, a distinct form of diadromy in aquatic organisms. Fish and Fisheries 8: 1–13.

Miwa, S. and Y. Inui. 1987. Effects of various doses of thyroxine and triiodothyronine on the metamorphosis of flounder (*Paralichthys olivaceus*). General and Comparative Endocrinology 67: 356–363.

Miwa, S. and Y. Inui. 1988. Thyroxine surge in metamorphosing flounder larvae. General and Comparative Endocrinology 70: 158–163.

Murphy, C.A. and J.H. Cowan. 2007. Production, marine larval retention or dispersal, and recruitment of amphidromous Hawaiian Gobioids: issues and implications. Bishop Museum Bulletin of Cultural and Environmental Studies 3: 63–74.

Nelson, S.G., J.E. Parham, R.B. Tibatts, F.A. Camacho, T. Leberer and B.D. Smith. 1997. Distribution and microhabitats of the amphidromous gobies in streams of Micronesia. Micronesica 30: 83–91.

Okada, N., T. Morita, M. Tanaka and M. Tagawa. 2005. Thyroid hormone deficiency in abnormal larvae of the Japanese flounder *Paralichtys olivaceus*. Fisheries Science 71: 107–114.

Pallas P.S. 1767. Spicilegia zoologica quibus novae imprimis et obscurae animalium species iconibus, descriptionibus atque commentariis illustrantur Fasciculus octavus. Lange, Berlin.

Schoenfuss, H.L. and R.W. Blob. 2007. The importance of functional morphology for fishery conservation and management: applications to Hawaiian amphidromous fishes. Bishop Museum Bulletin of Cultural and Environmental Studies 3: 125–141.

Schoenfuss, H.L., T.A. Blanchard and D.G. Kuamo'o. 1997. Metamorphosis in the cranium of postlarval *Sicyopterus lagocephalus* an endemic Hawaiian stream goby. Micronesica 30: 93–104.

Schreiber, A.M. and J.L. Specker. 1998. Metamorphosis in the summer flounder (*Paralichthys dentatus*): stage-specific developmental response to altered thyroid status. General and Comparative Endocrinology 11: 156–166.

Shiao, J-C. and P-P. Hwang. 2006. Thyroid hormones are necessary for the metamorphosis of tarpon *Megalops cyprinoides* leptocephali. Journal of Experimental Marine Biology and Ecology 331: 121–132.

Solbakken, J.S., B. Norberg, K. Watanabe and K. Pittman. 1999. Thyroxine as a mediator of metamorphosis of Atlantic halibut, Hippoglossus hippoglossus. Environmental Biology of Fishes 55: 53–65.

Tagawa, M. and M. Aritaki. 2005. Production of symmetrical flatfish by controlling the timing of thyroid hormone treatment in spotted halibut Verasper variegatus. General and Comparative Endocrinology 141: 184–189.

Taillebois, L., P. Keith, P. Valade, P. Torres, S. Baloche, S. Dufour and K. Rousseau. 2011. Involvement of thyroid hormones in the control of metamorphosis in *Sicyopterus lagocephalus* (Teleostei: Gobioidei) at the time of river recruitment. General and Comparative Endocrinology 173: 281–288.

Trijuno, D.D., K. Yosed, J. Hirokawa, M. Tagawa and M. Tanaka. 2002. Effects of thyroxine and thiourea on the metamorphosis of coral trout grouper Plectropomus leopardus. Fisheries Science 68: 282–289.

Valade, P., C. Lord, H. Grondin, P. Bosc, L. Taillebois, M. Iida, K. Tsukamoto and P. Keith. 2009. Early life history and description of larval stages of an amphidromous goby, *Sicyopterus lagocephalus* (Pallas, 1767) (Teleostei: Gobiidae: Sicydiinae). Cybium 33: 309–319.

Yamano, K., M. Tagawa, E.G. de Jesus, T. Hirano, S. Miwa and Y. Inui. 1991. Changes in whole body concentration of thyroid hormones and cortisol in metamorphosing conger eel. Journal of Comparative Physiology B 161: 371–375.

Yamano, K., K. Nomura and H. Tanaka. 2007. Development of thyroid gland and changes in thyroid hormone levels in leptocephali of japanese eel (*Anguilla japonica*). Aquaculture 270: 499–504.

# Salmonid Smoltification

*Karine Rousseau,*[1,a,][*] *Patrick Martin,*[2] *Gilles Boeuf*[3] and
*Sylvie Dufour*[1,b]

## 6.1 Introduction

The secondary metamorphosis is a non-classical metamorphosis occurring during the juvenile period in opposition to the first or "true" metamorphosis, which is a typical indirect development involving a larval metamorphosis leading to the juvenile period (Youson, 1988). Secondary metamorphosis, which occurs after a juvenile period, involves various morphological, physiological and behavioural modifications that pre-adapts the animal to life in the next environment, but that is less drastic than that observed during the larval metamorphosis. Two examples are found in Teleosts: smoltification (also called smolting or parr-smolt transformation) in salmon (this Chapter) and silvering in eel (Chapter 7). They both concern extreme migratory Teleosts and are related to their complex life cycles.

The Atlantic salmon (*Salmo salar*) spawns in freshwater (Fig. 1). The embryos hatch and develop into alevins (yolk-sac fry), which live off their nutrient rich yolk-sac that is attached to their underside. The young alevins will emerge as fry from the gravel when they absorb their nutrient rich yolk-sac and start actively searching for food. The small vulnerable

---

[1]Research Unit BOREA "Biology of Aquatic Organisms and Ecosystems" Muséum National d'Histoire Naturelle, CNRS 7208, IRD 207, UPMC, 7 rue Cuvier, CP32, 75231 Paris Cedex 05, France.
[a]E-mail: rousse@mnhn.fr
[b]E-mail: dufour@mnhn.fr
[2]Conservatoire National du Saumon Sauvage, 43 300 Chanteuges, France.
E-mail: pmartin@cnss.fr
[3]Laboratoire Arago, Research Team « Models in cell and evolutionary biology", University Pierre and Marie Curie- Paris 6/CNRS, Banyuls-sur-mer and Muséum national d'Histoire naturelle, 57 rue Cuvier, 75231 Paris Cedex 05, Paris, France.
E-mail: boeuf@mnhn.fr
[*]Corresponding author

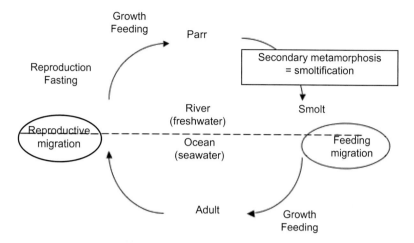

**Figure 1**. Biological life cycle of anadromous Atlantic salmon, *Salmo salar*. The secondary metamorphosis, smoltification occurrs between two periods of growth, one in freshwater and the other in seawater. It enables the fish to preadapt to a new environment, seawater, and to a perform a long feeding migration.

fries will start to develop parr marks on their sides as they start to feed and grow. The parr marks generally last a few months to years depending on the species of salmonid. After a period of growth in freshwater (FW), parrs will undergo a serie of physiological and morphological changes, called smoltification, that allow them to acclimate (adjust) to the salt water conditions of the new marine environment and start their downstream migration to the sea. Once in the ocean, smolts will feed and grow to fully achieved adult salmon. The adults, which start to starve, will migrate back to their natal streams to reproduce. Most Atlantic salmons, exhausted, will try to get back to the ocean, but only a few will succeed.

Smoltification occurs gradually over several months during the spring. One of the most pronounced changes in physiology of the parr is the gradual but consistent acquisition of the SW-tolerance (and increasing gill $Na+,K+$-ATPase activity) during the early spring, influenced by local latitude (photocycle) and riverine temperature. The climax smolt stage is reached a couple of months later and is characterised by fish having full tolerance as well as preference for SW, when given the challenge. At this stage (or slightly prior to), downstream movement is elicited, conditioned by the physiological development and triggered by environmental factors such as temperature and water discharge.

The morphological changes observed during the parr-smolt transformation (smoltification) will allow the fish to change environment (change from freshwater to seawater) and migrate for a long period in seawater; they thus involve modifications in organs involved in living in oceanic environment (skin, eye), in imprinting (olfactory system), in swimming (metabolism) and in osmoregulation (skin, gills, intestine and kidney).

## 6.2 Morphological, Behavioural and Physiological Changes Occurring During Smoltification

### 6.2.1 Skin

Gorbman et al. (1982) defined a parr as a young salmonid with vertical dark stripes, while a smolt is a slightly older and larger fish in which the dark stripes are obscured by a deposition of guanine and hypoxanthine in the scales and the skin, giving a silvery color (Fig. 2). This silvering of the skin would correspond to an adaptation to the oceanic pelagic adaptation, limiting the visibility of the fish for their predators. Guanine levels correlated well with changes in common physiological indices of smoltification (Staley and Ewing, 1992). As they reached maximum level 1 month earlier than development of seawater tolerance and transfer of juveniles coho salmon did not cause significant alterations in guanine concentration, it was concluded that guanine deposition in the skin during smolting may not represent an adaptation in preparation for an oceanic life (Staley and Ewing, 1992). In masu salmon (*O. masou*), it was reported that skin and serum guanine levels increased as smoltification progressed (Kazuhiro et al., 1994).

**Figure 2**. Atlantic salmon parr and smolt. Note the silvery color of smolt compared to the dark stripes of parr.

*Color image of this figure appears in the color plate section at the end of the book.*

In wild masu salmon, black pigmentation/darkening of dorsal, pectoral and caudal fin margins due to diffusion of melanin granules in melanophores, follows with development of seawater adaptability (Ura et al., 1997; Mizuno et al., 2001b). The level of dorsal fin pigmentation and ATPAse activity increased significantly from January or March to May during smoltification (Mizuno et al., 2004).

Epidermis is thinner in May (at the smoltification period) than in November (at the breeding season) (Rydevik, 1988). The increases in epidermal and dermal thickness in mature salmonids are correlated with androgen levels (Pottinger and Pickering, 1985; Rydevik, 1988).

The biological interface between fish and their aqueous environment is a mucus coat composed of biochemically diverse secretions, such as lysozyme (Schrock et al., 2001), which activity and levels peak 1 month before T4 surge during smoltification (Fagan et al., 2003). A sharp decrease in the numbers of mucus cells was observed in the skin of Atlantic salmon at the beginning of smoltification (O'Byrne-Ring et al., 2003).

## 6.2.2 Eye

Visual pigments (VP) comprise two components: an opsin protein and a chromophore. Some fishes have recently been found to express different subtypes of the five classes of vertebrate opsins and changes in expression levels of these subtypes have been associated with ontogenetic changes and metamorphic transitions. The other component of the VP, the chromophore can also be varied in some species. In order to compensate for the changes in spectral transmittance of water, salmonids are able to use mechanisms that alter spectral sensitivity, including gain or loss of a photoreceptor class (Allison et al., 2003, 2006a), changes in chromophore class [retinal (A1) or 3,4-dehydroretinal (A2)] (reviewed by Beatty, 1984; Temple et al., 2006) and expression of different opsin classes or subtypes within a photoreceptor class (reviewed by Bowmaker and Loew, 2008).

Salmonids have a paired-pigment visual system with photoreceptors possibly containing rhodopsin or porphyropsin. During smoltification, retina exhibits a shift from parr-stage porphyropsin (3-dehydroretinal; vitamin-A2-based visual pigment) dominance to a dominance in rhodopsin (retinal; vitamin-A1-based visual pigment) at smolt-stage. This is related to migration from FW, a medium that favours longer wavelengths of light, to marine (seawater) environment (Bridges and Delisle, 1974). Coho salmon possess rod photoreceptors and four classes of cone photoreceptors: ultraviolet wavelength-sensitive (UVS), short wavelength-sensitive (SWS), medium wavelength-sensitive (MWS) and long wavelength-sensitive (LWS). These photoreceptors express all five vertebrate opsin classes: RH1, UVS, SWS, RH2 and LWS, respectively (Dann et al., 2004). In coho salmon, the spectral absorbance of MWS and LWS cones differ among freshwater (alevin and parr) and ocean (smolt) phases (Temple et al., 2008a).

Ultraviolet wavelength-sensitive (UVS) cones are present throughout the retina of parrs in the land-locked rainbow trout (Allison et al., 2003), in migratory sockeye salmon (Novales Flamarique, 2000), as well as salmonids from the genus *Salmo* (Kunz, 1987; Kunz et al., 1994). In larger postmetamorphic rainbow trout, a decrease in UVS cone distribution is observed and is correlated with the reduced sensitivity to UV light (Allison et al., 2003) and during smoltification, salmonids lose most of their UVS through programmed cell death (Allison et al., 2006a). In sockeye salmon, UVS cones may disappear from the entire retina at the smolt stage (Novales Flamarique, 2000). On their return migration back to fresh water, some of these UVS cones are regenerated (Beaudet et al., 1997; Allison et al., 2006a).

Sea trout smolts, compared to parr, had downregulated ultraviolet-sensitive opsin (SWS1) expression in all quadrants of retina, lower long wavelength-sensitive opsin (LWS) expression dorsally, higher rod opsin (RH1) expression nasally and higher middle wavelength-sensitive opsin (RH2) expression dorsally (Veldhoen et al., 2006).

The change in visual pigment composition was speculated (Jacquest and Beatty, 1972; Evans and Fernald, 1990) to be a feature of smoltification process. Alexander and collaborators (1994) confirmed experimentally that the change from porphyropsin- to a rhodopsin-dominated visual system in coho salmon (*O. kisutch*) was one of the developmental processes of smoltification and not simply responses to the light environment being coincidental to the parr-smolt transformation (« migration-metamorphosis » hypothesis). The authors concluded that this shift of visual pigments could be used as an index of smoltification. However, later on, Temple and collaborators (2006) reported a shift in A1/A2 ratio correlated with season in both 0+ (<12 months old) coho parr that remained in fresh water for another year and in oceanic juvenile coho. They concluded that their findings support the hypothesis that the A1/A2 pigment pair system in coho salmon is an adaptation to seasonal variations in environmental variables (« seasonal » hypothesis) rather than to a change associated with migration or metamorphosis (Temple et al., 2006).

The shift in opsin expression for MWS and LWS cones occurs between parr and smolt stages, prior to seaward migration, as a means to prepare the visual system for a different photic environment and visual tasks. The flexibility of the spectral tuning mechanisms in salmonids might permit precise spectral tuning to the variable spectral environments, which they inhabit, while maintaining some optimum signal-to-noise ratio in the face of seasonal variation in temperature and light conditions (Temple et al., 2008a, 2008b).

### 6.2.3 Olfactory system

Salmon have long been known to imprint (olfactory imprinting) and home to natal stream odors. Olfactory imprinting in salmonids occurs during the smolt stage. Evidence came from studies in which coho salmon, brown trout (*Salmo trutta*) and steelhead trout (*Salmo gairdneri*) exposed to synthetic chemicals, either morpholine or phenethyl alcohol, during the smolt stage were attracted as adults to rivers scented with the chemicals (Cooper et al., 1976; Scholz et al., 1976; Johnsen and Hasler, 1980; Hasler and Scholz, 1983). If exposure to the chemicals occurs during the parr stage, no imprinting is observed (Hasler and Scholz, 1983). A sensitive period for olfactory imprinting (SPOI) has been identified in Atlantic salmon undergoing smoltification (Morin et al., 1989a), using measurement of the magnitude of a conditioned cardiac response to L-cysteine in fish during six 8-d intervals (Morin et al., 1989b). They concluded that there are two distinct optimal intervals for olfactory learning during smoltification but only one « sensitive » period for olfactory imprinting, occurring between 21 and 28 days after the onset of smoltification induced in the laboratory; fish tested for odor recognition exhibited a greater unconditioned cardiac deceleration to L-cysteine if they had been exposed during the third age interval of smoltification.

During induced smoltification (by a regime of increasing temperature and photoperiod), two peaks of olfactory activity occurred: the first peak occurred at the beginning of smoltification (the « acute phase ») coinciding with the previously described sensitive period for olfactory imprinting; the second peak occurring after the acute phase of smoltification coincided with a free thyroxine peak (Morin et al., 1994).

Studies in Atlantic salmon suggest a quadrupling of olfactory receptor cell number, as well as specific changes in the relative composition of the olfactory bulb neuropil during the parr-smolt transition (Bowers, 1988). More extensive investigation of Chinook salmon (*O. tshawytscha*) confirmed these findings suggesting growth in the input layer of the olfactory bulb coincident with smolting (Jarrard, 1997).

### 6.2.4 Osmoregulatory organs

#### 6.2.4.1 Gills

Apart from its role in respiration, nitrogenous excretion, thermal exchange and mucus production, the gill plays a major role in ionic and osmotic regulation through the action of chloride cells (mitochondria-rich cells or ionocytes). These specific cells increase considerably in number and activity in SW compared with FW fish (Conte and Lin, 1967) and during smoltification (Richman III et al., 1987). Chloride cells from the opercular membrane (an epithelium lining the branchial side of the operculum) increased nearly two-fold in late May, concurrent with the second increase in gill Na+/K+-ATPase activity and SW tolerance. The gill filament surface, which is rough initially, became smoother during smoltification and rough again toward the end of smoltification (Richman III et al., 1987). Two mitochondrion-rich (chloride) cell types were present in the gill epithelium during smoltification: the electron-lucent type I cell contained large, circular mitochondria, while the electron-dense type II cell contained thin, elongate mitochondria (Richman III et al., 1987). Chloride cells are the sites of active electrolyte transport (Payan et al., 1984). Chloride cell totally changes at the end of smolting with the development of a specific associated accessory cell (Pisam et al., 1988).

An increase in gill Na+/K+-ATPase activity during smoltification was first demonstrated in coho salmon by Zaugg and McLain (1970) and later reported in other salmonids (Zaugg and McLain, 1972; Zaugg and Wagner, 1973; McCartney, 1976; Lasserre et al., 1978; Boeuf and Harache, 1982). McCormick et al. (1989b) found that gill Na+/K+-ATPase activity increased along with the mitochondrial enzymes (citrate synthase and cytochrome C oxidase) between March and May during smoltification. The Na+/K+-ATPase activity remained high until August, while mitochondrial activity had declined. A further study reported a fivefold increase in gill Na+/K+-ATPase activity following transfer of Atlantic salmon to seawater, but a decrease in citrate synthase and cytochrome C oxidase activities in the gill (McCormick et al., 1989a). The enzyme Na+/K+-ATPase being composed of α and β subunits, an increase in the expression of α subunit mRNA in gill tissue was shown during smoltification in Atlantic salmon (D'Cotta et al., 1996, 2000; Seear et al., 2010).

Recently, five isoforms of claudins, tight junction proteins, were identified in gill libraries of Atlantic salmon and only the expression of branchial isoform 10e varied during smoltification, peaking in May at optimal seawater tolerance (Tipsmark et al., 2008).

### 6.2.4.2 Intestine

In salmonids, the intestine is the most important part of the digestive tract for osmoregulation, notably for water absorption. Indeed, whereas FW fish do not drink water and their gut is virtually watertight, in SW fish drink the ambient water continually and salt water is absorbed across the gut to reach the blood. During smoltification, the intestinal function must change from its FW role of preventing water inflow, to that of actively absorbing ions and water. Using an *in vitro* sac preparation, measurements of intestinal net fluid absorption (Jv) showed that FW coho salmon parr had significantly less Jv than FW or SW-adapted smolts (Collie and Bern, 1982). From 2 to 4 weeks following the springtime T4 surge, Jv increased in FW coho to a level comparable with that observed for SW wmolts; Jv remained elevated throughout most of the summer and then decreased in the autumn; this timing of the increase in Jv suggested a phase relationship to the T4 surge (Collie and bern, 1982). Rate of mucosal to serosal water movement, measured *in vitro* in non-everted midgut segments of Atlantic salmon, increased significantly during smoltification, but was not changed following transfer to SW (Usher et al., 1991). Veillette et al. (1993) reported, in Atlantic salmon, using *in vitro* intestinal sac preparations, a functional regionalization during smoltification with the posterior intestine taking on increased importance in osmoregulation in SW: the middle intestine of smolting fish underwent a significant decrease in fluid transport during the springtime, while posterior intestinal Jv significantly decreased and increased around the peak smolt period compared to parr. SW-adapted smolts generally exhibited posterior Jv approximately double that of FW cohorts. Sundell et al. (2003) demonstrated that, in Atlantic salmon, this preadaptative elevation in Jv during smoltification was at least due to an increase in intestinal Na+,K+-ATPase activity in the anterior intestine and a decrease of paracellular permeability (as judged by transepithelial resistance) in the posterior intestine. Nonnotte et al. (1995) demonstrated the influence of growth hormone on intestine physiology at the end of smolting in the Atlantic salmon.

### 6.2.4.3 *Kidney and urinary bladder*

The FW teleost produces a large volume of dilute urine, while in SW urine is sparse and relatively concentrated (for review: Boeuf, 1993). In FW, the kidney and urinary bladder are of considerable importance, their action consisting of eliminating excess water without losing ions; the glomerular filtration rate is very high and tubular reabsorption is low. In SW, the role of these organs is quite different. Ford (1958) showed a reduction in the number of functional *glomeruli* in SW salmon. There is a major change in kidney anatomy and physiology before migration in smoltifying salmonids, A decrease in electrolyte absorption in the urinary bladder of terminal coho smolts occurs in FW (Loretz et al., 1982), predisposing the fish to SW entry. Eddy and Talbot (1985) found an increase in urine production in silvery Atlantic salmon, which they interpreted as a consequence of pre-adaptation to increase the water permeability of the gut of the fish in FW. Mizuno et al. (2001c) demonstrated an increase in number and size of kidney juxtaglomerular cells (thought to produce renin) in smolting masu salmon, coincidentally with the increase in gill Na+,K+-ATPase.

## 6.2.5 Metabolism

A profound increase in oxygen consumption in smolts compared to parr has been reported (Baraduc and Fontaine, 1956; Higgins, 1985; Wiggs et al., 1989; Seddiki et al., 1996). Maxime et al. (1989) have reported an increase in oxygen consumption in juvenile Atlantic salmon during smolting. Changes in blood nucleoside triphosphate levels, hemoglobin concentrations and hematocrits (Zaugg and McLain, 1986) prepare for the increased oxygen demand during migration and greater energy requirements by erythrocytes during smoltification and seawater adaptation. A number of studies have reported that the hemoglobin system of salmon increases in complexity during smoltification (Vanstone et al., 1964; Wilkins, 1968; Giles and Vanstone, 1976; Sullivan et al., 1985). Adult hemoglobins are thought to pre-adapt the smolt to a marine environment where the oxygen tensions are lower than those experienced by fry and parr in freshwater (Giles and Randall, 1980). In parr, steady aerobic swimming occurs at a higher tailbeat frequency and with lower relative tailbeat amplitude, but during parr-smolt transformation, there is a developmental reduction in the myosin heavy chain composition of the red muscle (trout: Coughlin et al., 2001; Atlantic salmon: Martinez et al., 1993), which directly affects swimming behavior (Coughlin et al., 2001).

Smoltification is often characterized by a depletion of whole-body lipids (Fessler and Wagner, 1969; Farmer et al., 1978; Sheridan, 1989), especially of muscle (Woo et al., 1978; Sheridan, 1989) and liver lipid reservoirs (Sheridan, 1989), which is a process of natural energy depletion during a period of high energy requirements. During smoltification, lipids are depleted from depot sites primarily from the triacylglycerol fraction. Increases in tissue lipolytic rates and decreases in rates of lipid synthesis are among the factors responsible for lipid depletion (Sheridan, 1989). Lipolysis of depot fats and decrease in liver and muscle cholesterol during parr-smolt transformation are required by the increased energy demand (Bell et al., 1997; Sheridan, 1988; Sheridan et al., 1983, 1985 a and b).

A decrease in smolt liver glycogen was also reported (Fontaine and Hatey, 1950; Sweeting and McKeown, 1989). Sheridan et al. (1985a) reported a drop of 54% in glycogen synthesis in liver and an increase of 66% in glycogen phosphorylase activity. Hemre et al. (2002) observed redistribution of glycogen reservoirs from liver to muscle in Atlantic salmon as a consequence of parr-smolt transformation. Blood glucose increases in Atlantic salmon (Wendt and Saunders, 1973) during smoltification, while it decreases in coho salmon (Woo et al., 1978).

Protein catabolism is accelerated (Malikova, 1959). Depletion of protein reservoirs was observed, both as depleted whole-body protein (Fessler and Wagner, 1969) and depletion of protein in muscles and liver (see in Nordgarden et al., 2002). These results constrast with Woo et al. (1978) who reported equal muscle and liver protein levels in smolt and parr, and with Sweeting and McKeown (1989) who reported stable muscle protein levels throughout the smoltification process.

## 6.2.6 Migratory behaviour

Timing of downstream migration varies among species. Smolts usually start their downstream migration at the beginning of the spring for the individuals from South latitudes (for example, France and Spain: Boeuf, 1993; Utrilla and Lobon-Cervia, 1999), and in August for the individuals from the highest latitudes (for example, Iceland: Antonsson and Gudjonsson, 2002).

The salmon, which at the parr stage is sedentary, territorial and presents an aggressive behaviour, becomes a gregarious and migratory fish as a smolt (Jonsson and Jonsson, 1993; Thorpe, 1994). Behavioural changes include a decrease in territoriality and aggression and the subsequent formation of aggregations for the downstream migration (Iwata, 1995). There is a gradual decrease in aggressive behaviour as smolting progresses and reach minimum levels at the peak of smolting. Fish become aggressive again after the peak of smolting. The reduction of aggressive behaviour is a key factor in downstream migration, as it allows fish encounters to lead to schooling formations.

Downstream movement takes place preferentially in groups from the same parents and is stimulated by smolts coming from upstream areas (Hansen and Jonsson, 1985; Hvidsten et al., 1995; Olsen et al., 2004). A recent study indicated that solitary movements may initiate the downstream migration, followed by the schooling behaviour (Riley, 2007).

Downstream migration in freshwater is believed, by some authors, to be a passive displacement with the current (Huntsman, 1939; McCleave, 1978; Tytler et al., 1978; Thorpe et al., 1981). However, observations of smolt swimming actively have been reported (Kallenberg, 1958; Solomon, 1978; Hansen and Jonsson, 1985; Fängstam, 1994), namely with the smolts actively seeking areas of high velocity (Jonsson et al., 1991).

Generally, the downstream migration of salmonids occurs during the night (Neave, 1955; Aarestrup et al., 2002; Riley et al., 2002; Carlsen et al., 2004). The chum salmon fry forms school during daytime; after sunset, the schools break up, and the fry move to the water surface possibly because of poor vision in darkness and are carried downstream by the surface flow (Hoar, 1951).

## 6.3 Endocrine/internal Factors Involved in the Control of Smoltification

The involvement of pituitary in the smoltification process has been suggested by the fact that hypophysectomized fish are not able to smoltify (Nishioka et al., 1987) and that adenohypophysis cytology is clearly stimulated during this period (Olivereau, 1954; Komourdjian et al., 1976a, b; Nishioka et al., 1982).

### 6.3.1 Thyrotropic axis

The thyrotropic axis is mainly involved in the control of development and metabolism in vertebrates. Thyroid hormones (thyroxine, T4 and triiodothyronine, T3) are produced and secreted by the thyroid gland under the control of pituitary thyrotropin (TSH). The hypothalamic control of TSH is variable during evolution and according to developmental stages. For example, in amphibians, TSH is under the stimulatory control of Thyrotropin-Relasing Hormone (TRH) at adult stage, but of Corticotropin-Relasing Hormone (CRH) at larval stage. In teleosts, whatever the stage is, CRH seems to be the hypothalamic releasing factor for TSH.

#### 6.3.1.1 Hormonal changes

##### 6.3.1.1.1 Thyroid hormones

Thyroid involvement in smoltification was originally suggested by Hoar (1939) who observed histological activation of thyroid tissue of the Atlantic salmon (*Salmo salar*). The availability of RIA procedures has enabled Dickhoff et al. (1978) to demonstrate a T4 surge in coho salmon during the smoltification process. Different studies then determined that both plasma thyroid hormone levels and gill Na+, K+-ATPase activity were possible indicators of smolt stage (Folmar and Dickhoff, 1981; Langdon and Thorpe, 1984; Boeuf and Prunet, 1985). Many authors have demonstrated a dramatic surge of T4 at the end of smolting in freshwater, a few weeks before the highest level of gill ATPase activity is reached in both Pacific and Atlantic salmon (reviews: Boeuf, 1987; Hoar, 1988). In Atlantic salmon, sometimes T3 also peaks before and after the T4 surge (Boeuf and Prunet, 1985; Virtanen and Soivio, 1985; Boeuf et al., 1989; Prunet et al., 1989; Young et al., 1989). The T4 surge occurs much earlier in Atlantic salmon in long rivers than in those originating in short streams (Boeuf and Le Bail, 1990). This fact supports the concept of the role of TH in triggering migratory behaviour and memorizing the environment (i.e., imprinting).

### 6.3.1.1.2 TSH

Fridberg et al. (1981) reported that TSH-producing cells in the Atlantic salmon were more numerous in presmolts and smolts than in parr salmons, and had cytological features indicating an increased activity. However, Nishioka et al. (1982) showed no gross or ultrastructural changes of these cells during smoltification in coho salmon.

Martin et al. (1999) found that FW smolts had a lower concentration of TSH β subunit than parrs. Following a cohort of Atlantic salmon parrs in aquaculture at the Conservatoire National du Saumon Sauvage (Chanteuges, France), we could reveal, by quantitative real-time PCR, that TSHβ mRNA levels reached a peak in April at the time of parr-to-smolt transformation, i.e., when Atlantic salmons switched from positive to negative rheotactic behaviour (Fig. 3).

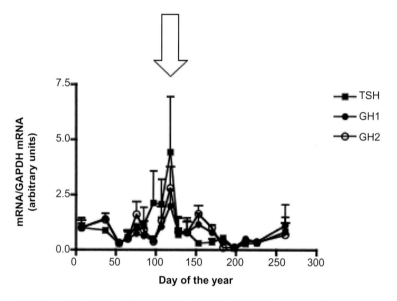

**Figure 3**. Changes in pituitary TSHβ, GH1 and GH2 mRNA during smoltification of Atlantic salmon. Arrows indicate the time when salmons shift from positive to negative rheotaxism in April. GH=growth hormone; TSH=thyrotropin.

### 6.3.1.1.3 CRH

CRH was shown to be potent stimulator of TSH secretion by coho salmon pituitary cells in culture, whereas TRH was not (Larsen et al., 1998). CRH being the stimulatory hypothalamic factor of the corticotropic axis, its involvement in smoltification is discussed in 2.1.3 part.

## 6.3.1.2 Experimental induction

The administration of exogenous thyroid hormones to juvenile parr-status salmonids results in morphological and physiological changes consistent with the parr-smolt transformation (for reviews: Robertson, 1949; Fontaine, 1975; Donaldson et al., 1979; Higgs et al., 1982; McBride et al., 1982; Sullivan et al., 1987). In addition, the use of anti-thyroid drugs such as propylthiouracil (PTU) in drinking water or food, as well as by immersion (Ebbesson et al., 1998) or intraperitoneal implants, in smolting salmon induced hypothyroidism by inhibiting the naturally occurring increases of plasma total T4 and total T3 and by affecting the conversion of T4 to T3. Because T4-treated salmonids do not undergo all the metamorphic changes necessary to pre-adapt them to saltwater, Eales (1979) referred to them as "pseudosmolts". Nevertheless, they remain a useful model for studying metamorphic physiology, for specific changes can be observed in isolation from the many other physiological factors that characterize true smoltification.

### 6.3.1.2.1 Body silvering/Fin darkening

Pioneer study by Robertson (1949) showed that intramuscular injection of mammalian thyroid extract and TSH produced silvery smolt stage in rainbow trout. Following studies confirmed that exogenous thyroid hormones (amago salmon: Miwa and Inui, 1983, 1985) and TSH (Premdas and Eales, 1976) induced silvery body color or guanine and hypoxanthine deposition in the skin of salmonids. Diet T4 for 72 days to underyearling amago salmon resulted in body silvering (Miwa and Inui, 1985). Administration of T4 from July to December induced silvery skin in underyearling masu salmon, but failed to induce seawater adaptability, darkening of dorsal and caudal fin margins and slimness of the body (Ikuta et al., 1985). T4 treatment, which can induce silvering of the body surface, failed to induce darkening of dorsal fin margins in masu and amago salmon (Miwa and Inui, 1983; Ikuta et al., 1985). T4 treatment increased the percentage of fish, which showed body silvering (amago salmon: Miwa and Inui, 1985; rainbow trout: Coughlin et al., 2001). T4 treatment caused increase in guanine levels in the skin and serum in masu salmon (Kazuhiro et al., 1994).

### 6.3.1.2.2 Metabolism

Treatment of coho salmon with T3 in the diet accelerated the increase in concentration of adult forms of blood hemoglobin, while dietary PTU reduced this increase during smoltification (Sullivan et al., 1985; for review: Hoar, 1988). Thyroid hormones induce a shift towards slower isoforms of the muscle protein myosin heavy chain (Coughlin et al., 2001). In T3-treated juvenile coho salmon, mean values of hematocrit were significantly decreased and the prolonged contraction (tetani) and twitch rates of contraction, relaxation and maximum force were significantly increased (Katzman and Cech, 2001).

T4 treatment stimulated lipid mobilization in parr, by decreasing total lipids and increasing lipolytic enzyme activity in the liver and dark muscle, as well as in mesenteric fat by decreasing total tissue mass and increasing lipase activity (Sheridan, 1986). Few reports suggested that T3 enhances lipid metabolism including synthesis, mobilization and degradation (Farbridge and Leatherland, 1988). A study in trout demonstrated that T3 could induce the liver (Na+K+)-ATPase activity (Pirini et al., 2002). Using dispersed salmon hepatocytes, Bhattacharya et al (1985) showed a stimulation of protein metabolism *in vitro* by T3.

### 6.3.1.2.3 Eye

Pioneer works demonstrated that administration of T4 favored proportions of porphyropsin in rods of rainbow trout (Munz and Swanson, 1965; Cristy, 1974; Allen, 1977). Similarly, Beatty (1969, 1972) showed in the kokanee salmon (*O. nerka*), a species of salmon, which does not normally exhibit any significant porphyropsin at any time of its life cycle that treatment with thyroid hormones or TSH induced the rhodopsin to porphyropsin conversion. Moreover, when PTU (goitrogen; Sullivan et al., 1987) or methimazole (deiodinase inhibitor; Alexander et al., 1998) has been administered to smoltifying coho salmon, there was impairment to the retinae pigmentary changes. A recent study by quantitative real-time PCR showed that TH-treated parr of rainbow trout had downregulated ultraviolet-sensitive (SWS1) opsin expression in all quadrants of retina and upregulated short wavelength-sensitive (SWS2) and middle wavelength-sensitive (RH2) opsin expressions in the temporal quadrants (Veldhoen et al., 2006). Proteomic analysis of opsins and thyroid hormone-induced retinal development using rainbow trout as a model of cone apoptosis and cone regeneration also demonstrated these changes in opsins (Allison et al., 2006b). In coho salmon, Temple and collaborators (2008a, b) showed that at least two RH2 opsin subtypes were expressed in MWS cones and that they were differentially expressed among alevin, parr and TH-treated alevin groups. They also demonstrated that, concerning the spectral absorbance of rods, TH-treated alevins were long wavelength shifted, consistent with an increase in A2 (Temple et al., 2008a). Exogenous TH induced an increase in spectral absorbance of all photoreceptors, consistent with an increase in vitamin A2 (Temple et al., 2008b), which is agreement with the seasonal hypothesis concerning opsin shift, proposed by Temple and collaborators (see paragraph 1.2).

In rainbow trout, Hawryshyn and collaborators have done many studies on ultraviolet photosensitivity and the role of thyroid hormones. They showed that T4 could induce a precocious loss of UV photosensitivity and an associated change in the retinal photoreceptor cell mosaic in small rainbow trout (Browman and Hawryshyn, 1992, 1994; Deutschlander et al., 2001). Six weeks after termination of hormone treatment, the same individuals once again possessed UV sensitivity due to UVS cone reappearance (Browman and Hawryshyn, 1994). T4 treatment to larger trout, which had lost their UV photoreceptor mechanism during normal development, were once again UV photosensitive and UVS cones were found in their retinae (Browman and Hawryshyn, 1994). These data suggest that thyroid hormones are involved in both the loss and reappearance of UV photosensitivity. Deutschlander and collaborators (2001) demonstrated that the reduction of UV sensitivity in T4-treated juvenile *O. mykiss* and anadromous steelhead *O. mykiss* smolts occurred primarily in the ventral retina. The remodelling of the smolt cone mosaic in response to T4 treatment is marked by localized proliferation or neuroregeneration of UVS cones in the ventral and nasal retinal areas (Hawryshyn et al., 2003). The optical quality of trout ocular lens was found to be decreased after T4 treatment; however, T3 treatment had no effect on trout lenses in culture (van Doorn et al., 2005).

*In vitro*, T3 is able to directly act on isolated coho salmon retinal pigment epithelial cells in order to alter the production of didehydro-derivatives of retinoids (Alexander et al., 2001).

### 6.3.1.2.4 Olfaction

An early study showed that artificially elevating thyroid hormone induced parr to imprint to artificial odorants, while parr with unaltered hormone levels did not. Moreover, pre-smolt coho salmon injected with TSH in doses sufficient to increase the T3 and T4 concentrations to smolt levels 5 months before smolt transformation would normally occurs, and simultaneously exposed to synthetic chemicals, became imprinted to the chemicals, whereas control fish receiving either ACTH or solvent did not (Hasler and Scholz, 1983). More recently, using intraperitoneal implants of T3 for 16–20 days to mimic smolting, Lema and Nevitt (2004) demonstrated that T3 could induce olfactory cellular proliferation in juvenile coho salmon. Scholz et al. (1985) demonstrated ontogenetic change in receptor binding kinetics during smolt transformation with all the putative target tissues for TH containing consistently more radioactivity in the TSH-induced smolts than in saline treated presmolts. In accordance with this result, it was shown that the olfactory epithelium of masu salmon (*O. masou*) became enriched with TH receptors during smolting, suggesting that olfactory tissues may be particularly sensitive to effects of TH at this time (Kudo et al., 1994).

The timing of olfactory imprinting is associated with elevations in plasma thyroid hormone levels (Hasler and Scholz, 1983; Morin et al., 1989a, 1994) and over the course of the parr-smolt transformation, changes in the density of cells in proliferation showed a positive relationship with natural fluctuations in plasma T4 (Lema and Nevitt, 2004). The olfactory epithelium may thus be better able to mirror subtle changes in plasma T4 triggered by changes in the fish's immediate environment. Whether this process extends to other parts of the brain potentially involved in imprinting (e.g., olfactory bulb and telencephalon) will be an interesting area for further research. T4 treatment depressed olfactory bulb responses to L-alanine (Morin et al., 1995) and lowered dopaminergic and serotonergic activity in the olfactory system (Morin et al., 1997).

### 6.3.1.2.5 Osmoregulation

Thyroid hormones do not seem to have a major role in osmoregulation and seawater adaptation in salmonids. Even if such action is feasible (for review: Fontaine, 1975; Donaldson et al., 1979; Higgs et al., 1982; Dickhoff et al., 1982; Refstie, 1982; Sullivan et al., 1987), many studies did not demonstrate improved SW adaptability following treatment with thyroid hormones (Miwa and Inui, 1983, 1985; Ikuta et al., 1985; Saunders et al., 1985; Omeljaniuk and Eales, 1986; Iwata et al., 1987; Sullivan et al., 1987; Madsen, 1990a; Boeuf et al., 1994).

### 6.3.1.2.6 Migratory behaviour/downstream migration

TH treatment decreased the aggressive behaviour of pre-smolts (Pacific and Atlantic salmon: Hoar, 1951; Godin et al., 1974; masu salmon: Iwata, 1995). Before the migratory period, chum salmon fry and coho salmon yearlings preferred shaded areas more than open areas. When immersed in T3 for 4 days, 80–90% preferred open water to shaded areas (the effect remaining for a week after treatment), while the controls and more than 78% treated with thiourea preferred to be in shade or to be in shelter (Iwata et al., 1989). This suggests that TH caused the fish to move to open water in daytime where it may be advantageous for the fish to form schools and migrate seaward (Iwata, 1995). During the pre-migratory season, chum salmon fry swam against the flow of current. T3 treatment changed their swimming direction to downstream (with the flow of current).

The involvement of TH in stimulation of downstream migratory behaviour has been well-documented (Baggerman, 1963; Godin et al., 1974; Fontaine, 1975; Youngson et al., 1985; Iwata et al., 1989; Boeuf and Le

Bail, 1990; Iwata, 1995; Munakata et al., 2000, 2001). Plasma T4 levels were shown to be higher in migrating fish than in non-migrating conspecifics (Biwa salmon, *Oncorhynchus rhodurus*: Fujioka et al., 1990; chum salmon: Ojima and Iwata, 2007). However, using exogenous treatments with T4 and/or thiourea, Birks et al. (1985) showed that TH were antagonistic to mechanism underlying seaward migration of steelhead trout.

### 6.3.2 Somatotropic axis

The somatotropic axis is classicaly involved in the control of growth, with the pituitary growth hormone, which can act by itself directly on target/ peripheral tissue or indirectly via the production and release of insulin-like growth factor 1 (IGF1) by the liver. The stimulatory hypothalamic control of GH production and release is multifactorial and varied during evolution, while the inhibitory action of somatostatin is conserved among vertebrates.

### 6.3.2.1 Hormonal changes

#### 6.3.2.1.1 GH

Hypertrophy and hyperplasia of somatotropic cells have been observed during smoltification of Pacific salmon (Clarke and Nagahama, 1977; for review: Donaldson et al., 1979). A rise in plasma GH during smoltification of coho salmon has been reported (Sweeting et al., 1985). In freshwater, during Atlantic salmon smoltification, plasma GH levels rose sharply concomitant with the T3 peak, 2 weeks before the peak of gill ATPase activity (Boeuf et al., 1989). After transfer to seawater, GH increased significantly remaining high for 7–10 days, and returning to basal levels after 14 days (Boeuf et al., 1989). Two peaks in plasma GH levels were observed in 1986, one in mid-April which coincided with a peak of plasma T3 and the second one in mid-May which lasted for 1 month, preceded T4 peak by 1–2 weeks and coincided with maximal gill Na+, K+ -ATPase activity (Prunet et al., 1989). More recently, in Atlantic salmon, Agustsson et al. (2003) showed that pituitary GH mRNA expression increased from early-April to reach a peak in mid-May, in agreement with increased GH secretion and plasma GH levels during this period (Agustsson et al., 2001).

Due to tetraploidy, two GH (GH1 and GH2) genes were evidenced in rainbow trout (Agellon et al., 1988; Mori et al., 2001). In Atlantic salmon, we measured, by quantitative real-time PCR, GH1 and GH2 mRNA levels during parr-to-smolt transformation and observed a peak in both hormones concomitant to that of TSH (Fig. 3).

Gill GHR mRNA levels peaked in early June, in parallel to branchial Na+, K+-ATPase activity in Atlantic salmon (Kiilerich et al., 2007).

### 6.3.2.1.2 IGF1

During salmon smoltification, liver and plasma IGF-1 levels increase (Duan et al., 1995; Agustsson et al., 2001) and hepatic mRNA levels are significantly correlated with plasma T4 levels in coho salmon (Duguay et al., 1994). The levels of gill IGF1 mRNA increase during smoltification of coho salmon (Duguay et al., 1994) and in response to GH (Sakamoto and Hirano, 1993). Beckman et al. (1999) reported that plasma IGF1 concentrations were good indicators of smolt quality and had significant relationships with smolt-to-adult returns (SAR) in hatchery reared Chinook salmon. Smolts of poor quality (i.e., behaviourally dysfunctional, physiologically compromised and disease prone) were found to have poor post-release survival and lower plasma IGF1 concentrations than higher quality smolts (Beckman et al., 1999).

### 6.3.2.1.3 GHRH

In chum salmon fry, the expression of GHRH in the preoptic area of the brain is not seen until the dowstream migratory period and is associated with the peak in plasma T4 levels (Parhar and Iwata, 1996).

Increased GH and IGF1 may also influence osmoregulatory capacities through their stimulation of cortisol production. Indeed, Young (1988) has shown that injection of coho salmon with GH enhances the sensitivity of the interregnal to ACTH. GH may also act via TH as it stimulates hepatic T4 5'monodeiodinase activity and T3 levels in rainbow trout (*Salmo gairdneri*) (MacLatchy and Eales, 1990).

### *6.3.2.2 Experimental induction*

Administration of GH to Pacific salmon induces smoltification-related changes in condition factor and skin pigment, and improves hypo-osmoregulatory ability and seawater survival of parr (for review: Donaldson et al., 1979; Wedemeyer et al., 1980).

### 6.3.2.2.1 Body silvering

Intramuscular injection of ovine GH for 72 days to underyearling amago salmon resulted in body silvering (Miwa and Inui, 1985). Komourdjian et al. (1976b) have reported that porcine GH administration induced marginal black color of the dorsal fin in Atlantic salmon.

## 6.3.2.2.2 Metabolism

Treatment with trout recombinant GH in FW resulted in a significant increase in gill (Na+K+)ATPase activity and in standard metabolism, and after direct transfer from FW to SW, attenuated the decrease in O2 affinity of hemoglobin foreseeable from the metabolic acidosis (Seddiki et al., 1995).

Preparations of GH from mammals and fish appear to be growth-promoting in a number of different teleosts (Donaldson et al., 1979). Free fatty acids are mobilized by GH in coho (Markert et al., 1977) and sockeye (Leatherland et al., 1974; McKeown et al., 1975) salmon. GH stimulated lipid mobilization from coho salmon parr with depletion of liver total lipids, increased lipolytic enzyme (triacylglycerol lipase) activity, while it had no effect in smolts (Sheridan, 1986). GH may also be diabetogenic (rainbow trout: Enomoto, 1964; sockeye salmon: McKeown et al., 1975). Injections of GH in juvenile coho salmon resulted in increases in plasma glucose levels but did not change plasma amino acid nitrogen levels (Sweeting et al., 1985). This hyperglycaemic effect of GH appeared to be dependent on the degree of smoltification of the coho juveniles (Sweeting et al., 1985). GH also appears to have a protein anabolic effect (*Salmo irideus*: Enomoto, 1964).

## 6.3.2.2.3 Osmoregulation

## 6.3.2.2.3.1 GH

In salmonids, many studies demonstrated that osmoregulation and SW adaptability is not regulated by thyroid hormones (see section 2.1.2.5.) but by GH. The effect of GH treatment on smoltification and seawater adaptability is well known (Smith, 1956; Komourdjian et al., 1976b; Clarke et al., 1977; Miwa and Inui, 1985; Bolton et al., 1987; Collie et al., 1989). The treatment with GH has been shown to have a positive influence on growth, smoltification and SW adaptability in juvenile Atlantic (Boeuf et al., 1990, 1994), coho salmon (Richman and Zaugg, 1987), in rainbow trout (Madsen, 1990b; Seddiki et al., 1995) and sea trout (Madsen, 1990c).

Treatment with GH (ovine or recombinant trout) of Atlantic salmon presmolts prior to smoltification (November, December and February) increased gill Na+, K+-ATPase activity and salinity tolerance, while treatment during April, May or June induced no difference compared to controls (Boeuf et al., 1994). Injections of juvenile coho salmon with bovine GH significantly increased corticosteroid receptor (CR) concentration in the gills, which may increase responsiveness of the gills to cortisol (Shrimpton et al., 1995). In addition, Shrimpton and McCormick (1998a) showed, in Atlantic salmon, that injection of ovine GH significantly increased gill

CR $B_{max}$ and $K_d$ while increasing gill Na+,K+-ATPase activity, which may regulate gill responsiveness to cortisol.

Treatment with oGH caused a greater increase in oxygen consumption in pre-smolts than in parr, and an increase of gill Na+, K+-ATPase activity in parr and pre-smolts, underlining the positive effect of oGH on SW adaptability in par and especially pre-smolts (Seddiki et al., 1996).

### 6.3.2.2.3.2 IGF1

Injection of rainbow trout with IGF1 enhances seawater tolerance by at least improving the ion transport capabilities of the gill (McCormick et al., 1991a). The osmoregulatory effect of GH may be mediated by hepatic production of IGF1 or by direct action on the gill with or without IGF1 involvement.

### 6.3.2.2.4 Migratory behaviour/downstream migration

Recently, using an artificial stream, Ojima and Iwata (2009) showed in chum salmon that icv injection of growth hormone-releasing hormone (GHRH) could stimulate both downstream movement (negative rheotaxis) and schooling behavior, whereas in the same species, an exogenous T4 treatment did not stimulate downstream migratory behavior even though plasma T4 level increased (Ojima and Iwata, 2007). Icv injection of GHRH also stimulated T4 secretion in chum, coho and sockeye salmon during the downstream migratory period (cited in Ojima and Iwata, 2009; Ojima and Iwata, 2010). In contrast, Munakata et al (2007) reported that administration of ovine GH in masu salmon did not stimulate downstream movement in artificial raceways during the migratory period. However, Iwata et al. (1990) demonstrated that GH stimulated movement from freshwater to brakish water in underyearling coho salmon held in a salinity preference tank.

### 6.3.3 Corticotropic axis

The corticotropic axis is commonly named stress axis. It is constituted of an hypothalamic factor, CRH, which stimulates the production and release of adrenocorticotropin (ACTH) from the pituitary. ACTH acts on the interrenal to stimulate the production and release of corticosteroids, mainly cortisol. As early as the 50s, activation and hypertrophy of the interrenal was demonstrated during smoltification (Fontaine and Olivereau, 1957; Olivereau, 1960).

## 6.3.3.1 Hormonal changes

### 6.3.3.1.1 Cortisol

The cells of interrenal tissue undergo hypertrophy during smoltification (Specker, 1982). This change coincides with the elevation of plasma cortisol in coho salmon (Specker, 1982); between January and April, plasma corticosteroid levels decrease and by the end of May in coho salmon smolts retained in fresh water, levels increase (Specker and Schreck, 1982). Young et al. (1989) showed, in coho salmon, increase in plasma cortisol from February to late March, a plateau from mid-April to July and a decline in September and October. Plasma cortisol levels increased during spring in both wild and hatchery-reared juvenile coho salmon (Shrimpton et al., 1994) and plasma clearance rate of cortisol is also elevated during the spring in this species (Patino et al., 1985). Using *in vitro* system (head kidney fragments), Young (1986) examined changes in interrenal sensitivity to ACTH during smoltification of coho salmon. He showed maximal *in vitro* responsiveness of interrenal tissue to ACTH in April, correlated with peak plasma T4 and enhanced hypoosmoregulatory ability, but not with plasma cortisol, which occurred in May.

In Atlantic salmon, plasma cortisol was constant from March to early May, and then started to increase, peaking in late June (Virtanen and Soivio, 1985). The rise of plasma cortisol concentration took place at the same time as the darkening of dorsal, pectoral and caudal fins (Virtanen and Soivio, 1985). Serum cortisol levels became elevated during smoltification, but after SW transfer, cortisol titers fell sharply (Langhorne and Simpson, 1986); these authors suggested that high serum cortisol levels represent a secondary response caused by the development of hypoosmoregulatory ability while still resident in FW. Comparing Atlantic salmon parrs separated by size into upper and lower mode, Shrimpton and McCormick (1998b) demonstrated that plasma cortisol levels increased significantly in upper mode fish in May, concurrently with increased smolt characteristics, but not in lower mode fish.

Recently, Kiilerich et al. (2007) reported a 2-fold increase in gill glucocorticoid receptor (GR) transcript levels during smoltification of Atlantic salmon. These data are consistent with the increased GR concentration and expression during smolting (coho salmon: Shrimpton et al., 1994; steelhead trout: McLeese et al., 1994; Atlantic salmon: Mazurais et al., 1998; Shrimpton and McCormick, 1998b; masu salmon: Mizuno et al., 2001b). This increase occurred before the increase in plasma cortisol, possibly explaining the increased responsiveness of gill tissue to cortisol observed in early spring (coho and Atlantic salmon: McCormick et al., 1991b) and demonstrated *in vitro* (rainbow trout: Shrimpton and McCormick,

1999). Changes in gill corticosteroid receptor (CR) concentration ($B_{max}$) and dissociation constant (kd) were observed when fish smolt during the spring (*O. kisutch*: Shrimpton, 1996; *S. salar*: Shrimpton and McCormick, 2003). Elevated GR expression suggests a need for increased cortisol signalling during smolting when SW-tolerance develops, and seems to occur prior to/or concurrent with elevated plasma cortisol levels (Shrimpton and McCormick, 2003).

### 6.3.3.1.2 ACTH

Measurement of the cell surface and nuclear area in the ACTH cells in Atlantic salmon showed an activation during smoltification prior to catadromous migration (Olivereau, 1975). Similarly, in coho salmon, McLeay (1975) reported a positive correlation between interenal nuclear diameters, interregnal cell size and ACTH cell nuclear diameters during smoltification. However, in the same species, Nishioka et al., (1982) demonstrated that ACTH-producing cells showed only moderate cytological changes indicative of increased synthetic cytological activity in the smolt.

### 6.3.3.1.3 CRH

CRH-like activity appears to increase temporally during the smoltification period in discrete brain regions (cited in Clements and Schreck, 2004).

### *6.3.3.2 Experimental induction*

### 6.3.3.2.1 Body silvering

The darkening of dorsal, pectoral and caudal fins can be induced by ACTH injection, but neither ACTH nor cortisol induced body silvering (Langdon et al., 1984).

### 6.3.3.2.2 Metabolism

Specker (1982) suggested that increased cortisol secretion may be responsible for the decrease in body lipid content characteristic of smoltification. Sheridan (1986) demonstrated that coho salmons with cortisol implant reduced total lipid concentration and triacylglycerol content of the liver and dark muscle, but increased mesenteric fat lipase activity. Cortisol replacement to hypophysectomized smolts restored liver lipase activity to approximately the same levels as those of controls (Sheridan, 1986).

## 6.3.3.2.3 Osmoregulation

Prolonged cortisol treatment in pre-smolt coho salmon caused an increase in gill Na+/K+-ATPase activity, while treatment of smolts had no effect (Richman et al., 1985). In young Atlantic salmon, cortisol treatment seems not to stimulate gill Na+/K+-ATPase activity, while ACTH does, but both ACTH and cortisol increased the SDH activity in gill cells from pre-smolts; neither hormone increased the size or abundance of gill chloride cells (Langdon et al., 1984). Cortisol treatment improved the development of hypoosmoregulatory mechanisms in rainbow trout (Madsen, 1990d) and enhanced the hypoosmoregulatory response to GH in rainbow and sea trouts (Madsen, 1990b, c). Intraperitoneal injections of cortisol were able to increase juxtaglomerular cell number and size during smoltification of masu salmon (Mizuno et al., 2001a).

In coho salmon, acute administration of cortisol resulted in a reduction in corticosteroid receptor numbers for 72h with no change in affinity, while chronic treatment resulted in a decrease in both concentration and affinity (Maule and Schreck, 1991; Shrimpton and Randall, 1994). In Atlantic salmon, Mazurais et al. (1998) also demonstrated that short-term treatment with cortisol (injections for up to 24 hours) induced a significant decrease of gill GR transcripts within 12 hours, but long-term treatment (implants for up to 26 days) did not induce any changes.

Using primary cultures of gill tissue, Tipsmark et al. (2009) demonstrated the stimulatory effect of cortisol on the expression of FW- (27a and 30) and SW-induced (10e) claudins (tight junction proteins); this effect was blocked by RU486, suggesting the involvement of a glucocorticoid type receptor. Similarly, injections of FW salmon with cortisol increased the expression of claudin 10e, 27a and 30.

Using tissue culture of FW-adapted sockeye salmon intestine, Veillette and Young (2005) were able to show that cortisol maintained Na+/K+-ATPase activity, which declined in controls. Slow-release implants of cortisol during 7 days stimulated Na+/K+-ATPase activity in pyloric coeca and posterior region of the intestine (Veillette and Young, 2005). In Atlantic salmon, cortisol implants stimulated the intestinal fluid absorption (Jv) during the parr and postsmolt stages to a rate comparable to that of control in the smolt stage, when Jv of controls was`low; conversely, RU 486 (corticosteroid antagonist) implants inhibited Jv only during the peak smolt period, when Jv of controls was elevated (Veillette et al., 1995).

## 6.3.3.2.4 Migratory behaviour/downstream migration

In juvenile Chinook salmon, central administration of CRH caused a dose-dependent increase in locomotor activity, which was prevented by the use of a peptide antagonist of CRH (Clements et al., 2002). It also increased the ability of juvenile salmonids to find cover in a novel environment (Clements et al., 2002). In a simulated stream environment, central administration of CRH was able to increase the proportion of juvenile chinook salmon that were distributed downstream of a mid-stream release site (Clements and Schreck, 2004). This alteration of downstream movement was similar to that of wild juvenile salmonids that were stressed (central administration of CRH) during their downstream migration (Clements and Schreck, 2004). Recently, icv injections of coho salmon held in circular-shaped channel tanks, with CRH showed that CRH stimulated both downstream movement and plasma T4 level (Ojima and Iwata, 2010). This last result is in agreement with the fact that ovine CRH was able to stimulate TSH release from coho salmon pituitary *in vitro* (Larsen et al., 1998) and that CRH-immunoreactive fibers terminate in pituitary regions containing TSH immunopositive cells in Chinook salmon (Matz and Hofeldt, 1999).

Similarly, an ip injection of cortisol stimulated downstream movement in masu salmon held in artificial raceways during the migratory period (Munakata et al., 2007). Furthermore, Ojima et al. (2007) reported that cortisol concentrations were higher in downstream migrating chum salmon fry than in pre-migrating fry (Ojima et al., 2007).

### 6.3.4 Gonadotropic axis

#### 6.3.4.1 Hormonal changes

The gonadotropic axis does not seem to be involved in the onset of smoltification. Indeed, in coho salmon, no changes in plasma levels of E2 in males and females, or in levels of 11-KT in males, were evident during spring when plasma T4 and cortisol were markedly elevated, indicating that the fish were undergoing smoltification (Patino and Schreck, 1986). Similarly, plasma sex steroids remain relatively constant throughout smoltification (chum salmon: Parhar and Iwata, 1996).

However, two other studies reported peaks of E2 and T at the same time as T4 (coho salmon: Sower et al., 1992; masu salmon: Yamada et al., 1993).

#### 6.3.4.2 Experimental induction

Early sexual maturation and sex steroid administration were able to inhibit smolt development and downstream migration (masu salmon: Ikuta et al., 1987; Munakata et al., 2001; Atlantic salmon: Madsen et al., 2004). Sex steroids inhibit smoltification (Aida et al., 1984; Ikuta et al., 1985, 1987; Miwa and Inui, 1986).

In masu salmon, methyltestosterone treatment was able to induce desmoltification (Yamazaki et al., 1973). Moreover, administration of methyltestosterone in early spring to yearling masu salmon parr inhibited natural smoltification; fish resembled parr or dark parr and seawater tolerance did not develop (Ikuta et al., 1985).

Aida et al. (1984) demonstrated that precocious gonadal maturation of masu salmon inhibited smoltification in the following spring since smoltification was induced in the completely castrated precocious male fish, but not in the fish, which had a small piece of remaining testis.

In masu salmon, sex steroids were shown to inhibit springtime smoltification (Ikuta et al., 1987). Indeed, administration of E2, T or 11-KT to masu salmon from February to May inhibited smoltification, but had no effect from September to December. Moreover, T and 11-KT were able to inhibit smoltification of masu salmon reared under artificially increased daylength.

Using artificial raceways and masu salmon with implants containing sex steroids, Munakata and collaborators (2001) showed that downstream migration of masu salmon smolts was inhibited significantly by T, E2 and 11-KT. Using very few fish in artificial stream channel, Munakata et al. (2001) found evidence that E2 (and androgens) inhibited downstream movement in smolts after 2 weeks of treatment. An inhibitory effect of downstream smolt migration by androgens was also reported in Atlantic salmon by Berglund et al. (1994).

Sex steroids may inhibit the normal body silvering of the smolting salmon (Madsen and Korsgaard, 1989), inhibit the smolt-associated increase in gill Na+,K+-ATPase activity (Madsen et al., 1997, 2004), negatively affect smolt growth rate (Arsenault et al., 2004) and impair the development of SW-tolerance during smolting (McCormick et al., 2005).

Ikuta et al. (1985, 1987) in masu salmon showed that treatment with sex steroids usually lowered plasma TH levels. These data suggest that sex steroids may inhibit smoltification by modulating thyrotropic axis.

## 6.4 Environmental/external Factors Involved in the Control of Smoltification

Among the external factors implicated, temperature and photoperiod are reported as the major actors in the initiation and the control of downstream migration (Boeuf, 1993; McCormick et al., 1998; Björnsson et al., 2011). Photoperiod is very active on the control of growth in fish (Boeuf and Le Bail, 1999) and parr-smolt transformation and largely used by farmers to stimulate it in using treatments consisting in increasing day length.

### *6.4.1 Photoperiod/melatonin*

Photoperiod is the most important environmental factor modulating the timing of smoltification in Atlantic and coho salmon (McCormick, 1994).

#### *6.4.1.1 Manipulation of photoperiod*

ATPase activity was decreased and migration reduced, when the length of increasing photoperiods approximated that of the summer solstice in steelhead trout (Zaugg and Wagner, 1973). An artificial shift in photoperiod from short to long days advances the onset of smoltification, whereas short-day exposure delays it (Saunders and Henderson, 1970). Smoltification of masu salmon (*O. masou*) can also be experimentally induced by artificial manipulation of photoperiod from short to long days (Ikuta et al., 1987). In addition, the larger the difference in day length after increasing photoperiod, the earlier smolts appear (Okumoto et al., 1989). Atlantic salmon exposed to continuous light into fall and winter showed a delayed parr-smolt transformation (delayed increases of branchial Na+,K+-ATPase activity and seawater tolerance: Saunders et al., 1985; McCormick et al., 1987; Saunders and Harmon, 1990).

#### *6.4.1.2 Melatonin treatment*

The onset of smoltification in pinealectomised Atlantic salmon was delayed by 3 weeks and melatonin implants, which elevate circulating melatonin to constant high levels, advanced the timing smoltification in pinealectomised Atlantic salmon (Porter et al., 1998). In contrast, in masu salmon, melatonin feeding, which reproduces melatonin profiles under short photoperiod, delays (but did not completely suppress) smoltification induced by long-day photoperiodic treatment (Iigo et al., 2005).

#### *6.4.1.3 Possible mechanisms of action*

Atlantic salmon juveniles exposed to abrupt increases in daylength (LD 15:9) in February or March exhibited earlier increase in plasma GH, with no changes in plasma TH (McCormick et al., 1987, 1995, 2007). When exposed to short daylength (LD 9:15) from January to May, fish had delayed increase in plasma GH (McCormick et al., 1995). The spring rise in plasma GH levels observed in Atlantic salmon in natural conditions or in simulated natural photoperiod was absent in salmon exposed to continuous light (Björnsson et al., 1995). Increasing daylength also increased plasma GH levels in masu salmon, while not affecting plasma T4 and T3 levels (Okumoto et al., 1989).

McCormick et al. (2007) demonstrated that smolt GH/IGF1 and cortisol axes had greater capacity to respond to increased daylength than parr ones.

## 6.4.2 Temperature

Temperature is less important than photoperiod in regulating physiological smolt development, but still has substantial impact (Boeuf, 1993; McCormick et al., 2000, 2002). According to Björnsson et al. (2011), temperature is likely to control ecological conditions that determine smolt survival during downstream migration and early ocean entry (an "ecological smolt window"). In at least some species of anadromous salmonids, temperature regulates the length of time during which the smolts are capable of good survival after ocean entry, with elevated temperatures narrowing this "physiological smolt window" (Björnsson et al., 2011).

ATPase activity was decreased and migration reduced, when steelhead trouts were subjected to temperatures of about 13°C or greater (Zaugg and Wagner, 1973). In Atlantic salmon smolts, an early increase in temperature from 5 to 12°C has been shown to advance the development of seawater tolerance compared with controls raised at ambient water temperature (5–6°C) (Solbakken et al., 1994). Similarly, the development of hypo-osmoregulatory ability and increase in gill Na+, K+-ATPase activity during smolting were significantly influenced by freshwater temperature, with earlier development of seawater tolerance in salmon smolts at higher temperatures: in smolts raised at 12.0°C, maximum gill Na+, K+-ATPase activity was reached in late April, compared to late May and mid-June in the 8.9°C and ambient groups (Handeland et al., 2004). Staurnes et al. (1994) even observed the development of typical smolt characters (marked silvering, high seawater tolerance and high Na+, K+-ATPase activity) in groups of Atlantic salmon reared at constant long day and seasonally changing water temperature that increased during spring, suggesting temperature can act as a cue for smolt development in the absence of an appropriate photoperiod stimulus. McCormick et al. (2000) were the first to demonstrate that low temperature limits the ability of photoperiod to advance the physiological aspects of smolting (condition factor, gill Na+, K+-ATPase). They clearly showed that smolting of Atlantic salmon was advanced by increased day length when fish were held at elevated temperatures (10°C) through the winter, but not when fish were held at cooler, ambient temperatures of 2°C.

Similar effects of temperature were reported for coho and chinook salmon by Zaugg and McLain (1976) and Clarke et al. (1981). In contrast, no effect of a rise in temperature from ambient to 10–11°C during spring on smolt development was observed by Dickhoff et al. (1989) and Duston and Saunders (1995) in these species.

McCormick et al. (2000) demonstrated that temperature could influence smolt development, most likely by affecting the rate of response to an endogenous circannual rhythm cued by changing photoperiod. They also showed that plasma T4 concentrations decreased following increased temperatures, whereas plasma IGF1, T3 and cortisol levels were only moderated affected. In contrast, changes of GH preceded and were strongly correlated with the physiological changes that resulted from the manipulations of both temperature and photoperiod.

## 6.5 Anthropic Factors Impairing Smoltification

### 6.5.1 Endocrine disruptors

Major declines in Atlantic salmon populations observed in Atlantic Canada were strongly associated with events of aerial forest spraying with insecticides containing 4-nonylphenol as primary solvent (Fairchild et al., 1999). Xenobiotics, such as alkylphenols, may interact negatively with physiological and biochemical aspects of smoltification in Atlantic salmon (Madsen et al., 1997), presumably due to their interaction with the estrogen receptor (White et al., 1994).

In Atlantic salmon, treatment with E2 and 4-nonylphenol (4-NP) impaired smolting as judged by elevated condition factor, reduced gill Na+,K+-ATPase activity and α-subunit Na+,K+-ATPase mRNA level, reduced muscle water content (Madsen et al., 2004). The inhibition of gill Na+,K+-ATPase activity and α-subunit Na+,K+-ATPase mRNA level by E2 treatment is in accordance with other studies in salmonids (Miwa and Inui, 1986; Madsen and Korsgaard, 1989; 1991; Madsen et al., 1997; McCormick et al., 2005). SW-transfer of 4-NP-treated fish resulted in 90% mortality within 24h, while no mortality was observed in FW-transferred controls; this was likely due to hypo-osmoregulatory failure after 4-NP treatment impairing SW-tolerance (Madsen et al., 2004). Moreover, treated fish initiated downstream migration with a delay of 6 (E2) and 8 (4-nonylphenol) days, and electro-fishing of the stream after 9 days revealed a more upstream position of E2-treated fish compared to controls (Madsen et al., 2004). Using waterborne exposure (30 days) during the peak migration period, Moore et al. (2003) were unable to confirm the negative effect of 4-NP on SW-tolerance in salmon smolts. It was suggested that 4-NP treatment in Madsen paper was initiated prior to migration, a period during which the developing smolts may be more sensitive to disruption. In another study, Fairchild et al. (2000) reported that a portion of Atlantic salmon smolts treated with E2 and 4-NP did not survive the transition from FW to SW.

Combined with impairment of smolt quality, a significant delay in migration may add to the risk of temperature-controlled desmoltification before reaching the ocean (Duston et al., 1990) and thus impair the chance of post-smolt survival in the ocean.

All these data show that disruption of sex hormones is subject of intensive research in vertebrates, but thyroid hormone synthesis and signalling are now also recognized as important targets of endocrine disruptors such as PCBs, perchlorates and brominated flame-retardants (for review: Brucker-Davis, 1998; Jugan et al., 2010; Zoeller, 2010). Thyroid-disrupting chemicals can target the thyrotropic axis at various levels, including synthesis of TH by thyroid gland and its regulation by hypothalamic-pituitary hormones, the catabolism and clearance of circulating TH by the liver and kidneys, the binding to transport proteins in the circulation, the cellular uptake of TH, the peripheral activating/inactivating metabolism of TH by iodothyronine deiodinases, the transcriptional activity of TH receptors and the expression of TH-regulated genes (for review: Brucker-Davis, 1998; Jugan et al., 2010; Zoeller, 2010). Even if data are available on thyroid-disrupting chemicals in teleosts (for review: Brown et al., 2004; Carr and Patino, 2011), studies in salmonids should aim at investigating their effects and mechanisms of action during smoltification. A pellet containing glucosinolate rapeseed used as diet in the rainbow trout triggers antithyroid and anti-growth effects (Burel et al., 2000, 2001).

Many studies showed that xenoestrogens likely exert some of their effects on smoltification indirectly by disturbing general hormonal processes such as the thyrotropic and somatotropic axes. Sex steroids depressed plasma thyroid hormone levels (Ikuta et al., 1987; Munakata et al., 2001), which are involved in the stimulation of downstream migration (Iwata, 1995; Munakata et al., 2000, 2001). Plasma TH levels were clearly depressed by 4-NP and E2 (McCormick et al., 2005). E2 has been shown to modulate somatotroph function and/or GH dynamics in amago salmon (Miwa an Inui, 1986) and rainbow trout (Holloway and Leatherland, 1997). During the early temporal window within the final stages of smoltification, short-term water-borne E2 and 4-NP exposures resulted in depressed growth and plasma IGF1 concentrations (Arsenault et al., 2004). McCormick et al. (2005), also reported decreased circulating IGF1, and as no changes in plasma GH levels were observed, the authors concluded that the effects of E2 and 4-NP were not through their impact at the pituitary level, but rather at the hepatic level. Similarly, treatment of juvenile Atlantic salmon with 4-NP and E2 did not affect pituitary GH mRNA levels (Yadetie and Male, 2002).

### 6.5.2 Anthropic disruptors of migration

During downstream migration, the timing between arrival in various environments (freshwater, estuary and finally seawater) and smolt physiological status is crucial for survival. For example, the period during which smolt is able to adapt to seawater is limited to several weeks ("physiological smolt window") (Boeuf et al., 1985; Berglund et al., 1992; Boeuf, 1993; McCormick et al., 1998). Therefore, any disturbance during the phase of downstream movement in freshwater resulting in a late estuarine arrival could be lethal due, firstly, to the loss of physiological capacities to seawater adaptation ("physiological smolt window") and secondly, to unfavourable local environmental conditions ("ecological smolt window") such as high temperature, low dissolved oxygen or presence of pollutants (McCormick et al., 1998; Hvidsten et al., 1998, Antonsson and Gudjonsson, 2002; Jutila et al., 2003). This problem is much more significant in the case of very long rivers where climatic and anthropic conditions (temperature, flow, pollution and dams) can considerably influence and delay downstream migration (McCormick et al., 2003).

Disruption to migratory pathways of both juveniles and adults is contributing o the decline of salmonid populations (Raymond, 1988; Rivinoja et al., 2001; Wilson, 2003). For instance, in the Pacific Northwest (USA), a large number of man made obstacles are in place on rivers, including hydroelectric and irrigation dams and passage through such structures can induce a severe stress response in juvenile salmonids (Congleton et al., 2000) and delay fish arrival at the ocean.

## Conclusions

All these data clearly show the involvement of the thyrotropic axis in smoltification (Fig. 4). Such a major stimulatory control by thyroid hormones has been well described for amphibian metamorphosis (Tata, 2006) and teleost first/larval metamorphosis. Nevertheless, two other axes, the somatotropic axis (with GH and IGF1) and the corticotropic axis (with cortisol) are also necessary for the control of some aspects of smoltification (Fig. 4). GH is very important as the hormone responsible for osmoregulation and SW adaptability. At the hypothalamic level, CRH may represent a coordinator between thyrotropic and somatotropic axes (Fig. 4), as it is able to stimulate in fish both TSH (salmon: Larsen et al., 1998) and GH (eel: Rousseau et al., 1999) *in vitro*. In addition, strong interactions exist between these different neuroendocrine axes. Indeed, cortisol seems to have a synergizing action with GH and TH in the control of osmoregulation and metabolism. In contrast, the gonadotropic axis is likely inhibitory on smoltification.

Data show that the various aspects of smolting process are likely to be under the control of different environmental cues. For example, physiological changes appear to precede the switch to migratory behavior, and it has been suggested that changes in behavior require environmental (e.g., photoperiod) and endocrine 'priming' factors that may be similar to

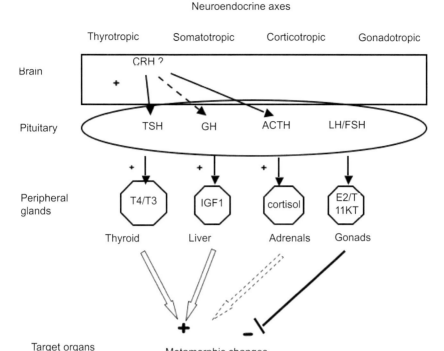

**Figure 4**. Neuroendocrine axes involved in the control of secondary metamorphosis (smoltification) in salmonid. Smoltification is triggered mainly by thyrotropic and somatotropoic axes with some synergism of corticotropic axis. Gonadotropic axis is likely inhibitory. CRH=corticotropin-releasing hormone; TSH=thyrotropin; T4=thyroxine; T3=triiodothyronine; GH=growth hormone; IGF=insulin-like growth factor; ACTH=corticotropin; LH=luteinizing hormone; FSH=follicle-stimulating hormone; E2=estradiol; T=testosterone; 11KT=11-ketotestosterone.

those that control physiological development, followed by 'releasing' factors (e.g., water temperature or flow) that initiate behaviours (McCormick et al., 1998). Many studies provided evidence that GH was the major endocrine mediator of the effects of environmental factors (such photoperiod and temperature) on the timing of smolt development.

## Acknowledgements

We thank Nicole Dunn and Sarah Harris (MNHN/University of Keele, UK; ERASMUS European Program) for their work on the assays of pituitary hormone mRNAs. Thanks to Jocelyn Rancon (Conservatoire National du Saumon Sauvage) for providing pictures of salmon parr and smolt.

# References

Aarestrup, K., C. Nielsen and A. Koed. 2002. Net ground speed of downstream migrating radio-tagged Atlantic salmon (*Salmo salar* L.) and brown trout (*Salmo trutta* L.) smolts in relation to environmental factors. Hydrobiologia 483: 95–102.

Agellon, B., S.L. Davies, C-M. Lin, T. Chen and D.A. Powers. 1988. Rainbow trout has two genes for growth hormone. Molecular Reproduction and Development 1: 11–17.

Agustsson, T., K. Sundell, T. Sakamoto, V. Johansson, M. Ando and B. Th. Björnsson. 2001. Growth hormone endocrinology of Atlantic salmon (*Salmo salar*): pituitary gene expression, hormone storage, secretion and plasma levels during parr-smolt transformation. Journal of Endocrinology 270: 227–234.

Agustsson, T., K. Sundell, T. Sakamoto, M. Ando and B. Th. Björnsson. 2003. Pituitary gene expression of somatolactin, prolactin and growth hormone during Atlantic salmon parr-smolt transformation. Aquaculture 222: 229–238.

Aida, K., T. Kato and M. Awaji. 1984. Effects of castration on the smoltification of precocious male masu salmon Oncorhynchus masou. Bulletin of the Japanese Society of Scientific Fisheries 50: 565–571.

Alexander, G., R. Sweeting and B. McKeown. 1994. The shift in visual pigment dominance in the retinae of juvenile coho salmon (*Oncorhynchus kisutch*): an indicator of smolt status. Journal of Experimental Biology 195: 185–197.

Alexander, G., R. Sweeting and B. McKeown. 1998. The effect of thyroid hormone and thyroid hormone blocker on visual pigment shifting in juvenile coho salmon (*Oncorhynchus kisutch*). Aquaculture 168: 157–168.

Alexander, G., R. Sweeting and B.A. McKeown. 2001. The effects of 3,4,3'-triiodo-L-thyronine on didehydroretinol synthesis by isolated coho salmon retinal pigment epithelial cells. General and Comparative Endocrinology 123: 192–202.

Allen, D.M. 1977. Measurement of serum thyroxine and the proportions of rhodopsin and porphyropsin in rainbow trout. Canadian Journal of Zoology 55: 836–842.

Allison, W.T., S.G. Dann, J.V. Helvik, C. Bradley, H.D. Moyer and C.W. Hawryshyn. 2003. Ontogeny of ultraviolet-sensitive cones in the retina of rainbow trout (*Oncorhynchus mykiss*). The Journal of Comparative Neurology 461: 294–306.

Allison, W.T., S.G. Dann, K.M. Veldhoen and C.W. Hawryshyn. 2006a. Degeneration and regeneration of ultraviolet cone photoreceptors during development in rainbow trout. The Journal of Comparative Neurology 499: 702–715.

Allison, W.T., K.M. Veldhoen and C.W. Hawryshyn. 2006b. Proteomic analysis of opsins and thyroid hormone-induced retinal development using isotope-coded affinity tags (ICAT) and mass spectrometry. Molecular Vision 12: 655–672.

Antonsson, T. and S. Gudjonsson. 2002. Variability in timing and characteristics of Atlantic salmon smolt in Icelandic rivers. Transactions of the American Fisheries Society 131: 643–655.

Arsenault, J.T.M., W.L. Fairchild, D.L. MacLatchy, L. Burridge, K. Haya and S.B. Brown. 2004. Effects of water-borne 4-nonylphenol and 17β-estradiol exposures during parr-smolt transformation on growth and plasma IGF-I of Atlantic salmon (*Salmo salar* L.). Aquatic Toxicology 66: 255–265.

Baggerman, B. 1963. The effect of TSH and anti-thyroid substances on salinity preference and thyroid activity in juvenile Pacific salmon. Canadian Journal of Zoology 41: 307–319.

Balon, E.K. 1985. Saltatory ontogeny and life history models revisited. In: Early life history of fishes (Balon EK, Ed.). Nijhoff/Junk, Dordrecht, pp. 13–30.

Baraduc, M.M. and M. Fontaine. 1956. Etude comparée du métabolisme respiratoire du jeune saumon sédentaire (parr) et migrateur (smolt). Comptes Rendus des Séances de la Société de Biologie Paris 149: 1327–1329.

Beatty, D.D. 1969. Visual pigment changes in juvenile kokanee salmon in response to thyroid hormones. Vision Research 9: 855–864.

Beatty, D.D. 1972. Visual pigment changes in salmonid fishes in response to exogenous L-thyroxine, bovine TSH and 3-dehydroretinol. Vision Research 12: 1947–1960.

Beatty, D.D. 1984. Visual pigments and the labile scotopic visual system of fish. Vision Research 24: 1563–1573.

Beaudet, L., I. Novales Flamarique and C.W. Hawryshyn. 1997. Cone photoreceptor topography in the retina of sexually mature Pacific salmonid fishes. Journal of Comparative Neurology 383: 49–59.

Beckman, B.R., W.W. Dickhoff, W.S. Zaugg, C. Sharpe, S. Hirtzel, R. Schrock, D.A. Larsen, R.D. Ewing, A. Palmisano, C.B. Schreck and C.V.W. Mahnken. 1999. Growth, smoltification, and smolt-to adult return of spring chinook salmon from hatcheries on the Deshutes River Oregon. Transactions of the American Fisheries Society 128:1125–1150.

Bell, G., D.R. Tocher, B.M. Farndale, D.I. Cox, R.W. McKinney and J.R. Sargent. 1997. The effects of dietary lipid on polyunsaturated fatty acid metabolism in Atlantic salmon (*Salmo salar*) undergoing parr-smolt transformation. Lipids 32: 515–525.

Berglund, I., M. Schmitz and H. Lundqvist. 1992. Seawater adaptability in Baltic salmon (*Salmo salar*): a bimodal smoltification pattern in previously mature males. Canadian Journal of Fisheries and Aquatic Sciences 49: 1097–1106.

Berglund, I., H. Lundqvist and H. Fänstam. 1994. Downstream migration of immature salmon (*Salmo salar*) smolts blocked by implantation of the androgen 11-ketoandrostenedione. Aquaculture 121: 269–276.

Bhattacharya, S., E. Plistetskaya, W.W. Dickhoff and A. Gorbman. 1985. The effects of estradiol and triiodothyronine on protein synthesis by hepatocytes of juvenile coho salmon (*Oncorhynchus kisutch*). General and Comparative Endocrinology 57: 103–109.

Birks, E.K., R.D. Ewing and A.R. Hemmingsen. 1985. Migration tendency in juvenile steelhead trout, *Salmo gairdneri* Richardson, injected with thyroxine and thiourea. Journal of Fish Biology 26: 291–300.

Björnsson, B.T., T. Ogasawara, T. Hirano, J.P. Bolton and H.A. Bern. 1988. Elevated growth hormone levels in stunted Atlantic salmon, *Salmo salar*. Aquaculture 73: 275–281.

Björnsson, B.T., S.O. Stefansson and T. Hansen. 1995. Photoperiod regulation of plasma growth hormone levels during parr-smolt transformation of Atlantic salmon: implications for hypoosmoregulatory ability and growth. General and Comparative Endocrinology 100: 73–82.

Björnsson, B.T., S.O. Stefansson and S.D. McCormick. 2011. Environmental endocrinology of salmon smoltification. General and Comparative Endocrinology 170: 290–298.

Boeuf, G. 1987. Contribution à l'étude de l'adaptation à l'eau de mer chez les poissons salmonidés. Détermination de critères de smoltification par measures de l'activité (Na+-K+)-ATPasiques des microsomes de la branchie et des hormones thyroïdiennes plasmatiques. PhD Thesis, Université de Bretagne Occidentale, Brest.

Boeuf, G. 1993. Salmonid smolting: a pre-adaptation to the oceanic environment, in: Fish Ecophysiology (Rankin JC and Jensen FB, Eds). Chapman and Hall: London, pp. 105–135.

Boeuf, G. and Y. Harache. 1982. Criteria for adaptation of salmonids to high salinity seawater Aquaculture 28: 163–176.

Boeuf, G. and P. Prunet. 1985. Measurements of gill (Na+-K+)-ATPase activity and plasma thyroid hormones during smoltification in Atlantic salmon, *Salmo salar* L. Aquaculture 45: 111–119.

Boeuf, G. and P.Y. Le Bail. 1990. Growth hormone and thyroid hormones levels during smolting in different populations of Atlantic salmon. In: Progress in Clinical Research (Epple A, Scanes CG and Steson MH, Eds.). Wiley-Liss Inc, New York, 342: 193–197.

Boeuf, G. and P.Y. Le Bail. 1999. Does light have an influence on fish growth? Aquaculture 177: 129–152.

Boeuf, G., A. Le Roux, J.L. Gaignon and Y. Harache. 1985. Gill (Na+-K+)-ATPase activity and smolting in Atlantic salmon *Salmo salar* L. in France. Aquaculture 45: 73–81.

Boeuf, G., P.Y. Le Bail and P. Prunet. 1989. Growth hormone and thyroid hormones during Atlantic salmon, *Salmo salar* L., smolting and after transfer to seawater. Aquaculture 82: 257–268.

Boeuf, G., P. Prunet and P.Y. Le Bail. 1990. Un traitement à l'hormone de croissance peut-il stimuler la smoltification du saumon atlantique ? Comptes Rendus de l'Académie des Sciences série III, 310: 75–80.

Boeuf, G., A-M. Marc, P. Prunet, P.Y. Le Bail and J. Smal. 1994. Stimulation of parr-smolt transformation by hormonal treatment in Atlantic salmon (*Salmo salar* L.). Aquaculture 121: 195–208.

Bolton, J.P., G. Young, R.S. Nishioka, T. Hirano and H.A. Bern. 1987. Plasma growth hormone levels in normal and stunted yearling coho salmon, *Oncorhynchus kisutch*. Journal of Experimental Zoology 242: 379–382.

Bowers S.M. 1988. Morphologic differences in the Atlantic salmon olfactory bulb between the parr and smolt stages of development. Yale University.

Bowmaker, J.K. and E. Loew. 2008. Vision in fish. In: The Senses: A comprehensive Reference. Vol. 1, A.I. Basbaum, A. Kaneko, G.M. Sheperd and G. Westhiemer (Eds.). Academic Press, San Diego, pp. 53–76

Bridges, C.D.B. and C.E. Delisle. 1974. Evolution of visual pigments. Experimental Eye Research 18: 323–332.

Browman, H.I. and C.W. Hawryshyn. 1992. Thyroxine induces a precocial loss of ultraviolet photosensitivity in rainbow trout (*Oncorhynchus mykiss*, Teleostei). Vision Research 32: 2303–2312.

Browman, H.I. and C.W. Hawryshyn. 1994. The development trajectory of ultraviolet photosensitivity in rainbow trout is latered by thyroxine. Vision Research 34: 1397–1406.

Brown, S.B., E.E., Robert, L. Vandenbyllardt, K.W. Finnson, V.P. Palace, A.S. Kane, A.Y. Yarechewski and D.C.G. Muir. 2004. Altered thyroid status in lake trout (Salvelinus namaycush) exposed to co-planar 3, 3', 4, 4', 5-pentachlorobiphenyl. Aquatic Toxicology 67: 75–85.

Brucker-Davis, F. 1998. Effects of environmental synthetic chemicals on thyroid function. Thyroid 8: 827–856.

Burel, C., T. Boujard, A.M. Escaffre, S.J. Kaushik, G. Boeuf, K.A. Mol, S. Van Der Geyten and E.R. Kühn. 2000. Dietary low-glucosinolate rapeseed meal affects thyroid status and nutrient utilisation in rainbow trout (*Oncorhynchus mykiss*). British Journal of Nutrition 83: 653–664.

Burel, C., T. Boujard, S.J. Kaushik, G. Boeuf, K.A. Mol, S. Van Der Geyten, V.M. Darras, E.R. Kühn, B. Pradet-Balade, B. Quérat, A. Quinsac, M. Krouti and D. Ribaillier. 2001. Effects of rapeseed meal-glucosinates on thyroid metabolism and feed utilization in rainbow trout. General and Comparative Endocrinology 124: 343–358.

Carlsen, K.T., O.K. Berg, B. Fingstad and T.G. Heggberget. 2004. Diel periodicity and environmental influence on the smolt migration of Arctic charr, *Salvelinus alpinus*, Atlantic salmon, *Salmo salar*, and Brown Trout, *Salmo trutta*, in Northern Norway. Environmental biology of fishes 70: 403–413.

Carr, J.A. and R. Patino. 2011. The hypothalamus-pituitary-thyroid axis in teleosts and amphibians: endocrine disruption and its consequences to natural populations. General and Comparative Endocrinology 170: 299–312.

Clarke, W.C. and Y. Nagahama. 1977. Effect of premature transfer to seawater on growth and morphology of the pituitary, thyroid, pancreas and interregnal in juvenile coho salmon *Oncorhynchus kisutch*. Canadian Journal of Zoology 55: 1620–1630.

Clarke, W.C., S.W. Farmer and K.M. Hartwell. 1977. Effect of teleost pituitary growth hormone on growth of Tilapia mossambica and on growth and seawater adaptation of sockeye salmon *O. nerka*. General and Comparative Endocrinology 33: 174–178.

Clarke, W.C., J.E. Shelbourn and J.R. Brett. 1981. Effect of artificial photoperiod cycles, temperature and salinity on growth and smolting in underyearling coho (*Oncorhynchus kisutch*), Chinook (*O. tshawytscha*) and sockeye (*O. nerka*) salmon. Aquaculture 22: 105–116.

Clements, S. and C.B. Schreck. 2004. Central administration of corticotropin-releasing hormone alters downstream movement in an artificial stream in juvenile Chinook salmon (*Oncorhynchus tshawytscha*). General and Comparative Endocrinology 137: 1–8.

Clements, S., C.B. Schreck, D.A. Larsen and W.W. Dickhoff. 2002. Central administration of corticotropin-releasing hormone stimulates locomotor activity in juvenile Chinook salmon (*Oncorhynchus tshawytscha*). General and Comparative Endocrinology 125: 319–327.

Collie N.L. and H.A. Bern. 1982. Changes in intestinal fluid transport associated with smoltification and seawater adaptation in coho salmon, *Oncorhynchus kisutch* (Walbaum). Journal of Fish Biology 21: 337–348.

Collie, N.L., J.P. Bolton, H. Kawauchi and T. Hirano. 1989. Survival of salmonids in seawater and the time-frame of growth hormone. Fish Physiology and Biochemistry 7: 315–321.

Congleton, J.L., W.J. LaVoie, C.B. Schreck and L.E. Davis. 2000. Stress indices in migrating juvenile Chinook salmon and steelhead of wild and hatchery origin before and after barge transportation. Transactions of the American Fisheries Society 129: 946–961.

Conte, F.P. and D.H. Lin. 1967. Kinetics of cellular morphogenesis in gill epithelium during salt water adaptation of Oncorhynchus Walbaum. Comparative Biochemistry and Physiology 23: 945–957.

Cooper, J.C., A.T. Scholz, R.M. Horrall, A.D. Hasler and D.M. Madison. 1976. Experimental confirmation of the olfactory hypothesis with artificially imprinted homing coho salmon (*Oncorhynchus kisutch*). Journal of the Fisheries Research Board of Canada 33: 703–710.

Coughlin, D.J., J.A. Forry, S.M. McGlinchey, J. Mitchell, K.A. Saporetti and K.A. Stauffer. 2001. Thyroxine induces transitions in red muscle kinetics and steady swimming kinematics in rainbow trout (*Oncorhynchus mykiss*). Journal of Experimental Zoology 290: 115–124.

Cristy, M. 1974. Effects of prolactin and thyroxine on the visual pigments of trout, *Salmo gairdneri*. General and Comparative Endocrinology 23: 58–62.

D'Cotta, H., C. Gallais, B. Saulier and P. Prunet. 1996. Comparison between parr and smolt Atlantic salmon (*Salmo salar*) α subunit gene expression of Na+/K+-ATPase in gill tissue. Fish Physiology and Biochemistry 15: 29–39.

D'Cotta, H., C. Valotaire, F. Le Gac and P. Prunet. 2000. Synthesis of gill Na+/K+-ATPase in Atlantic salmon smolts: differences in α-mRNA and α-protein levels. American Journal of Physiology 278: R101–110.

Dann, S.G., W.T. Allison, D.B. Levin, J.S. Taylor and C.W. Hawryshyn. 2004. Salmonid opsin sequences undergo positive selection and indicate an alternative evolutionary relationship in Oncorhynchus. Journal of Molecular Evolution 58: 400–412.

Deutschlander, M.E., D.K. Greaves, T.J. Haimberger and G.W. Hawryshyn. 2001. Functional mapping of ultraviolet photosensitivity during metamorphic transitions in a salmonid fish, *Oncorhynchus mykiss*. Journal of Experimental Biology 204: 2401–2413.

Dickhoff, W.W., L.C. Folmar and A. Gorbman. 1978. Changes in plasma thyroxine during smoltification of coho salmon, *Oncorhynchus kisutch*. General and Comparative Endocrinology 36: 229–232.

Dickhoff, W.W., L.C. Folmar, J.L. Mighell and C.V.W. Mahnken. 1982. Plasma thyroid hormones during smoltification of yearling and underyearling coho salmon and yearling Chinook salmon and steelhead trout. Aquaculture 28: 39–48.

Dickhoff, W.W., C.V.W. Mahnken, W.S. Zaugg, F.W. Waknitz, M.G. Bernard and C.V. Sullivan. 1989. Effects of temperature and feeding on smolting and seawater survival of Atlantic salmon (*Salmo salar*). Aquaculture 82: 93–103.

Donaldson, E., U. Fagerlund, D. Higgs and J. McBride. 1979. Hormonal enhancement of growth. In: Fish Physiology vol VIII, W.S. Hoar, D.J. Randall and J.R. Brett (Eds.). Academic Press, New York, pp. 456–597.

Duan, C., E.M. Plisetskaya and W.W. Dickhoff. 1995. Expression of insulin-like growth factor I in normally and abnormally developing coho salmon (*Oncorhynchus kisutch*). Endocrinology 136: 446–52.

Duguay, S.J., P. Swanson and W.W. Dickhoff. 1994. Differential expression and hormonal regulation of alternatively spliced IGF1-mRNA transcripts in salmon. Journal of Molecular Endocrinology 12: 25–37.

Duston, J. and R.L. Saunders. 1995. Increased winter temperature did not affect completion of smolting in Atlantic salmon. Aquaculture International 3: 196–204.

Duston, J., R.L. Saunders and D.E. Knox. 1990. Effects of increases in freshwater temperature on loss of smolt characteristics in Atlantic salmon (*Salmo salar*). Canadian Journal of Fisheries and Aquatic Sciences 48: 164–169.

Eales, J.G. 1979. Thyroid functions in cyclostomes and fishes. In: Hormones and Evolution vol 1, E.J.W. Barrington (Ed.). Academic Press, London/New York, pp. 341–436.

Ebbesson, L.O.E., B.T. Björnsson, S.O. Stefansson and P. Ekström. 1998. Propylthiouracil-induced hypothyroidism in coho salmon, *Oncorhynchus kisutch*: effects on plasma total thyroxine, total triiodothyronine, free thyroxine, and growth hormone. Fish Physiology and Biochemistry 19: 305–314.

Eddy, F.B. and C. Talbot. 1985. Urine production in smolting Atlantic salmon *Salmo salar* L. Aquaculture 45: 67–72.

Enomoto, Y. 1964. A preliminary experiment on the growth promoting effect of growth hormone with thyroid-stimulating hormone and prolactin to the young rainbow trout (*Salmo irideus*). Bulletin of the Japanese Society of Scientific Fisheries 30: 537–541.

Evans, B.I. and R.D. Fernald. 1990. Metamorphosis and fish vision. Journal of Neurobiology 21: 1037–1052.

Fagan, M.S., N. O'Byrne-Ring, R. Ryan, D. Cotter, K. Whelan and U. Mac Evilly. 2003. A biochemical study of mucus lyzozyme proteins and plasma thyroxine of Atlantic salmon (*Salmo salar*) during smoltification. Aquaculture 222: 287–300.

Fairchild, W.L., E.O. Swansburg, J.T. Arsenault and S.B. Brown. 1999. Does an association between pesticide use and subsequent declines in catch of Atlantic salmon (*Salmo salar*) represent a case of endocrine disruption? Environmental Health Perspectives 107: 349–358.

Fairchild, W.L., J.T. Arsenault, K. Haya, L.E. Burridge, J.G. Eales, D.L. MacLatchy, R.E. Evans, B.k. Burnison, J.P. Sherry, D.T. Bennie and S.B. Brown. 2000. Effects of water-borne 4-nonylphenol and estrogen on the growth, survival and physiology of Atlantic salmon (*Salmo salar*) smolts. In: Proceedings of the 27th Annual Aquatic Toxicity Workshop, St. John's. Newfounland, 1–4 October 2000. Canadian Technical Report of Fisheries and Aquatic Sciences No. 2331, p. 58.

Fängstam, H. 1994. Individual swimming speed and time allocation during smolt migration in salmon. Nordic Journal of Freshwater Research 69, 99.

Farbridge, K.J. and J.F. Leatherland. 1988. Interaction between ovine growth hormone and triiodo-L-thyronine on metabolic reserves of rainbow trout, *Salmo gairdneri*. Fish Physiology and Biochemistry 5: 141–151.

Farmer, G.J., J.A. Ritter and D. Ashfield. 1978. Seawater adaptation and parr smolt transformation of juvenile Atlantic salmon *Salmo salar*. Journal of the Fisheries Research Board of Canada 35: 93–100.

Fessler, J.L. and H.H. Wagner. 1969. Some morphological and biochemical changes in steelhead trout *Salmo gairdneri* during parr-smolt transformation. Journal of the Fisheries Research Board of Canada 26: 2823–2841.

Folmar, L.C. and W.W. Dickhoff. 1981. Evaluation of some physiological parameters as predictive indices of smoltification. Aquaculture 23: 309–324.

Fontaine, M. 1975. Physiological mechanisms in the migration of marine and amphihaline fish. Advances in Marine Biology 13: 241–255.

Fontaine, M. and J. Hatey. 1950. Variations de la teneur du foie en glycogène chez le jeune saumon *Salmo salar* L. au cours de la smoltification. Comptes Rendus de la Société de Biologie 144 : 953–955.

Fontaine, M. and M. Olivereau. 1957. Interénal antérieur et smoltification chez *Salmo salar* L. Journal of Physiology 49: 174–176.

Ford, P. 1958. Studies on the development of the kidney of the Pacific pink salmon Oncorhynchus gorbuscha Walbaum II. Variation in glomerular count of the kidney of the Pacific pink salmon. Canadian Journal of Zoology 36: 45–47.

Fridberg, G., K. Lindahl and B. Ekengren. 1981. The thyrotropic cell in the Atlantic salmon, *Salmo salar*. Acta Zoologica 62: 43–51.

Fujioka, Y., S. Tushiki, M. Tagawa, T. Ogasawara and T. Hirano. 1990. Downstream migratory behavior and plasma thyroxine levels of Biwa salmon Oncorhynchus rhodurus. Nippon Suisan Gakkaishi 56: 1773–1779.

Giles, M.A. and D.J. Randall. 1980. Oxygenation characteristics if the polymorphic hemoglobins of coho salmon (*Oncorhynchus kisutch*) at different developmental stages. Comparative Biochemistry and Physiology A 65: 265–271.

Giles, M.A. and W.E. Vanstone. 1976. Ontogenetic variation in the multiple hemoglobins of coho salmon (*Oncorhynchus kisutch*) and effect of environmental factors on their expression. Journal of the Fisheries Research Board of Canada 33: 1144–1149.

Godin, J.G., P.A. Dill and D.E. Drury. 1974. Effects of thyroid hormones on behaviour of yearling Atlantic salmon *Salmo salar* L. Journal of the Fisheries Research Board of Canada 31: 1787–1790.

Gorbman, A., W.W. Dickhoff, J.L. Mighell, E.F. Prentice and F.W. Waknitz. 1982. Morphological indices of developmental progress in the parr-smolt coho salmon, *Oncorhynchus kisutch*. Aquaculture 28: 1–19.

Handeland, S.O., E. Wilkinson, B. Sveinsbo, S.D. McCormick and S.O. Stefansson. 2004. Temperature influence the development and loss of seawater tolerance in two fast-growing strains of Atlantic salmon. Aquaculture 233: 513–529.

Hansen, L.P. and B. Jonsson. 1985. Downstream migration of hatchery-reared smolts of Atlantic salmon (*Salmo salar* L.) in the River Imsa, Norway. Aquaculture 45: 237–248.

Hasler, A.D. and A.T. Scholz. 1983. Olfactory imprinting and homing in salmon. Springer-Verlag, Berlin New York.

Hawryshyn, C.W., G. Martens, W.T. Allison and B.R. Anholt. 2003. Regeneration of ultraviolet-sensitive cones in the retinal cone mosaic of thyroxin-challenged post-juvenile rainbow trout (*Oncorhynchus mykiss*). The Journal of Experimental Biology 206: 2665–2673.

Hemre, G.-I., M. Bjørnevik, C. Beattie, B.T. Björnsson and T. Hansen. 2002. Growth and salt-water tolerance of juvenile Atlantic salmon, *Salmo salar*, reared under different combinations of dietary carbohydrate and photoperiod regimes. Aquaculture 8: 23–32.

Higgins, P.J. 1985. Metabolic differences between Atlantic salmon, *Salmo salar*, parr and smolts. Aquaculture 45: 33–53.

Higgs, D.A., U.H.M. Fagerland, J.G. Eales and R.E. McBride. 1982. Application of thyroid and steroid hormones as anabolic agents in fish culture. Comparative Biochemistry and Physiology 73B: 143–176.

Hoar, W.S. 1939. The thyroid gland of the Atlantic salmon. J. Morph 65: 257–295.

Hoar, W.S. 1951. The behaviour of chum, pink and coho salmon in relation to their seaward migration. Journal of the Fisheries Research Board of Canada 8: 241–263.

Hoar, W.S. 1976. Smolt transformation: evolution, behaviour and physiology. Journal of the Fisheries Research Board of Canada 33: 1233–1252.

Hoar, W.S. 1988. The physiology of smolting salmonids. In: Fish Physiology, vol 11B, W.S> Hoar and D.J. Randall (Eds). Academic Press, New York, pp 275–343.

Holloway, A.C. and J.F. Leatherland. 1997. Effect of gonadal steroid hormones on plasma growth hormone concentrations in sexually immature rainbow trout, *Oncorhynchus mykiss*. General and Comparative Endocrinology 105: 246–254.

Huntsman, A.G. 1939. Migration and conservation of Atlantic salmon for Canada's maritime provinces. In: The Migration and Conservation of Salmon, F.R. Moulton (Ed.). American Association for the Advancement of Science Publisher 8, 32–44.

Hvidsten, N.A., J.A. Jensen, H. Vivaas, O. Bakke and T.G. Heggberget. 1995. Downstream migration of Atlantic salmon smolts in relation to water flow, water temperature, moon phase and social interaction. Nordic Journal of Freshwater Research 70: 38–48.

Hvidsten, N.A., T.G. Heggberget and A.J. Jensen. 1998. Sea Water Temperatures at Atlantic Salmon Smolt Enterance. Nordic Journal of Freshwater Research 74: 79–86.

Iigo, M., K. Ikuta, S. Kitamura, M. Tabata and K. Aida. 2005. Effects of melatonin feeding on smoltification in masu salmon (Oncorhynchus masou). Zoological Science 22: 1191–1196.

Ikuta, K., K. Aida, N. Okumoto and I. Hanyu. 1985. Effects of thyroxine and methyltestosterone on smoltification of masu salmon Oncorhynchus masou. Aquaculture 45: 289–303.

Ikuta, K., K. Aida, N. Okumoto and I. Hanyu. 1987. Effects of sex steroids on the smoltification of masu salmon, Oncorhynchus masou. General and Comparative Endocrinology 65: 99–110.

Iwata, M. 1995. Downstream migratory behavior of salmonids and its relationship with cortisol and thyroid hormones: a review. Aquaculture 135: 131–139.

Iwata, M., R.S. Nishioka and H.A. Bern. 1987. Whole animal transepithelial potential (TEP) of coho salmon during the parr-smolt transformation and effects of thyroxine, prolactin and hypophysectomy. Fish Physiology and Biochemistry 3: 25–38.

Iwata, M., T. Yamanome, M. Tagawa, H. Ida and T. Hirano. 1989. Effects of thyroid hormones on phototaxis of chum and coho salmon juveniles. Aquaculture 82: 329–338.

Iwata, M., K. Yamauchi, R.S. Nishioka, R. Lin and H.A. Bern. 1990. Effects of thyroxine, growth hormone and cortisol on salinity preference of juvenile coho salmon (*Oncorhynchus kisutch*). Marine Behavior and Physiology 17: 191–201.

Jacquest, W.L. and D.D. Beatty. 1972. Visual pigment changes in the rainbow trout, *Salmo gairdneri*. Canadian Journal of Zoology 50: 1117–1126.

Jarrard, H.E. 1997. Postembryonic changes in the structure of the olfactory bulb of the chinook salmon (*Oncorhynchus tshawytscha*) across its life history. Brain Behavior and Evolution 49: 249–260.

Johnsen, P.B. and A.D. Hasler. 1980. The use of chemical cues in the upstream migration of coho salmon, *Oncorhynchus kisutch* Walbaum. Journal of Fish Biology 17: 67–73.

Jonsson, B. and N. Jonsson. 1993. Partial migration: niche shift versus sexual maturation in fishes. Reviews of Fish Biology and Fisheries 3: 348–365.

Jonsson, B., N. Jonsson and L.P. Hansen. 1991. Differences in life history and migratory behaviour between wild and hatchery reared Atlantic salmon in nature. Aquaculture 98: 69–78.

Jugan, M.L, Y. Levi and J.P. Blondeau. 2010. Endocrine disruptors and thyroid hormone physiology. Biochemical Pharmacology 79: 939–947.

Jutila, E., E. Jokikokko, I. Kallio-Nyberg, I. Saloniemi and P. Pasanen. 2003. Differences in sea migration between wild and reared Atlantic salmon (*Salmo salar* L.) in the Baltic sea. Fisheries Research 60: 333–343.

Kalleberg, H. 1958. Observations in a stream tank of territoriality and competition in juvenile salmon and trout (*Salmo salar* L. and *Salmo trutta* L.). Report from the Institute of Freshwater Research 39: 55–98.

Katzman, S. and J.J. Jr Cech. 2001. Juvenile coho salmon locomotion and mosaic muscle are modified by 3′, 3′, 5′-tri-iodo-L-thyronine (T3). Journal of Experimental Biology 204: 1711–1717.

Kazuhiro, U., H. Akihiko and Y. Kohei. 1994. Serum thyroid hormone, guanine and protein profiles during smoltification and after thyroxine treatment in the masu salmon, Oncorhynchus masou. Comparative Biochemistry and Physiology A 107: 607–612.

Kiilerich, P., K. Kristiansen and S.S. Madsen. 2007. Hormone receptors in gill of smolting Atlantic salmon, *Salmo salar* : expression of growth hormone, prolactin, mineralocorticoid and glucocorticoid receptors and 11 β-hydroxysteroid dehydrogenase type 2. General and Comparative Endocrinology 152: 295–303.

Komourdjian, M.P., R.L. Saunders and J.C. Fenwick. 1976a. Evidence for the role of growth hormone as a part of a « light pituitary axis » in growth and smoltification of Atlantic salmon, *Salmo salar* L. Canadian Journal of Zoology 54: 544–551.

Komourdjian, M.P., R.L. Saunders and J.C. Fenwick. 1976b. The effect of porcine somatotropin on growth and survival in seawater of Atlantic salmon, *Salmo salar* L. Canadian Journal of Zoology 54: 531–535.

Kudo, H., Y. Tsuneyoshi, M. Nagae, S. Adachi, K. Yamauchi, U. Ueda and H. Kawamura. 1994. Detection of thyroid hormone receptors in the olfactory system and brain of wild masu salmon, Oncorhynchus masou (Brevoort) during smolting by *in vitro* autoradiography. Aquaculture and Fisheries Management 25: 171–182.

Kunz, Y.W. 1987. Tracts of putative ultraviolet receptors in the retina of the two-year-old brown trout (*Salmo trutta*) and the Atlantic salmon (*Salmo salar*). Experientia 43: 2102–2104.

Kunz, Y.W., G. Wildenburg, L. Goodrich and E. Callaghan. 1994. The fate of ultraviolet receptors in the retina of the Atlantic salmon (*Salmo salar*). Vision Research 11: 1375–1383.

Langdon, J.S. and J.E. Thorpe. 1984. Responses of gill Na+,K+ATPase activity, succinic dehydrogenase activity and chloride cells to seawater adaptation in Atlantic salmon, *Salmo salar* L., parr and smolt. Journal of Fish Biology 24: 323–331.

Langdon, J.S., J.E. Thorpe and R.J. Roberts. 1984. Effects of cortisol and acth on gill Na+/K+-ATPase, SDH and chloride cells in juvenile atlantic salmon *Salmo salar* L. Comparative Biochemistry and Physiology A 77: 9–12.

Langhorne, P. and T.H. Simpson. 1986. The interrelationship of cortisol, gill (Na+KATPase), and homeostasis during the parr-smolt transformation of atlantic salmon (*Salmo salar* L.). General and Comparative Endocrinology 61: 203–213.

Larsen, D.A., P. Swanson, J.T. Dickey, J. River and W.W. Dickhoff. 1998. *In vitro* thyrotropin-releasing activity of corticotropin-releasing hormone-family peptides in coho salmon, *Oncorhynchus kisutch*. General and Comparative Endocrinology 109: 276–285.

Lasserre, P., G. Boeuf and Y. Harache. 1978. Osmotic adaptation of *Oncorhynchus kisutch* Walbaum. I. Seasonal variations of the gill Na+-K+ ATPase activity in coho salmon, 0+ age and yearling, reared in freshwater. Aquaculture 14: 365–382.

Leatherland, J.F., B.A. McKeown and T.M. John. 1974. Circadian rhythm of plasma prolactin, growth hormone, glucose and free fatty acid in juvenile kokanee salmon, Oncorhynchus nerka. Comparative Biochemistry and Physiology A 47: 821–828.

Lema, S.C. and G.A. Nevitt. 2004. Evidence that thyroid hormone induces olfactory cellular proliferation in salmon during a sensitive period for imprinting. Journal of Experimental Biology 207: 3317–3327.

Loretz, C.A., N.L. Collie, N.H. Richman and H.A. Bern. 1982. Osmoregulatory changes accompanying smoltification in coho salmon. Aquaculture 28: 67–74.

MacLatchy, D.L. and J.G. Eales. 1990. Growth hormone stimulates hepatic thyroxine 5'monodeiodinase activity and 3, 5, 3'-triiodothyronine levels in rainbow trout (*Salmo gairdneri*). General and Comparative Endocrinology 78: 164–172.

Madsen, S.S. 1990a. Effect of repetitive cortisol and thyroxine injections on chloride cell number and Na+-K+ ATPase activity in gills of freshwater acclimated rainbow trout, *Salmo gairdneri*. Comparative Biochemistry and Physiology A 95: 171–175.

Madsen, S.S. 1990b. Enhanced hypoosmoregulatory response to growth hormone after cortisol treatment in immature rainbow trout, *Salmo gairdneri*. Fish physiology and Biochemistry 8: 271–279.

Madsen, S.S. 1990c. The role of cortisol and growth hormone in seawater adaptation and development of hypoosmoregulatory mechanisms in sea trout parr (*Salmo trutta trutta*). General and Comparative Endocrinology 79: 1–11.

Madsen, S.S. 1990d. Cortisol treatment improves the development of hypoosmoregulatory mechanisms in sea trout parr (*Salmo trutta trutta*). Fish Physiology and Biochemistry 8: 45–52.

Madsen, S.S. and B. Korsgaard. 1989. Time-course effects of repetitive estradiol–17β and thyroxine injections on the natural spring smolting of Atlantic salmon, *Salmo salar* L. Journal of Fish Biology 35 : 119–128.

Madsen, S.S. and B. Korsgaard. 1991. Opposite effects of 17β-estradiol and growth hormone-cortisol treatment on hypo-osmoregulatory performance in sea trout presmolts, *Salmo trutta*. General and Comparative Endocrinology 83: 276–282.

Madsen, S.S., A.B. Mathiesen and B. Korsgaard. 1997. Effects of 17β-estradiol and 4-nonylphenol on smoltification and vitellogenesis in Atlantic salmon (*Salmo salar*). Fish Physiology and Biochemistry 17: 303–312.

Madsen, S.S., S. Skovbølling, C. Nielsen and B. Korsgaard. 2004. 17-β estradiol and 4-nonylphenol delay smolt development and downstream migration in Atlantic salmon, *Salmo salar*. Aquatic Toxicology 68: 109–120.

Matz, S.P. and G.T. Hofeldt. 1999. Immunohistochemical localization of corticotrophin-releasing factor in the brain and corticotrophin-releasing hormone in the pituitary of Chinook salmon (*Oncorhynchus tshawytscha*). General and Comparative Endocrinology 114: 151–160.

Malikova, E.M. 1959. Biochemical analyses of young salmon at the time of their transformation to a condition close to the smolt stage and during retention of smolts in fresh water. Fish. Res. Board Can. Trans. Ser. 232: 1–19.

Markert, J.R., D.A Higgs, H.M. Dye and D.W. MacQuarrie. 1977. Influence of bovine growth hormone on growth rate, appetite, and food conversion of yearling coho salmon (*Oncorhynchus kisutch*) fed two diets of different composition. Canadian Journal of Zoology 55: 74–83.

Martin, S.A.M., W. Wallner, A.F. Youngson and T. Smith. 1999. Differential expression of Atlantic salmon thyrotropin β subunit mRNA and its cDNA sequence. Journal of Fish Biology 54: 757–766.

Martinez, I., B. Bang, B. Hatlen and P. Blix. 1993. Myofibrillar proteins in skeletal muscles of parr, smolt and adult atlantic salmon (*Salmo salar* L.). Comparison with another salmonid, the arctic charr *Salvelinus alpinus* (L.). Comparative Biochemistry and Physiology B 106: 1021–1028.

Maule, A.G. and C.B. Schreck. 1991. Stress and cortisol treatment changed affinity and number of glucocorticoid receptors in leukocytes and gill of coho salmon. General and Comparative Endocrinology 84: 83–93.

Maxime, V., G. Boeuf, J.P. Pennec and C. Peyraud. 1989. Comparative study of the energetic metabolism of Atlantic salmon (*Salmo salar*) parr and smolts. Aquaculture 82: 163–171.

Mazurais, D., B. Ducouret, M. Tujague, Y. Valotaire, H. D'Cotta, C. Gallais and P. Prunet. 1998. Regulation of the glucocorticoid receptor mRNA levels in the gills of Atlantic salmon (*Salmo salar*) during smoltification. Bulletin Français de la Pêche et de la Pisciculture 350–351: 499–510.

McBride, J.R., D.A. Higgs, U.H.M. Fagerlund and J.T. Buckley. 1982. Thyroid hormones and steroid hormones: potential for control of growth and smoltification of salmonids. Aquaculture 28: 201–210.

McCartney, T.H. 1976. Sodium-potassium dependent adenosine triphosphate activity in gills and kidneys of Atlantic salmon (*Salmo salar*). Comparative Biochemistry and Physiology 53: 351–353.

McCleave, J.D. 1978. Rhythmic aspects of estuarine migration of hatchery-reared Atlantic salmon (*Salmo salar*) smolts. Journal of Fish Biology 12: 559–570.

McCormick, S.D. 1994. Ontogeny and evolution of salinity tolerance in anadromous salmonids: hormones and heterochrony. Estuaries 17: 26–33.

McCormick, S.D., R.L. Saunders, E.B. Henderson and P.R. Harmon. 1987. Photoperiod control of parr smolt transformation in Atlantic salmon (*Salmo salar*): changes in salinity tolerance, gill Na+-K+ ATPase activity and plasma thyroid hormones. Canadian Journal of Fisheries and Aquatic Sciences 44: 1462–1468.

McCormick, S.D., C.D. Moyes and J.S. Ballantyne. 1989a. Influence of salinity on the energetics of gill and kidney of Atlantic salmon (*Salmo salar*). Fish Physiology and Biochemistry 6: 243–254.

McCormick, S.D., R.L. Saunders and A.D. MacIntyre. 1989b. Mitochondrial enzyme and Na+, K+-ATPase activity, and ion regulation during parr-smolt transformation of Atlantic salmon (*Salmo salar*). Fish Physiology and Biochemistry 6: 231–241.

McCormick, S.D., T. Kakamoto, S. Hasegawa, I. Hirano. 1991a. Osmoregulatory actions of insulin-like growth factor-I in rainbow trout (*Oncorhynchus mykiss*). Journal of Endocrinology 130: 87–92.

McCormick, S.D., W.W. Dickhoff, J. Duston, R.S. Nishioka and H.A. Bern. 1991b. Developmental differences in the responsiveness of gill Na+, K+-ATPase to cortisol in salmonids. General and Comparative Endocrinology 84: 308–317.

McCormick, S.D., B. Th. Björnsson, M. Sg-heridan, C. Eilertson, J.B. Carey and M. O'Dea. 1995. Increased daylength stimulates plasma growth hormone and gill Na+,K+-ATPase in Atlantic salmon (*Salmo salar*). Journal of Comparative Physiology B 165: 245–254.

McCormick, S.D., L.P. Hansen, T.P. Quinn and R.L. Saunders. 1998. Movement, migration, and smelting of Atlantic salmon (*Salmo salar*). Canadian Journal of Fisheries and Aquatic Sciences 55: 77–92.

McCormick, S.D., S. Moriyama and B.T. Björnsson. 2000. Low temperature limits photoperiod control of smolting in Atlantic salmon through endocrine mechanisms. American Journal of Physiology 278: R1352–1361.

McCormick, S.D., J.M. Shrimpton, S. Moriyama, B.T. Björnsson. 2002. Effects of an advanced temperature cycle on smolt development and endocrinology indicate that temperature is not a zeitgeber for smolting in Atlantic salmon. The Journal of Experimental Biology 205: 3553–3560.

McCormick, S.D., M.F. O'Dea, A.M. Moeckel and B.T. Björnsson. 2003. Endocrine and physiological changes in Atlantic salmon smolts following hatchery release. Aquaculture 222: 45–57.

McCormick, S.D., M.F. O'Dea, A.M. Moeckel, D.T. Lerner and B.T. Björnsson. 2005. Endocrine disruption of parr-smolt transformation and seawater tolerance of Atlantic salmon by 4-nonylphenol and 17β-estradiol. General and Comparative Endocrinology 142: 280–288.

McCormick, S.D., J.M. Shrimpton, S. Moriyama and B.T. Björnsson. 2007. Differential hormonal responses of Atlantic salmon parr and smolt to increased daylength: a possible developemtal basis for smolting. Aquaculture 273: 337–344.

McFarland, W.F. and D.M. Allen. 1977. The effect of extrinsic factors on two distinctive rhodopsin-porphyropsin systems. Canadian Journal of Zoology 55: 1000–1009.

McKeown, B.A., J.F. Leatherland and T.M. John. 1975. The effect of growth hormone and prolactin on the mobilization of free fatty acids and glucose in the kokanee salmon, Oncorhynchus nerka. Comparative Biochemistry and Physiology B 50: 425–430.

McLeay, D.J. 1975. Variations in the pituitary-interreanl axis and the abundance of circulating blood-cell types in juvenile coho salmon, *Oncorhynchus kisutch*, during stream residence. Canadian Journal of Zoology 53: 1882–1891.

McLeese, J.M., J. Johnsson, F.M. Huntley, W.C. Clarke and M. Weisbart. 1994. Seasonal changes in osmoregulation, cortisol, and cortisol receptor activity in the gills of parr/smolt of steelhead trout and steelhead-rainbow trout hybrids, *Oncorhynchus mykiss*. General and Comparative Endocrinology 93: 103–113.

Miwa, S. and Y. Inui. 1983. Effects of thyroxine and thiourea on the parr smolt transformation of amago salmon Oncorhynchus rhodurus. Bulletin of national Research Institute of Aquaculture 4: 41–52.

Miwa, S. and Y. Inui. 1985. Effects of L-thyroxine and ovine growth hormone on smoltification of amago salmon (Oncorhynchus rhodurus). General and Comparative Endocrinology 58: 436–442.

Miwa, S. and Y. Inui. 1986. Inhibitory effect of 17α methyltestosterone and estradiol 17β on smoltification of sterilized amago salmon Oncorhynchus rhodurus. Aquaculture 53: 21–39.

Mizuno, S., K. Ura, Y. Onodera, H. Fukada, N. Misaka, A. Hara, S. Adachi and K. Yamauchi. 2001a. Changes in transcript levels of gill cortisol receptor during smoltification in wild masu salmon, Oncorhynchus masou. Zoological Science 18: 853–860.

Mizuno, S., N. Misaka and N. Kasahara. 2001b. Effects of cortisol and angiotensin II on the number and size of juxtaglomerular cells in masu salmon, Oncorhynchus masou. Fish Physiology and Biochemistry 25: 249–254.

Mizuno, S., N. Misaka and N. Kasahara. 2001c. Morphological changes in juxtaglomerular cells of the kidney during smoltification in masu salmon Oncorhynchus masou. Fisheries Science 67: 538–540.

Mizuno, S., N. Misaka, D. Ando and T. Kitamura. 2004. Quantitative changes of black pigmentation in the dorsal fin margin during smoltification in masu salmon, Oncorhynchus masou. Aquaculture 229: 433–450.

Moore, A., A.P. Scott, N. Lower, I. Katsiadaki and L. Greenwood. 2003. The effects of 4-nonylphenol and atrazine on Atlantic salmon (*Salmo salar* L) smolts. Aquaculture 222: 253–263.

Mori, T., F. Deguchi and K. Ueno. 2001. Differential expression of GH1 and GH2 genes by competitive RT-PCR in rainbow trout pituitary. General and Comparative Endocrinology 123: 137–143.

Morin, P.P., J.J. Dodson and F.Y. Doré. 1989a. Thyroid activity concomitant with olfactory learning and heart rate changes in Atlantic salmon, *Salmo salar*, during smoltification. Canadian Journal of Fisheries and aquatic Sciences 46: 131–136.

Morin, P.P., J.J. Dodson and F.Y. Doré. 1989b. Cardiac responses to a natural odorant as evidence of a sensitive period for olfactory imprinting in young Atlantic salmon, *Salmo salar*. Canadian Journal of Fisheries and aquatic Sciences 46: 122–130.

Morin, P.P., O. Andersen, E. Haug and K.B. Doving. 1994. Changes in serum free thyroxine, prolactin and olfactory activity during induced smoltification in Atlantic salmon (*Salmo salar*). Canadian Journal of Fisheries and aquatic Sciences 51: 1985–1992.

Morin, P.P., T.J. Hara and J.G. Eales. 1995. T4 depresses olfactory responses to L-alanine and plasma T3 and T3 production in smoltifying Atlantic salmon. Americam Journal of Physiology 269: R1434–1440.

Morin, P.P., T.J. Hara and J.G. Eales. 1997. Thyroid function and olfactory responses to L-alanine during induced smoltification in Atlantic salmon, *Salmo salar*. Canadian Journal of Fisheries and Aquatic Sciences 54: 596–602.

Munakata, A., M. Amano, K. Ikuta, S. Kitamura and K. Aida. 2000. Inhibitory effects of testosterone on downstream migratory behavior in masu salmon, Oncorhynchus masou. Zoological Science 17: 863–870.

Munakata, A., M. Amano, K. Ikuta, S. Kitamura and K. Aida. 2001. The involvement of sex steroid hormones in downstream and upstream migratory behavior of masu salmon. Comparative Biochemistry and Physiology B 129: 661–669.

Munakata, A., M. Amano, K. Ikuta, S. Kitamura and K. Aida. 2007. Effects of growth hormone and cortisol on the downstream migratory behavior in masu salmon, Oncorhynchus masou. General and Comparative Endocrinology 150: 12–17.

Munz, F.W. and R.T. Swanson. 1965. Thyroxine-induced changes in the proportions of visual pigments. American Zoologist 5: 583.

Neave, F. 1955. Notes on the seaward migration of pink and chum salmon fry. Journal of the Fisheries Research Board of Canada 12: 369–374.

Nishioka, R.S., H.A. Bern, K.V. Lai, Y. Nagahama and E.G. Grau. 1982. Changes in the endocrine organs of coho salmon during normal and abnormal smoltification. An electron microscope study. Aquaculture 28: 21–38.

Nishioka, R.S., N.H. Richman, G. Young, P. Prunet and H.A. Bern. 1987. Hypophysectomy of coho salmon (*Oncorhynchus kisutch*) and survival in freshwater and seawater. Aquaculture 65: 343–352.

Nonnotte, L., G. Boeuf and G. Nonnotte. 1995. The role of growth hormone in the adaptability of Atlantic salmon (*Salmo salar* L.) to seawater: effects on the morphology of the mucosa of the middle intestine. Canadian Journal of Zoology 73: 2361–2374.

Nordgarden, U., G-I. Hemre and T. Hansen. 2002. Growth and body composition of Atlantic salmon (*Salmo salar* L.) parr and smolt fed diets varying in protein and lipid contents. Aquaculture 207: 65–78.

Novales Flamarique, I. 2005. Temporal shifts in visual pigment absorbance in the retina of Pacific salmon. Journal of Comparative Physiology A 91: 37–49.

O'Byrne-Ring, N., K. Dowling, D. Cotter, K. Whelan and U. MacEvilly. 2003. Changes in mucus cell numbers in the epidermis of the Atlantic salmon at the onset of smoltification. Journal of Fish Biology 63: 1625–1630.

Ojima, D. and M. Iwata. 2009. Central administration of growth hormone-releasing hormone triggers downstream movement and schooling behavior of chum salmon (*Oncorhynchus keta*) fry in an artificial stream. Comparative Biochemistry and Physiology A 152: 293–298.

Ojima, D. and M. Iwata. 2010. Central administration of growth hormone-releasing hormone and corticotropin-releasing hormone stimulate downstream movement and thyroxine secretion in fall-smolting coho salmon (*Oncorhynchus kisutch*). General and Comparative Endocrinology 168: 82–87.

Ojima, D., T. Yoshinaga, M. Harada and M. Iwata. 2007. Role of cortisol on the onset of downstream migration in hatchery reared chum salmon *Oncorhynchus keta* fry. Coastal and Marine Science 31: 47–52.

Okumoto, N., K. Ikuta, K. Aida, I. Hanyu and T. Hirano. 1989. Effects of photoperiod on smolting and hormonal secretion in masu salmon, Oncorhynchus masou. Aquaculture 82: 63–76.

Olivereau, M. 1954. Hypophyse et glande thyroïde chez les poissons. Etude histophysiologique de quelques corrélations endocriniennes, en particulier chez *Salmo salar*. Annales de l'Institut océanographique de Monaco 29: 95–296.

Olivereau, M. 1960. Etude volumétrique de l'interrénal antérieur au cours de la smoltification de *Salmo salar* L. Acta Endocrinologica 33: 142–156.

Olivereau, M. 1975. Histophysiologie de l'axe hypophyso-corticosurrénalien chez de saumon de l'Atlantique (cycle en eau douce, vie thalassique et reproduction). General and Comparative Endocrinology 27: 9–27.

Olsen, K.H., E. Petersson, B. Ragnarsson, H. Lundqvist and T. Järvi. 2004. Downstream migration in Atlantic salmon (*Salmo salar*) smolt sibling groups. Canadian Journal of Fisheries and Aquatic Sciences 61: 328–331.

Omeljaniuk, R.J. and J.G. Eales. 1986. The effect of 3, 5, 3'-triido-L-thyronine on gill Na+-K+-ATPase activity of rainbow trout *Salmo gairdneri* in freshwater. Comparative Biochemistry and Physiology A 84: 427–429.

Patino, R. and C.B. Schreck. 1986. Sexual dimorphism of plasma sex steroid levels in juvenile coho salmon, *Oncorhynchus kisutch*, during smoltification. General and Comparative Endocrinology 61: 127–133.

Patino, R., C.B. Schreck and J.M. Redding. 1985. Clearance of plasma corticosteroids during smoltification of coho salmon, *Oncorhynchus kisutch*. Comparative Biochemistry and Physiology A 82: 531–535.

Parhar, I.S. and M. Iwata. 1996. Intracerebral expression of gonadotropin-releasing hormone and growth hormone-releasing hormone is delayed until smoltification in the salmon. Neuroscience Research 26:299–308.

Payan, P., J.P. Girard and N. Mayer-Gostan. 1984. Branchial ion movements in teleost: the roles of respiratory and chloride cells. In: Fish Physiology, W.S. Hoar and D.J. Randall (Eds.). Academic Press, New York, pp. 39–63.

Pirini, M., A. Pagliarani, V. Ventrella, F. Trombetti, G. Trigari and A.R. Borgatti. 2002. Response to T3 treatment and changing environmental salinity of liver lipid composition, mitochondrial respiration and (Na+K+)-ATPase activity in rainbow trout *Oncorhynchus mykiss* Walbaum. Aquaculture Research 33: 891–905.

Pisam, M., P. Prunet, G. Boeuf and A. Rambourg. 1988. Ultrastructural features of chloride cells on the gill epithelium of the Atlantic salmon, *Salmo salar*, and their modifications during smoltification. American Journal of Anatomy 183: 235–244.

Porter, M.J.R., C.F. Randall, N.R. Bromage and J.E. Thorpe. 1998. The role of melatonin and the pineal gland on development and smoltification of Atlantic salmon (*Salmo salar*) parr. Aquaculture 168:139–155.

Pottinger, T.G. and A.D. Pickering. 1985. Changes in skin structure associated with elevated androgen levels in maturing male brown trout, *Salmo trutta* L. Journal of Fish Biology 26: 745–753.

Premdas, F.H. and J.G. Eales. 1976. The influence of TSH and ACTH on purine and pteridine deposition in the skin of rainbow trout (*Salmo gairdneri*). Canadian Journal of Zoology 54: 576–581.

Prunet, P., G. Boeuf, J.P. Bolton and G. Young. 1989. Smoltification and seawater adaptation in Atlantic salmon (*Salmo salar*): plasma prolactin, growth hormone and thyroid hormones. General and Comparative Endocrinology 74: 355–364.

Raymond, H.L. 1988. Effects of hydroelectric development and fisheries enhancement on spring and summer Chinook salmon and Steelhead in the Columbia River basin. North American Journal of Fisheries Management 8: 1–24.

Refstie, T. 1982. The effect of feeding thyroid hormones on saltwater tolerance and growth rate of Atlantic salmon. Canadian Journal of Zoology 60: 2706–2712.

Richman III, N.H. and W.S. Zaugg. 1987. Effects of cortisol and growth hormone on osmoregulation in pre- and desmoltified coho salmon (*Oncorhynchus kisutch*). General and Comparative Endocrinology 65: 189–198.

Richman, N.H., S.T. de Diaz, R.S. Nishioka and H.A. Bern. 1985. Developmental study of coho gill functional morphology and the effects of cortisol. Aquaculture 45: 386–387.

Riley, W.D. 2007. Seasonal downstream movements of juvenile Atlantic salmon, *Salmo salar* L., with evidence of solitary migration of smolts. Aquaculture 273: 194–199.

Riley, W.D., M.O. Eagle and S.J. Ives. 2002. The onset of downstream movement of juvenile Atlantic salmon, *Salmo salar* L., in a chalk stream. Fisheries Management and Ecology 9: 87–94.

Rivinoja, P., S. McKinell and H. Lundqvist. 2001. Hindrances to upstream migration of Atlantic salmon (*Salmo salar*) in anirthern Swedish River caused by a hydroelectric power-station. Regulated Rivers: Research and Management 17: 101–115.

Robertson, O.G. 1949. Production of silvery smolt stage in rainbow trout by intramuscular injection of mammalian thyroid extract and thyrotropic hormone. Journal of Experimental Zoology 110: 337–355.

Rousseau, K., N. Le Belle, J. Marchelidon and S. Dufour. 1999. Evidence that corticotropin-releasing hormone acts as a growth hormone-releasing factor in a primitive teleost, the European eel (Anguilla anguilla). Journal of Neuroendocrinology 11: 385–392.

Rydevik, M. 1988. Epidermis thickness and secondary sexual characters in mature male and immature Baltic salmon, *Salmo salar* L. parr: seasonal variations and effects of castration and androgen treatment. Journal of Fish Biology 33: 941–944.

Sakamoto, T. and T. Hirano. 1993. Expression of insulin-like growth factor I gene in osmoregulatory organs during seawater adaptation of the salmonid fish: possible mode of osmoregulatory action of growth hormone. Procedings of the National Academy of Sciences 90: 1912–1916.

Saunders, R.L. and E.B. Henderson. 1970. Influence of photoperiod on smolt development and growth of Atlantic salmon *Salmo salar*. Journal of the Fisheries Research Board of Canada 27: 1295–1311.

Saunders, R.L. and P.R. Harmon. 1990. Influence of photoperiod on growth of juvenile Atlantic salmon and development of salinity tolerance during winter-spring. Transactions of the American Fisheries Society 119: 689–697.

Saunders, R.L., E.B. Henderson and P.R. Harmon. 1985. Effects of photoperiod on juvenile growth and smolting of Atlantic salmon and subsequent survival and growth in sea cages. Aquaculture 45: 55–56.

Scholz, A.T., R.M. Horrall, J.C. Cooper and A.D. Hasler. 1976. Imprinting to chemical cues: the basis for home stream selection in salmon. Science 192: 1247–1249.

Scholz, A.T., R.J. White, M. Muzi and T. Smith. 1985. Uptake of radiolabelled triiodothyronine in the brain of steelhead trout (*Salmo gairdneri*) during parr-smolt transformation: implications for the mechanisms of thyroid activation of olfactory imprinting. Aquaculture 45: 199–214.

Schrock, R.M., S.D. Smith, A.G. Maule, S.K. Doulos and J.J. Rockowski. 2001. Mucous lysozyme levels in hatchery coho salmon (*Oncorhynchus kisutch*) and spring Chinook salmon (*O. tshawytscha*) early in the parr-smolt transformation. Aquaculture 198: 169–177.

Seddiki, H., V. Maxime, G. Boeuf and C. Peyraud. 1995. Effects of growth hormone on plasma ionic regulation, respiration and extracellular acid-base status in trout (*Oncorhynchus mykiss*) transferred to seawater. Fish Physiology and Biochemistry 14: 279–288.

Seddiki, H., G. Boeuf, V. Maxime and C. Peyraud. 1996. Effects of growth hormone treatment on oxygen consumption and sea water adaptability in Atlantic salmon parr and presmolts. Aquaculture 148: 49–62.

Seear, P.J., S.N. Carmichael, R. Talbot, J.B. Taggart, J.E. Bron and G.E. Sweeney. 2010. Differential gene expression during smoltification of Atlantic salmon (*Salmo salar* L.): a first-scale microaaray study. Marine Biotechnology 12: 126–140.

Sheridan, M.A. 1986. Effects of thyroxin, cortisol, growth hormone, and prolactin on lipid metabolism of coho salmon, *Oncorhynchus kisutch*, during smoltification. General and Comparative Endocrinology 64: 220–238.

Sheridan, M.A. 1988. Lipid dynamics in fish: aspects of absorption, transportation, deposition and mobilization. Comparative Biochemistry and Physiology 90: 679–690.

Sheridan M.A. 1989. Alterations of lipid metabolism accompanying smoltification and seawater adaptation of salmonid fish. Aquaculture 82: 191–204.

Sheridan, M.A., W.V. Allen and T.H. Kerstetter. 1983. Seasonal variation in the lipid composition of the steelhead trout *Salmo gairdneri* Richardson, associated with the parr-smolt transformation. Journal of Fish Biology 23: 125–134.

Sheridan M.A., N.Y.S. Woo and H.A. Bern. 1985a. Changes in the rates of glycogenesis, glycogenolysis, lipogenesis and lipolysis in selected tissues of the coho salmon *Oncorhynchus kisutch* associated with the parr-smolt transformation. Journal of Experimental Zoology 236: 35–44.

Sheridan M.A., W.V. Allen and T.H. Kerstetter. 1985b. Changes in the fatty acid composition of steelhead trout *Salmo gairdneri* Richardson associated with the parr-smolt transformation. Comparative Biochemistry and Physiology 80B: 671–676.

Shrimpton, J.M. 1996. Relationship between size, gill corticosteroid receptors, Na+-K+ ATPase activity and smolting in juvenile coho salmon (*Oncorhynchus kisutch*) in autum and spring. Aquaculture 147:127–140.

Shrimpton, J.M. and S.D. McCormick. 1998a. Regulation of gill cytosolic corticosteroid receptors in juvenile Atlantic salmon: interaction effects of growth hormone with prolactin and triiodothyronine. General and Comparative Endocrinology 112: 262–274.

Shrimpton, J.M. and S.D. McCormick. 1998b. Seasonal differences in plasma cortisol and gill corticosteroid receptors in upper and lower mode juvenile Atlantic salmon. Aquaculture 168: 205–219.

Shrimpton, J.M. and S.D. McCormick. 1999. Responsiveness of gill Na+/K+-ATPase to cortisol is related to gill corticosteroid receptor concentration in juvenile rainbow trout. The Journal of Experimental Biology 202: 987–995.

Shrimpton, J.M. and S.D. McCormick. 2003. Environmental and endocrine control of gill corticosteroid receptor number and affinity in Atlantic salmon (*Salmo salar*) during smolting. Aquaculture 222: 83–99.

Shrimpton, J.M. and D.J. Randall. 1994. Downregulation of corticosteroid receptors in gills of coho salmon due to stress and cortisol treatment. American Journal of Physiology 267: R432–438.

Shrimpton, J.M., N.J. Bernier and D.J. Randall. 1994. Changes in cortisol dynamics in wild and hatchery-reared juvenile coho salmon (*Oncorhynchus kisutch*) during smoltification. Canadian Journal of Fisheries and Aquatic Sciences 51: 2179–2187.

Shrimpton, J.M., R.H. Devlin, E. McLean, J.C. Byatt, E.M. Donaldson and D.J. Randall. 1995. Increases in gill cytosolic corticosteroid receptor abundance and saltwater tolerance in juvenile coho salmon (*Oncorhynchus kisutch*) treated with growth hormone and placental lactogen. General and Comparative Endocrinology 98: 1–15.

Smith, D.C.W. 1956. The role of the endocrine organs in the salinity tolerance of trout. Mem. Soc. Endocrinol 5: 83–101.

Solbakken, V.A., T. Hansen and S.O. Stefansson. 1994. Effects of photoperiod and temperature on growth and parr-smolt transformation in Atlantic salmon (*Salmo salar* L.) and subsequent performance in seawater. Aquaculture 121: 13–27.

Solomon, D.J. 1978. Some observations on salmon smolt migration in a chalkstream. Journal of Fish Biology 12: 571–574.

Sower, S.A, K.H. Karlson and R.S. Fawcett. 1992. Changes in plasma thyroxine, estradiol-17β, and 17α,20β-dihydroxy-4-pregnen-3-one during smoltification of coho salmon. General and Comparative Endocrinology 85: 278–285.

Specker, J.L. 1982. Interrenal function and smoltification. Aquaculture 28: 59–66.

Specker, J.L. and C.B. Schreck. 1982. Changes in plasma corticosteroids during smoltification of coho salmon, *Oncorhynchus kisutch*. General and Comparative Endocrinology 46: 53–58.

Staley, K.B. and R.D. Ewing. 1992. Purine levels in the skin of juvenile coho salmon (*Oncorhynchus kisutch*) during parr-smolt transformation and adaptation to seawater. Comparative Biochemistry and Physiology B 101: 447–452.

Staurnes, M., T. Sigholt and O.A. Gulseth. 1994. Effects of seasonal changes in water temperature on the parr-smolt transformation of Atlantic salmon and anadromous Arctic charr. Transactions of the American Fisheries Society 123: 408–415.

Sullivan, C.V., W.W. Dickhoff, C.V. Mahnken and W.K. Hershberger. 1985. Changes in hemoglobin system of the coho salmon *Oncorhynchus kisutch* during smoltification and triidothyronine and propylthiouracil treatment. Comparative Biochemistry and Physiology A 81: 807–813.

Sullivan, C.V., D.S. Darling and W.W. Dickhoff. 1987. Effects of triiodothyronine and propylthiouracil on thyroid function and smoltification of coho salmon (*Oncorhynchus kisutch*). Fish Physiology and Biochemistry 4: 121–135.

Sundell, K., F. Jutfelt, T. Agustsson, R-E. Olsen, E. Sandblom, T. Hansen and B.Th. Björnsson. 2003. Intestinal transport mechanisms and plasma cortisol levels during normal and out-of-season parr-smolt transformation of Atlantic salmon, *Salmo salar*. Aquaculture 222: 265–285.

Sweeting, R.M., G.F. Wagner and B.A. McKeown. 1985. Changes in plasma glucose, amino acid, nitrogen and growth hormone during smoltification and seawater adaptation in coho salmon, *Oncorhynchus kisutch*. Aquaculture 45: 185–197.

Sweeting, R.M. and B.A. McKeown. 1989. Changes in plasma growth hormone and various metabolic factors during smoltification of coho salmon, *Oncorhynchus kisutch*. Aquaculture 82: 279–295.

Tata, J.R. 2006. Amphibian metamorphosis as a model for the developmental actions of thyroid hormone. Molecular and Cellular Endocrinology 246: 10–20.

Temple, S.E., E.M. Plate, S. Ramsden, T.J. Haimberger, W.M. Roth and C.W. Hawryshyn. 2006. Seasonal cycle in vitamin A(1)/A(2)-based visual pigment composition during the life history of coho salmon (*Oncorhynchus kisutch*). Journal of Comparative Physiology A 192: 301–313.

Temple, S.E., K.M. Veldhoen, J.T. Phelan, N.J. Veldhoen and C.W. Hawryshyn. 2008a. Ontogenetic changes in photoreceptor opsin gene expression in coho salmon (*Oncorhynchus kisutch*, Walbaum). The Journal of Experimental Biology 211: 3879–3888.

Temple, S.E., S. Ramsden, T.J. Haimberger, K.M. Veldhoen, N.J. Veldhoen, N.L. Carter, W.M. Roth and C.W. Hawryshyn. 2008b. Effects of exogenous thyroid hormones on visual pigment composition in coho salmon (*Oncorhynchus kisutch*). The Journal of Experimental Biology 211: 2134–2143.

Thorpe, J.E. 1994. An alternative view of smelting in salmonids. Aquaculture 121: 105–113.

Thorpe, J.E., L.G. Ross, G. Struthers and W. Watts. 1981. Tracking Atlantic salmon smolts, *Salmo salar* L., through Loch Voil, Scotland. Journal of Fish Biology 19: 519–537.

Tipsmark, C.K., P. Kiilerich, T.O. Nilsen, L.O.E. Ebbesson, S.O. Stefansson and S.S Madsen. 2008. Branchial expression patterns of claudin isoforms in Atlantic salmon during seawater acclimation and smoltification. American Journal of Physiology 294: R1563–1574.

Tipsmark, C.K., C. Jorgensen, N. Brande-Lavridsen, M. Engelund, J.H. Olesen and S.S. Madsen. 2009. Effect of cortisol, growth hormone and prolactin on gill claudin expression in Atlantic salmon. General and Comparative Endocrinology 163: 270–277.

Tytler, P., J.E. Thorpe and W.M. Shearer. 1978. Ultrasonic tracking of the movements of atlantic salmon smolts (*Salmo salar* L.) in the estuaries of two Scottish rivers. Journal of Fish Biology 12: 575–586.

Ura, K., S. Mizuno, T. Okubo, Y. Chida, N. Misaka, S. Adachi and K. Yamauchi. 1997. Immunohistochemical study on changes in gill Na/K-ATPase α-subunit during smoltification in the wild masu salmon, Oncorhynchus masou. Fish Physiology and Biochemistry 17: 397–403.

Usher, M.L., C. Talbott and F.B. Eddy. 1991. Intestinal water transport in juvenile Atlantic salmon (*Salmo salar* L.) during smolting and following transfer to seawater. Comparative Biochemistry and Physiology A 100: 813–818.

Utrilla, C.G. and J. Lobon-Corvia. 1999. Life-history patterns in a southern population of Atlantic salmon. Journal of Fish Biology 55: 68–83.

Van Doorn, K.L.H., J.G. Sivak and M.M. Vijayan. 2005. Optical quality changes of the ocular lens during induced parr-to-smolt metamorphosis in rainbow trout (*Oncorhynchus mykiss*). Ocular lens optical quality during induced salmonid metamorphosis. Journal of Comparative Physiology A 191: 649–657.

Vanstone, W.E., E. Roberts and H. Tsuyuki. 1964. Changes in the multiple hemoglobin patterns of some pacific samon, genus Oncorhynchus, during the parr-smolt transformation. Canadian Journal of Physiology and Pharmacology 42: 697–703.

Veillette, P.A. and G. Young. 2005. Tissue culture of sockeye salmon intestine: functional response of Na+-K+-ATPase to cortisol. American Journal of Physiology 288: R1598–R1605.

Veillette, P.A., R.J. White and J.L. Specker. 1993. Changes in intestinal fluid transport in Atlantic salmon (*Salmo salar* L.) during parr-smolt transformation. Fish Physiology and Biochemistry 12: 193–202.

Veillette, P.A., K. Sundell and J.L. Specker. 1995. Cortisol mediates the increase in intestinal fluid absorption in Atlantic salmon during parr-smolt transformation. General and Comparative Endocrinology 97: 250–258.

Veldhoen, K., W.T. Allison, N. Veldhoen, B.R. Anholt, C.C. Helbing and C.W. Hawryshyn. 2006. Spatio-temporal characterization of retinal opsin gene expression during thyroid hormone-induced and natural development of rainbow trout. Visual Neuroscience 23: 169–179.

Virtanen, E. and A. Soivio. 1985. The patterns of T3, T4, cortisol and Na+-K+ ATPase during smoltification of hatchery reared *Salmo salar* and comparison with wild smolts. Aquaculture 45: 97–109.

Wedemeyer, A.H., R.L. Saunders and W.L. Clarke. 1980. Environmental factors affecting smoltification and early marine survival of anadromous salmonids. Marine Fisheries Review 42: 1–14.

Wendt, C.A.G. and R.L. Saunders. 1973. Changes in carbohydrate metabolism in young Atlantic salmon in response to various forms of stress. International Atlantic salmon Foundation Special Publication Series 4: 55–82.

White, R., S. Jobling, S.A. Hoare, J.P. Sumpter and M.G. Parker. 1994. Environmentally persistent alkylphenolic compounds are estrogenic. Endocrinology 135: 175–182.

Wiggs, A.J., E.B. Henderson, R.L. Saunders and M.N. Kutty. 1989. Activity, respiration and excretion of ammonia by Atlantic salmon (*Salmo salar*) smolt and postsmolt. Canadian Journal of Fisheries and Aquatic Sciences 46: 790–795.

Wilkins, N.P. 1968. Multiple hemoglobins of the Atlantic salmon (*Salmo salar*). Journal of the Fisheries Research Board of Canada 25: 2651–2653.

Wilson, P.H. 2003. Using population projection matrices to evaluate recovery strategies for snake river spring and summer Chinook salmon. Conservation Biology 17: 782–794.

Woo, N.Y.S., H.A. Bern and R.S. Nishioka. 1978. Changes in body composition associated with smoltification and premature transfer to seawater in coho salmon (*Oncorhynchus kisutch*) and king salmon (*O. tschawytscha*). Journal of Fish Biology 13: 421–428.

Yadetie, F. and R. Male. 2002. Effects of 4-nonyphenol on gene expression of pituitary hormones in juvenile Atlantic salmon (*Salmo salar*). Aquatic Toxicology 58: 113–129.

Yamada, H., H. Ohta and K. Yamauchi. 1993. Serum thyroxine, estradiol–17β, and testosterone profiles during the parr-smolt transformation of masu salmon, Oncorhynchus masou. Fish Physiology and Biochemistry 12: 1–9.

Yamazaki, F., T. Awakuwa, M. Atoda and S. Tanada. 1973. Scienfic Reports of the Hokkaido Fish Hatchery 29: 1–10.

Young, G. 1986. Cortisol secretion *in vitro* by the interrenal of coho salmon (*Oncorhynchus kisutch*) during smoltification: relationship with plasma thyroxine and plasma cortisol. General and Comparative Endocrinology 63: 191–200.

Young, G. 1988. Enhanced esponse of the interregnal of coho salmon (*Oncorhynchus kisutch*) to ACTH after growth hormone treatment *in vivo* and *in vitro*. General and Comparative Endocrinology 71: 85–92.

Young, G., B.Th. Björnsson, P. Prunet, R.J. Lin and H.A. Bern. 1989. Smoltification and seawater adaptation in coho salmon (*Oncorhynchus kisutch*): plasma prolactin, growth hormone, thyroid hormone and cortisol. General and Comparative Endocrinology 74: 335–345.

Young, G., S.D. McCormick, B.T. Björnsson and H.A. Bern. 1995. Circulating growth hormone, cortisol and thyroxine levels after 24h seawater challenge of yearling coho salmon at different developmental stages. Aquaculture 136: 371–384.

Youngson, A.F., D.C.B. Scott, R. Johnstone and D. Pretswell. 1985. The thyroid system's role in the downstream migration of Atlantic salmon (*Salmo salar* L.) smolts. Aquaculture 45: 392–393.

Youson, J.H. 1988. First metamorphosis. In: Fish Physiology Vol XI, Physiology of developing fish. Part B, Viviparity and posthatching juveniles, W.S. Hoar and D.J. Randall (Eds.). Academic Press, San Diego, pp. 135–196.

Zaugg, W.S. and L.R. McLain. 1970. Adenosine triphosphate activity in gills of salmonids: seasonal variations and salt water influence in coho salmon, *Oncorhynchus kisutch*. Comparative Biochemistry and Physiology 35: 587–596.

Zaugg, W.S. and L.R. McLain. 1972. Changes in gill adenosine-triphosphate activity associated with parr-smolt transformation in steelhead trout, coho and spring Chinook salmon. Journal of the Fisheries Research Board of Canada 29: 167–171.

Zaugg, W.S. and L.R. McLain. 1976. Influence of water temperature on gill sodium, potassium-stimulated ATPase activity in juvenile coho salmon (*Oncorhynchus kisutch*). Comparative Biochemistry and Physiology A 54: 419–421.

Zaugg, W.S. and L.R. McLain. 1986. Changes in blood levels of nucleoside triphosphates, haemoglobin and hematocrits during parr-smolt transformation of coho salmon (*Oncorhynchus kisutch*). Comparative Biochemistry and Physiology A 84: 487–493.

Zaugg, W.S. and H.H. Wagner. 1973. Gill ATPase activity related to parr-smolt transformation and migration in steelhead trout (*Salmo gairdneri*): influence of photoperiod and temperature. Comparative Biochemistry and Physiology B 45: 955–965.

Zoeller, T.R. 2010. Environmental chemicals targeting thyroid. Hormones 9: 28–40.

# CHAPTER 7

# Eel Secondary Metamorphosis: Silvering

*Karine Rousseau,*[1,a] *Salima Aroua*[1] *and Sylvie Dufour*[1,b,]*

## 7.1 Introduction

As mentioned before, secondary metamorphosis, which occurs in juveniles, involves various morphological, physiological and behavioral modifications that pre-adapt the animal to life in a new environment. These modifications are less drastic than those observed during larval/true/first metamorphosis. This type of metamorphosis in teleosts seems to be related to extreme migration and complex biological cycles, as observed in salmons (smoltification) and eels (silvering).

Eels have complex migratory life cycle with the occurrence of two metamorphoses (for reviews: Sinha and Jones, 1975; Tesch, 1977; Haro, 2003; Dufour and Rousseau, 2007; Rousseau and Dufour, 2008)(Fig. 1). They present a typical larval (first) metamorphosis, leptocephali larva being transformed into glass eels (See Chapter 3). After this drastic transformation, the growth phase starts in the continental waters and glass eels become elvers and "yellow" eels. After many years in freshwater, the yellow eels transform into "silver" eels which stop growing and start their downstream migration towards the ocean and the area of reproduction. To allow this transition from sedentary life in freshwater to migrant life in seawater, eels undergo their second metamorphosis, known as silvering. Silvering not only preadapts the eel to deep-sea conditions (Sébert, 2003), but also prepares the sexual maturation, which will only be completed during the

[1]Research Unit BOREA «Biology of Aquatic Organisms and Ecosystems» Muséum National d'Histoire Naturelle, CNRS 7208, IRD 207, UPMC, 7 rue Cuvier, CP32, 75231 Paris Cedex 05, France.
[a]E-mail: rousse@mnhn.fr
[b]E-mail: dufour@mnhn.fr
*Corresponding author

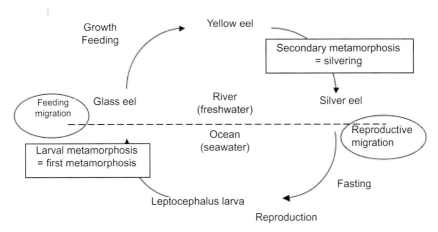

**Figure 1**. Biological life cycle of catadromous European eel, *Anguilla anguilla*. The secondary metamorphosis, silvering, occurs after a long period of growth and feeding in freswater. It prepares the eel for the long reproductive migration in seawater, during which the eel will not feed anymore.

oceanic migration towards the Sargasso sea (Tesch, 1982, 1989; Dufour and Fontaine, 1985; Fontaine, 1985; Dufour, 1994). At the silver stage, eels are blocked at a prepubertal stage (Dufour et al., 2003).

## 7.2 Morphological, Behavioural and Physiological Changes Occurring During Silvering

During the silvering process, in addition to important behavioral changes such as transition from a sedentary to a migrating behavior and start of starvation, drastic morphological changes are observed. The most spectacular modifications concern the tegument and the eyes but various other organs are also remodeled.

### 7.2.1 Changes linked to adaptation to seawater (Osmoregulation)

The primary gill epithelium of silver eels is thicker and contain larger and more numerous chloride cells with enlarged mitochondria, as compared to yellow European eels (Fontaine et al., 1995). More recently, Kaneko et al. (2003) demonstrated in Japanese eels that filament chloride cells were activated in saltwater-adapted or silver eels, and lamella chloride cells were mainly observed in freshwater-adapted eels or yellow eels, presumably acting as sites responsible for salt secretion in seawater and ion uptake in freshwater respectively. Moreover, toward the time of downstream migration of the eel to the sea, the ratio of filament chloride cells increased, while that of lamella chloride cells decreased. These observations indicate that the silvering process in eels corresponds to a preparation to seawater life.

And, in constrast to parr, eels at yellow stage have already the capacity to survive if transfered to seawater. In addition, during their biological cycle, eels are commonly going from freshwater to brackish water at yellow stage and they are living in coastal waters throughout their sedentary life (otolith microchemistry with wavelength dispersive X-ray spectrometry: Tzeng et al., 1997). However, Tsukamoto et al. (1998), as well as Tsukamoto and Arai (2001) reported, using Synchroton Radiation Induced X-ray Fluorescence analysis for quantifying strontium and calcium deposits in otoliths, populations of yellow and silver eels of both Japanese and European eels in marine areas adjacent to their typical freshwater habitats which have never migrated into freshwater and have spent their entire life history in the ocean.

Experiments of transfer to seawater of yellow or silver eels showed changes in the levels of chloride cells and (Na+, K+)-dependent ATPase in the gills (Thomson and Sargent, 1977; Cutler et al., 1995a and b), in the expression of guanylin (peptide which regulates electrolyte and water transport) in the intestine (Comrie et al., 1999) and aquaporins (water transporters) in the kidney (Martinez et al., 2005).

When yellow eels were transferred from FW to SW, percentage of chloride cells increased from about 2 to 5% after 6 days and to 6% from 13 to 21 days, and Na+, K+-ATPase activity increased 2.5 times after 2 weeks (Thomson and Sargent, 1977). In contrast, for silver eels, chloride cells formed about 6% in FW and SW and only a slight increase of Na+, K+-ATPase activity was observed (Thomson and Sargent, 1977).

In the Japanese eel, the ratio lamellar to filamentary chloride cells is much lower in yellow eels than in silver eels; silver eels caught in brackish water during downstream migration had very few chloride cells on the lamellae (Kaneko et al., 2003).

In summary, the silver eel does show pre-adaptations to marine life as far as gill chloride cells are concerned. There may be an advantage in having the machinery for marine osmoregulation in place before feeding stops preparatory to the spawning migration, during which all the energy reserves of the body will be required for swimming and maturation in addition to maintaining homeostasis. This energy required for the latter includes the cost of osmoregulation in a hyperosmotic environment, which the eel continues to carry out efficiently up to the time of maturation. The situation resembles that in pressure acclimation where the yellow eel can acclimatize to high pressure but the silver eel is already prepared.

## 7.2.2 Changes linked to adaptation to depth and changes of pressure

### 7.2.2.1 Skin

During the sedentary growth phase, the back of the eel is greenish brown and the ventral face is creamy white with yellow bands, while at the time of the oceanic migration of reproduction, they are respectively blackish brown and silvery white (Fig. 2). The silvering of the skin is thought to be the result of the increase of dorsal melanin and ventral purines (Bertin, 1951; Pankhurst and Lythgoe, 1982). In addition to this change in color, the skin becomes thicker (Bertin, 1951; Pankhurst, 1982a). Other skin changes also occur resulting in a greater resistance to abrasion, as increase in mucus production (Bertin, 1951; Saglio et al., 1988) and variation in scales (Pankhurst, 1982a).

**Figure 2**. European eel yellow and silver stages. Note the silvery color of the ventral side of the silver eel as compared to the yellow eel. The eye diameter is enlarged in silver eels compared to yellow eels.

*Color image of this figure appears in the color plate section at the end of the book.*

### 7.2.2.2 Eye

Eye is probably one of the most modified sensory organs during eel silvering. Indeed, during the sedentary phase, eel is a benthic fish and lives in rivers, where conditions for vision can be poor. In contrast, in the open ocean, conditions for vision are better even considering that in depth only few light goes through (Lythgoe, 1979). In deep sea, photosensitive pigments have maxima absorption displaced towards the blue end of the spectrum (Denton and Warren, 1957). A recent study using miniaturized pop-up satellite archival transmitter reported that when eels moved into the mesopelagic zones, they all undertook distinct diel vertical migrations, predominantly between depths of 200 and 1,000 m (Aarestrup et al., 2009).

During the transition from yellow to silver eel, increase in eye diameter and retina surface is observed (*A. anguilla*: Pankhurst, 1982b; *A. japonica*: Yamamoto and Yamauchi, 1974; *A. dieffenbachii* and *A. australis*: Todd, 1981a; Lokman et al., 1998). Structural modifications of the retina are also reported, with an increase in the number of rods and a decrease in the number of cones (Pankhurst and Lythgoe, 1983; Braekevelt, 1985, 1988). Pankhurst and Lythgoe (1983) suggest that these changes in retina surface and structure lead to an increase in sensitivity, even if the degeneration of cones will probably reduce the ability to perceive color.

Finally, the visual system also adapts from a freshwater duplex retina with rods and cones to a specialized deep-sea retina containing only rods. There is also a change in visual pigment at the chromophore level with predominance of rhodopsin (opsin linked to a vitamin A1 aldehyde chromophore) in silver eels compared to porphyropsin (opsin linked to a vitamin A2 aldehyde chromophore) in yellow eels (Carlisle and Denton, 1959; Beatty, 1975), which increases the blue light sensitivity of rods in silver eels (Archer et al., 1995). All of these modifications of the visual system are in agreement with a vision in deep sea (Partridge et al., 1989) and prepare the fish to its future reproductive migration. Using experimental maturation, Wood and Partridge (1993) demonstrated that the new 'deep-sea' opsin (*dso*) was expressed within existing rod photoreceptors and formed a shortwave sensitive visual pigment that could be detected at the base of existing rod outer segments after one week of hormone treatment. Hope et al. (1998) then found that the switch in opsin gene expression was fast, new gene being detectable within 6h of a single hormone injection. The fresh water rod opsin gene (*fwo*) and deep-sea rod opsin gene (*dso*) were cloned in *Anguilla japonica* and the expression of *fwo* was shown to decrease with sexual maturation while that of *dso* increased (Zhang et al., 2000). The fate of retinal cones during eel metamorphosis was reported by Bowmaker et al. (2008): yellow eels possess only two spectral classes of single cones, one sensitive in the green (presumably an RH2 opsin gene) and the second sensitive in the blue (expressing an SWS2 opsin gene); the fully FW yellow eels have cone pigments that are almost pure porphyropsin; during the early stages of metamorphosis, the pigments switch to rhodopsins; and then, the cones are almost completely lost. A recent study (Cottrill et al., 2009) investigated the developmental dynamics of cone photoreceptors in the European eel. They showed that only the Rh2 cone opsin was present in small yellow eels, while the sws2 opsin was additionally expressed in large yellow and silver eels.

## 7.2.2.3 Swimbladder

During the spawning migration, eels encountered two drastically different environments, the shallow fresh water and the deep sea. Thus, some modifications of swimbladder during silvering will permit the transition from the shallow to the deep environment, preparing the eels to hydrostatic pressure encountered in their future deep environment. First, an increase in the crystalline guanine content of the swimbladder in silver compared to yellow American eels was observed, permitting to reduce the gas loss (Kleckner, 1980a). An increase in gas deposition rates was also demonstrated (Kleckner, 1980b), which could result from the increase in length and in luminal diameter of the *rete mirabilis* capillaries (Kleckner and Krueger, 1981).

## 7.2.2.4 Cellular metabolism

Exposure of FW eels to high pressure (HP; 101 ATA equivanlent to a 1,000 m depth) for some hours induces a sharp increase in oxygen consumption and at the muscle level, a decrease in ATP and glycogen contents, in cytochrome oxidase activity and an increase in LDH activity, fatty acids stores and circulating lactate (for review: Sébert, 2002). Hydrostatic pressure induces a state rsembling histotoxic hypoxia via a decraese in membrane fluidity, thus aletring areobic metabolism (Sébert et al., 1993). When exposure to pressure is maintained for several days or weeks, eels are able to acclimatize to pressure effects. The pressure acclimatization process is the « normalization » of metabolism through membrane fluidification that allows normal functioning of the respiratory chain and oxidative phosphorylation (for review: Sébert, 2002).

Vettier and collaborators demonstrated that the silvering process appeared to mimic pressure acclimatization observed in yellow eels (Simon et al., 1989), allowing silver eels to cope with pressure effects with a minimal energy cost (Vettier et al., 2005). Indeed, the global sensitivity to compression measured by the muscle fibre oxygen consumption increased by 60% between before (1 ATA) and at the end of compression after 21 days at 101 ATA. At the end of the experiment, a significant decrease in HSI and a trend to increase GSI were observed (Vettier et al., 2005). In another paper, they demonstrated that yellow eels exhibited a higher pressure sensitivity (compression effects) compared to silver eels. The acclimatization period (21 days at 10.1 MPa) of yellow eels cancelled the differences in pressure sensitivity and in aerobic metabolism at 0.1 MPa observed between the two stages.this capacity to cope with the membrane pressure effects was due to the fact that silver eel exhibited higher membrane fluidity (Vettier et al., 2006). The mechanisms, which take place in yellow eels during acclimatization to high pressure, appear to be already present in silver eels before pressure exposure. They also showed that Lake Balaton eels, which never become silver and do not migrate (Biro, 1992), have an alteration in aerobic metabolism and muscle fibre maximal aerobic capacities after 3 days under pressure, suggesting lower pressure tolerance (Vettier et al., 2003). *In vitro* studies using permeabilised red muscle fibres of silver eels suggested that the best conditions to optimize energy metabolism were low temperature combined to high pressure (Scaion et al., 2008). High hydrostatic pressure (101 ATA) improves the swimming efficiency of European migrating silver eel by significantly lowering oxygen consumption (Sébert et al., 2009).

## 7.2.3 Changes linked to preparation to prolonged migration

### 7.2.3.1 Olfactory system

In contrast to vision, olfaction is important during the sedentary life, as a precious system of prey detection. In female *Anguilla rostrata*, Pankhurst and Lythgoe (1983) observed a decrease in the number of mucous cells in the olfactory epithelium of silver eels compared to yellow eels, but the structure of the lamellae remained the same. Moreover, in artificially matured eels, they reported a degeneration of the olfactory epithelium with a change in cell organization and a decrease in mucous cell number. Sorensen and Pankhurst (1988) reported the same modifications in male *Anguilla rostrata*. These modifications can be related to the fact that eel stops "hunting" and begins to starve.

Different studies also showed the importance of olfaction in the migration of eels. Tagged European eels were inhibited in either the visual, magnetic or olfactory sense, and the group that had been made anosmic by injection of elastomer into the nasal cavity showed irregular swimming behaviour, slower speed and no common direction (Westin, 1990; Tesch et al., 1991). Similarly, ultrasonic telemetry showed the importance of olfaction in the estuarine migration of silver-phase American eels as the anosmic eels with nares filled with petroleum jelly spent more time in the estuary, whereas the control eels moved upstream and downstream with the tides (Barbin, 1998).

Recently, it has been shown an androgen-dependent stimulation of brain dopaminergic system in the olfactory bulbs of European eel, suggesting that androgens, which increase during silvering, may enhance the central processing of olfactory cues, essential for navigation during eel catadromous migration toward the spawning grounds (Weltzien et al., 2006).

Olfaction is probably also necessary to pheromonal communication during the migration and in the spawning area as suggested by a recent study indicating that sexually mature males can stimulate spermatogenesis of their neighbors (Huertas et al., 2006).

In summary, during silvering, there is a switch from feeding related-olfaction to migration/reproduction related olfaction.

### 7.2.3.2 Metabolism

At the silver stage, eels starve (Fontaine and Olivereau, 1962) and this starvation is accompanied by a regression of the digestive tract (Han et al., 2003; Aroua et al., 2005; Durif et al., 2005). Eels need important metabolic changes to be able to accomplish their long oceanic migration, which is comprised between 4,000 km for Japanese eels (*A. japonica*: Tsukamoto, 1992) and American eel (*A. rostrata*: Tucker, 1959; Mc Cleave et al., 1987) and ≥ 6,000 km for European eel (*A. anguilla*: Schmidt, 1923). To permit this long oceanic migration, animals have to optimize the use of energy stores, for both swimming and gonadal maturation. Van Ginneken and van den Thillart (2000) demonstrated, using large swim-tunnels, that for their swim effort of 6,000 km, 40% of the European eels' energy reserves are needed while the remaining 60% of their energy stores can be used for gonad development. During silvering, changes in intermediary metabolism were observed (Boström and Johansson, 1972; Lewander et al., 1974; Dave et al., 1974; Barni et al., 1985; Eggington 1986; Van Ginneken et al., 2007a). Dave et al. (1974) reported a slightly higher amount of unsaturated and longer fatty acids in muscle and a significantly lower level of 18:0 and 20:4n-6 in the liver of silver compared to yellow European eels. Comparable trends were more recently observed in the shortfin eel *Anguilla australis* (De Silva et al., 2002). An increase in lipids (phospholipids, free fatty acids and cholesterol) content has also been reported recently at silvering in the blood of European eels (van Ginneken et al., 2007a). According to Lewander et al. (1974), a redistribution of cholesterol occurs from other tissues to the gonads in silver eel. These metabolic variations would be amplified by environmental conditions, encountered during the migration, such as depth and water temperature (Théron et al., 2000), as well as by locomotor activity (van Ginneken et al., 2007b).

During the silvering process, red muscle volume increases (Pankhurst, 1982c), probably due more to an increase in fat and mitochondria than to an increase in number of muscle fibers (Lewander et al., 1974; Pankhurst, 1982c). Activities of enzymes involved in the aerobic pathway are higher in silver than yellow eels (Boström and Johansson, 1972; Egginton, 1986) together with higher slow muscle output and power (Ellerby et al., 2001) and also a change in the main energy stores from glycogen in the yellow stage to fat in migrating fish (Lewander et al., 1974; Barni et al., 1985; Zara et al., 2000). Concomitantly, an increase in myoglobin content is observed

(Egginton, 1986). Silver eels have higher aerobic activities than yellow eels. The mass-specific power output of slow muscle is greater at silver stage than at yellow stage (Ellerby et al., 2001).

### 7.2.3.3 Migratory behaviour

Downstream/seaward migration is the last step/stage of eel silvering. It is a process sensitive to environmental factors such as flood/water discharge, water temperature, tubidity, photoperiod, light intensity and moon phase and/or atmospheric depression (Vollestad et al., 1986; Okamura et al., 2002a and b; Durif et al., 2008). At silvering, eels changed from sedentary fishes to migrant fishes, i.e., migration from brakish waters, rivers and streams to the sea. Silver eel migration behaviour is characterized by regular diurnal activity, active swimming very close to the surface at night and occasional dives of short duration (Westerberg et al., 2007). Using miniaturized pop-up staellite archival transmitter, Aarestrup et al. (2009) were able to follow silver European eels during their spawning migration up to 1,300 km from release and thus to provide unique behavioral insights. They showed that when eels moved into the mesopelagic zone they all undertook distinct vertical migrations (DVM) between 200 and 1,000m. During the night, they occupied shallow warm water; at dawn, they made a steep dive into the cool dsphotic zone and at night they ascended steeply back into the upper layer (Aarestrup et al., 2009). The authors hypothesize that DVM reflects thermoregulation in order to keep temperature below 11°C, delaying gonadal development.

### 7.2.4 Changes linked to preparation to reproduction (Changes in gonads)

The gonadosomatic index (GSI) increases progressively in female yellow eels from 0.3 to $\geq 1.5$ in silver eels with increase of follicular diameter, thickening of follicular wall and appearance of many lipidic vesicles (Fontaine et al., 1976; Lopez and Fontaine, 1990). This increase in gonad size was shown to be a good criterion to estimate the state of advancement of the silvering process in the different eels (Marchelidon et al., 1999; Durif et al., 2005). Durif et al. (2005) described five stages with physiological and morphological validation. In this study, a growth phase (stages I and II), a pre-migrating stage (stage III) and a migrating phase (stages IV and V) were defined. Stages I and II correspond to the previous "yellow" stage with a GSI<0.4%; the gonads show small primary, non-vitellogenic oocytes, with a dense ooplasma and a dense nucleus with a large nucleolus (Aroua et al., 2005). Stage III corresponds to the previous "intermediate" or "yellow/silver" stage with $0.4\% \leq GSI < 1.2\%$; oocytes are larger and a few lipidic vesicles are observed in the ooplasma, which indicates the initiation of the incorporation of lipidic stores in the oocytes, also referred as "endogenous vitellogenesis" (Aroua et al., 2005). Stages IV and V correspond to the previous "silver" stage with a $GSI \geq 1.2\%$; oocytes are further enlarged with a large nucleus and small nucleoli at a peripheral position and numerous lipidic vesicles in the ooplasma, with is the oil-droplet stage of early vitellogenesis (Aroua et al., 2005). In the most advanced stage of silvering, vitellogenin can be observed in the ooplasm, as well as in the plasma, which corresponds to the start of exogenous vitellogenesis.

Considering the start of gonadal maturation, silvering should be considered as an initiation of puberty. As the development of gonads and sexual maturity are blocked at this stage and until the occurrence of oceanic reproductive migration, our group defined eel silvering as a prepuberty. These data concerning gonadal maturation show that eel silvering is quite different from salmon smoltification, which occurs before a growth phase and is not associated with changes related to reproduction.

## 7.3 Endocrine/internal Factors Involved in the Control of Secondary Metamorphosis

As indicated previously, silvering consists of various morphological, physiological and behavioural changes. Among the modified organs, some are related to sensory organs, others to hydrostatic pressure or seawater adaptation, similarly to changes observed during smoltification, which traditionally led to eel silvering being defined as a metamorphosis. However, unlike smoltification, silvering also includes some changes related to an onset of sexual maturation such as gonad development, which led to the hypothesis that silvering corresponds to a pubertal event. Puberty, the major post-embryonic developmental event in the life cycle of all vertebrates, encompasses various morpho-physiological and behavioural changes, which unlike metamorphic changes are induced by sexual steroids (for review, see Romeo, 2003). While metamorphosis is mostly triggered in vertebrates by the activation of the thyrotropic axis, puberty is triggered by the activation of the gonadotropic axis. In order to assess which neuroendocrine axis may be involved in the induction of silvering, we analyzed the profiles of pituitary and peripheral hormones during the transition from yellow to silver eels.

### 7.3.1 Hormones

#### 7.3.1.1 Thyrotropic axis

Callamand and Fontaine (1942) observed a hyperactivity of the thyroid gland before and during the downstream migration. It was then thought that the thyroid activation was responsible for the important morphological modifications observed during silvering. Especially, concerning eye changes, some cases of exophthalmy can be observed in humans having hyperthyroidian pathology (for reviews: Bradley, 2001; Wiersinga and Bartalena, 2002), supporting the former hypothesis of a role of thyroid hormones in eel eye changes. Moreover, Fontaine (1953) suggested that the thyroid activation during eel silvering at the moment of the downstream migration could be responsible for the increased locomotor activity of the animals, as is observed during smoltification. In subadult American eels, elevated T4 plasma levels are correlated with increased locomotor activity under natural conditions (Castonguay et al., 1990).

In contrast with the previous hypothesis, the study of the expression profiles of TSH showed a non-significant or a weak increase in TSHβ mRNA between yellow and silver eels (*Anguilla anguilla*: Aroua et al., 2005; *Anguilla japonica*: Han et al., 2004) (Fig. 3a). Moreover, measurement of plasma levels of thyroid hormones in yellow and silver eels showed a moderate increase in thyroxine (T4) and no significant variations in triiodothyronine (T3) during silvering (*Anguilla anguilla*: Marchelidon et al., 1999; Aroua et al., 2005; *Anguilla japonica*: Han et al., 2004) (Fig. 3b). These recent results suggest that the thyrotropic axis is poorly implicated in the neuroendocrine control of the silvering process. The weak variations observed on TSHβ mRNA and T4 plasma level could be involved in the increased activity of eels related to their migratory behavior. This role in motility may not be specific to the silver stage as thyroid hormones can also induce an increase in locomotor activity in yellow (Castonguay et al., 1990) and glass eels (Edeline et al., 2005).

Future studies should aim at investigating the brain control of pituitary TSH. Interestingly, preliminary data on eel pituitary cells suggest a role for CRH in the control of TSH, rather than TRH (personal communication).

#### 7.3.1.2 Somatotropic axis

A recent study showed no significant differences in mRNA levels between yellow and silver stage, and a significant decrease of GH pituitary content (Aroua et al., 2005). These data suggest that, unlike in salmonid smoltification where GH has a strong role as a factor controlling osmoregulation and seawater adaptability (Björnsson, 1997), in eel silvering the GH role is less

a)

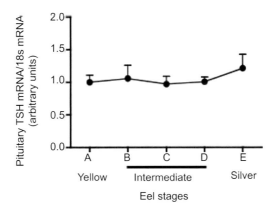

b)

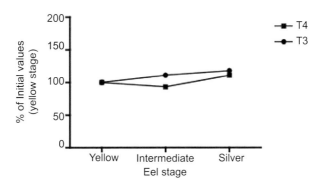

**Figure 3**. Lack of involvement of thyrotropic axis in the control of eel silvering. During the process of silvering, no activation of thyrotropic axis is observed; as there is no change of pituitary TSH β mRNA levels (a) nor plasma T4 and T3 levels (b). TSH=thyrotropin; T4=thyroxine ; T3=triiodothyronine.

critical, probably due to the fact that even at the yellow stage eels are able to pass into seawater.

Studies focusing on GH cell regulation have shown that basal release and synthesis of GH persist *in vitro*, in the absence of secretagogues or serum, using organ-cultured pituitaries (European eel: Baker and Ingleton, 1975; Japanese eel: Suzuki et al., 1990) or primary cultures of pituitary cells (European eel: Rousseau et al., 1998, 1999; rainbow trout: Yada et al., 1991; turbot: Rousseau et al., 2001). All these observations lead to the suggestion that the major control of these cells *in vivo* is an inhibitory control (Rousseau and Dufour, 2004). Our *in vitro* study in the European eel demonstrates that the brain inhibitory control of GH is exerted by somatostatin (SRIH) and that insulin-like growth factor 1 (IGF-1) exerts a negative feedback on GH. Thus, during silvering, the major regulation of the somatotroph cells, which are responsible of GH synthesis, would be an increase of their inhibitory control.

It is important to note that seasonal data suggest that a peak in GH and body growth during summer may occur (Durif et al., 2005). This suggests that while GH may not be involved in the control of the silvering process itself, the somatotropic axis may participate earlier in the initiation of the silvering. A similar growth surge is observed at puberty in mammals.

### 7.3.1.3 Corticotropic axis

Only a few studies have focused on the corticotropic axis during the transition from yellow to silver stage. It is probably because of the difficulty of sampling blood in order to measure plasma cortisol levels without stressing animals. Recently, Van Ginneken et al. (2007a) demonstrated elevated cortisol levels in silver eels prior to migration. This is in agreement with the fact that during the downstream migration, eels are fasting and it is well know that the production of cortisol is induced in response to starvation. A role of cortisol may be to permit the mobilization of energy stores needed by the fish at this critical period.

Forrest et al. (1973) found that Na+, K+-ATPase activity rose very slowly after transfer of yellow american eels (*A. rostrata*) to SW. If the eels were treated with cortisol before transfer, the enzyme activity increased to levels found in SW-acclimated eels and did not increase further following transfer to SW. Injection of yellow American eels with cortisol, in addition to increasing Na+, K+-ATPase of gill and intestine, caused their ventral surface to turn silver (Epstein et al., 1971).

In addition to the effect of cortisol on energy mobilization and seawater adaptation, it was previously demonstrated in the European eel that cortisol had also a strong positive effect on LH production *in vivo* as well as *in vitro* (Huang et al., 1999). This stimulation is stronger when eels are treated by a combination of cortisol and androgens, indicating synergistic action of these hormones on LH (Huang, 1998; Sbaihi, 2001). It is interesting to note that while in amphibians, cortisol has a synergistic effect with thyroid hormones on metamorphosis, a synergy between cortisol and sex steroids is observed in the control of eel silvering. In the eel, cortisol has also an inhibitory action on digestive tract somatic index and can mobilize mineral stores from vertebral skeleton (Sbaihi et al., 2009).

The various effects of cortisol demonstrated in the eel indicate that the corticotropic axis may play an important role throughout silvering by permitting energy mobilization and probably acting at the onset of puberty. Cortisol may therefore control the metabolic challenge occurring during both metamorphosis and puberty/reproduction in teleosts. It may also act in synergy with different other hormones to induce the various morphological changes observed: with sex steroids during eel silvering, with THs during larval metamorphosis and with GH and TH during salmonid smoltification.

### 7.3.1.4 Gonadotropic axis

As highlighted by several authors, the silvering process is also characterized by gonadal modifications (Lopez and Fontaine, 1990; Fontaine and Dufour, 1991). Field studies clearly demonstrated an increase in gonad weight and modifications of the oocyte structure throughout the silvering process (Marchelidon et al., 1999; Aroua et al., 2005; Durif et al., 2005). In addition, measurement of sexual steroids, estrogens (E2) and androgens (T and 11-KT), in the plasma showed an increase between yellow and silver stage (*A. australis and A. dieffenbachii*: Lokman et al., 1998; *A. anguilla*: Sbaihi et al., 2001; Aroua et al., 2005; *A. rostrata*: Cottrill et al., 2001; *A. japonica*: Han et al., 2003). A first increase in E2 levels at the early stages of the silvering process (in intermeditae eels), then a further increase of E2, accompanied by an increase in 11-KT and T in seilver eels (Aroua et al., 2005) (Fig. 4b). All these results suggest that the gonadotropic axis occupies an important position during the silvering process.

Recent studies have focused on the two pituitary hormones involved in the control of reproduction, the gonadotropins (luteinizing hormone, LH and follicle-stimulating hormone, FSH). In *A. japonica* (Han et al., 2003), as well as in *A. anguilla* (Aroua et al., 2005), variation on mRNA levels of the alpha and the beta subunits of the gonadotropins were observed throughout silvering. In *A. japonica*, authors observed a concomitant increase in mRNA of the different subunits, LHβ, FSHβ and the glycoprotein alpha (GPα) (Han et al., 2003). In *A. anguilla*, LH and FSH were shown to be differentially expressed during the silvering process, with an early increase in FSHβ expression and a late increase of LHβ expression (Aroua et al., 2005) (Fig. 4a). These data suggest that FSH could play an early role in the activation of gonads, while LH may have an important role later in the silvering process. Indeed, a concomitance exists between the increase in FSH expression and the start of lipid incorporation in oocytes (also called "endogenous vitellogenesis"), which suggests that FSH could be responsible for the initiation of lipidic vitellogenesis. The early increase in FSH may also be responsible for the first increase in steroid production (E2), observed in « intermediate » eels. In contrast, the late increase in vitellogenin (Vg) plasma levels, concomitant with the late increase in LH expression, and the

**a)**

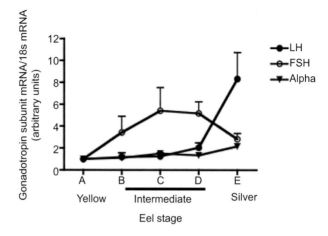

**b)**

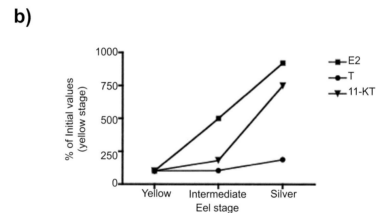

**Figure 4**. Involvement of gonadotropic axis in the control of eel silvering. During the process of silvering, there is an activation of gonadotropic axis as pituitary gonadotropin (LH β and FSH β) mRNA levels (a) and plasma sex steroids (E2, T and 11-KT) levels increase. LH=luteinizing hormone; FSH=follicle-stimulating hormone; E2=estradiol; T=testosterone; 11KT=11-ketotestosterone; alpha=GP alpha.

slight decrease in FSH, suggests that LH may participate in the induction of Vg production and initiation of the "exogenous vitellogenesis". Similarly, LH may also participate in the second increase in sex steroid levels, in silver eels.

In summary, among the different pituitary hormones that were studied, the gonadotropins, LH and FSH, showed the biggest variations during eel silvering.

## 7.3.1.5 Experimental induction of silvering changes

All these results clearly indicate that the gonadotropic axis is activated during silvering. The interesting question is to understand if this activation of the gonadotropic axis is related to the "metamorphic" changes observed during the transition from yellow to silver eel.

In 2008, the first discovery of mature freshwater eels in the open ocean, at the southern part of the West Mariana Ridge, was reported by Tsukamoto and collaborators (Chow et al., 2009). The three male eels, one giant mottled eel (*A. marmorata*) and two Japanese eels, presented dark brown or blackish gray body color, larger eye index than silver eels and bigger GSI compared to silver eels or even compared to fully matured male eels induced by hormonal injection (Chow et al., 2009).

Previous studies on the experimental induction of eel sexual maturation showed an amplification of anatomic changes observed during silvering (Pankhurst, 1982 a, b, c). Thus, morphological changes observed during silvering such as increase of eye diameter (Boëtius and Larsen, 1991), enhancement of silver-colour-body and decrease of gut weight (Pankhurst and Sorensen, 1984) could be further induced by gonadotropic treatments (with human chorionic gonadotropin in males or carp pituitary extracts in females).

Experimental data using exogenous sex steroid treatments are in agreement with this involvement of gonadotropic axis in the induction of silvering. Early study showed that injections of male silver European eels with 17α-methyltestosterone resulted in enlarged eye diameter, increased skin thickness and darkened head and fins (*A. anguilla*: Olivereau and Olivereau 1985). Similarly, implants of testosterone induced an increase of eye size in male silver eels (*A. anguilla*: Boëtius and Larsen, 1991). Moreover, immature female *Anguilla australis* which received implants of 11-KT, a non-aromatizable androgen, for 6 weeks presented the external morphological changes observed during silvering: increased eye diameter, larger gonads and thicker dermis, compared to controls (Rohr et al., 2001). Finally, recent studies in *A. anguilla* showed that treatment with testosterone induced a decrease in the digestive tract-somatic index (Vidal et al., 2004; Aroua et al., 2005) and an increase in ocular index (Aroua et al., 2005), while E2 has no effect (Aroua et al., 2005) (Fig. 5).

All these data suggest that the silvering changes observed during our experiments are androgen-dependent. In contrast, a 3-month-treatment of yellow eels with thyroid hormone (T3) did not induce any changes in ocular index and digestive tract-somatic index (Aroua et al., 2005) (Fig. 6). However, cortisol may have a synergistic action with steroids in this complex process of eel silvering, as we demonstrated that concomitant administration of E2, T and cortisol was most efficient in inducing the silvering of the skin in eels (Sbaihi, 2001).

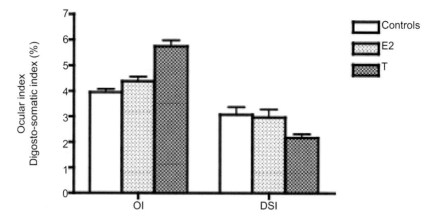

**Figure 5**. Involvement of gonadotropic axis in the control of eel silvering. Treatment of yellow eels with androgen (T) for 3 months induces an increase in ocular index (OI) and a decrease in digesto-somatic index (DSI).

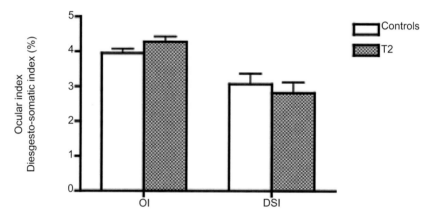

**Figure 6**. Lack of involvement of thyrotropic axis in the control of eel silvering. Treatment of yellow eels with thyroxine (T4) for 3 months does not induce any change in ocular index (OI) nor digesto-somatic index (DSI).

All these data demonstrate the involvement of androgens as crucial actors in the morpho-physiological changes of eel silvering. This is in contrast to smoltification, in which a precocious sexual maturation and sex steroid administration are able to inhibit/prevent smoltification and downstream migration (See Chapter 6).

### 7.3.2 Internal factors linked to fat

#### 7.3.2.1 Initiation of silvering/Age at silvering

##### 7.3.2.1.1 Fat stores

In general, a minimum, critical amount of fat is needed, as a trigger for metamorphosis or sexual maturation, in order to withstand the metabolic needs due to metamorphosis and after migration and reproduction. In the eel, there is a wide variability in the age at silvering (Rossi and Colombo, 1976; Vollestad and Jonsson, 1986; Poole and Reynolds, 1996; Svedäng et al., 1996). In 2006, Durif et al showed, in a study collecting European eels from 5 different locations in France, that age ranged from 5 to 24 years. The idea of a 'critical fat mass' is discussed in Larsson et al. (1990), in which the authors examined the percentage of lipid in "muscular tissue" (15–20 g tail, including skin) in yellow and silver eels (all females). They found that a fat % of 28 (in the tail) may be a prerequisite for the change from yellow to silver stage and under a critical fat mass, silvering may not even be initiated (Larsson et al., 1990). However, a number of large yellow eels had fat above 28 and a number of silver eels had fat below 28, indicating that a certain fat % is not the 'triggering factor', but may be necessary and need to be associated with other triggering factors (internal and external). Svedäng and Wickström (1997) suggested that silver eels with low fat concentrations may temporarily halt migration, revert to a feeding stage, and "bulk up" until fat reserves are sufficient to carry out successful migration to the spawning area. Combining both Boëtius and Boëtius (1985) and Larsson et al. (1990) studies, Larsen and Dufour (1993) found that silver eels (regardless of sex) rarely have a fat % lower than 20. Careful analysis of the energy budget of migratory eels have been made (Boëtius and Boëtius, 1980, 1985; van Ginneken and van den Thillart, 2000; van den Thillart et al., 2004) and showed that silver eels contain energy reserves and organic material sufficient for both the long migration and the extensive gonadal growth. By artificially inducing sexual maturation with CPE injections and monitoring individual characteristics, Durif et al. (2006) demonstrated that the initial state of the eel in terms of energy stores partly explained the variability of subsequent sexual maturation. Indeed, the best response to CPE treatment came from eels with the highest initial condition factor and largest initial body diameter.

##### 7.3.2.1.2 Growth

Comparing maturing European eel from 38 different locations throughout Europe and South Africa, Vollestad (1992) reported geographic variation in age at metamorphosis, which could be explained by variation in growth rate as growth rate varied with both latitude and longitude. Silvering seems to be triggered by a period of high growth. Indeed, an important increase in GH was observed in pre-migrant European eels (stage III = beginning of silvering) (Durif et al., 2005). Using primary cultures of eel pituitary cells, it was shown that insulin-like growth factor IGF-1 (produced by the liver under GH stimulatory control) was able to inhibit GH (Rousseau et al., 1998) and stimulate LH release (Huang et al., 1998), leading to the hypothesis that this growth factor could be the link between body growth and induction of puberty in the eel. The high GH levels observed during spring in the eel could stimulate liver secretion of IGF-1, which could trigger the initial LH synthesis and decrease in GH. In addition, various studies showed that favourable growth conditions cause eels to silver rapidly (Vollestad, 1988, 1992; De Leo and Gatto, 1995), such as is the case in aquaculture, under experimental conditions (Tesch, 1991; Fernandez-Delgado et al., 1989) or in brackish water and at low latitudes (Lee, 1979; Fernandez-Delgado et al., 1989), involving environmental factors as triggerers of silvering.

### 7.3.2.2 Swimming

Lipid mobilization and initiation of maturation that occur during silvering are linked to migration, as showed by the positive correlation between intermediate stages and migration (Durif et al., 2005) and the negative correlation between migration distance to the spawning rounds and GSI at the start of oceanic migration (Todd, 1981b).

The swimming physiology of European silver eels and the effects on sexual maturation and reproduction has been recently reviewed (Palstra and van den Thillart, 2010).

Swimming induced an increase of eye diameter in different swim trials (Palstra et al., 2006, 2007, 2008a). However, changes in the length and shape of the pectoral fins, which are indicative of the degree of silvering, were not reported in any of the swim-trials (Palstra et al., 2007).

Swimming induced changes in gonad histology of both females and males (Palstra et al., 2007, 2008b). In contrast to resting eels, swimming female eels from Lake Balaton had higher GSI and oocyte diameter and some of their oocytes were in the lipid droplet stage 3 (Palstra et al., 2007). Swimming male eels had also a higher GSI, reflecting thick layers of connective tissue and more stage 2 and late type b spermatogonia compared to resting males (Palstra et al., 2008b).

Swimming induced changes in pituitary and blood plasma/expression of maturation parameters. After swimming 5,500-km (173 days) in freshwater, young farmed female eels showed increased plasma levels of estradiol and 11-ketotestosterone, pituitary levels of LH (Van Ginneken et al., 2007a). In this study, no significant differences were measured in eye index, GSI, HSI and plasma levels of vitellogenin and cortisol. Female silver eels that swam for 1.5 or 3 months had reduced ERα, VTG1 and VTG2 liver expression and lower plasma calcium levels in their blood, indicative of suppressed hepatic vitellogenesis (Palstra et al., 2010).

Males stimulated by three month-swimming showed a similar two- to three- fold higher LHβ expression level than males injected with GnRH agonist, compared to controls (Palstra et al., 2008b). Both treatments also caused a three- to five-fold increase in GSI and an induced spermatogenesis (>80% presence of spermatogonia late type b); one male in the swimmer group even showed the formation of spermatocytes (Palstra et al., 2008b). However, in this study, females were not stimulated by SW-swimming nor by GnRH agonist, and even showed regression of maturation overt time as demonstrated by lower LHβ expression, GSI and oocyte diameters in all groups after 3 months (Palstra et al., 2008b). This unresponsiveness of females may be due to the fact that, in contrast to males, their pituitaries are not sensitized and still under dopaminergic control.

Different mechanisms have been suggested by van Thillart and collaborators. Swimming may activate lipid metabolism and/or inhibits vitellogenesis (Palstra et al., 2009a). A possible role of cortisol in this phenomenon has been hypothetized as silver eels have higher cortisol levels (Van Ginneken et al., 2007a) and higher cortisol levels have been measured in swimming eels of Lake Grevelingen and Lake Balaton (Palstra et al., 2009b).

In summary, swimming trials performed by van den Thillart's group did show different effects in males and females. In silver males, swimming trials continued sexual maturation. In contrast, suppressed gonadotropin expression and vitellogenesis, associated with stimulated lipid deposition in oocytes are observed after swimming trials in silver females. These latter observations in females suggest that in nature, a different sequence of events may occur compared to artificial maturation and eels may undergo vitellogenesis and final maturation near or at the spawning grounds.

## 7.4 Environmental Factors Involved in the Control of Eel Silvering

### 7.4.1 Photoperiod/melatonin

Melatonin is a hormone principally released by the pineal organ during the night known to act as a transmitter of environmental cues, notably photic information, and to play a role in synchronising various behaviours and physiological processes (for review: Zachmann et al., 1992; Pandi-Perumal et al., 2006).

In the eel, a long-term melatonin treatment increased brain tyrosine hydroxylase (TH, the rate limiting enzyme of DA synthesis) mRNA expression in a region dependent way (Sébert et al., 2008). Melatonin stimulated the dopaminergic system of the preoptic area, which is involved in the inhibitory control of gonadotrophin (LH and FSH) synthesis and release. Moreover, the increased TH expression was consistent with melatonin binding site distribution as shown by 2[$^{125}$I]-melatonin labelling studies. On the other hand, melatonin had no effects on the two eel native forms of GnRH (mGnRH and cGnRH-II) mRNA expression. Concerning the pituitary-gonad axis, melatonin treatment decreased both gonadotrophin β-subunit (LHβ, FSHβ) mRNA expression and reduced sexual steroid (11-ketotestosterone, oestradiol) plasma levels. This indicates that melatonin treatment had a negative effect on eel reproductive function. By this mechanism melatonin could represent one pathway by which environmental factors could modulate reproductive function in the eel.

### 7.4.2 Hydrostatic and atmospheric pressure

Professional fishermen and researchers have observed that the migration of silver eel often occurs during storms (i.e., atmospheri depression). Eels become more active (active swimming and escapement behaviour) during thunderstorms (Lowe, 1952). Hvidsten (1985) also indicated that atmospheric pressure is statistically correlated with migration peaks.

Preliminary experiments were performed using female silver eels kept in a cage 450 m in depth (about 46 atm) during 3 months (Dufour and Fontaine, 1985; Fontaine et al., 1985). Significantly higher ovarian development was observed (x1.39) and the pituitary LH content was 27 times higher in immersed eels. Serum LH concentrations, which were undetectable (≤0.2 ng/ml) in control eels, were significantly increased (x1.4) in immersed eels. In a more recent study, female silver eels were submitted to 101 ATA, in a hyperbaric chamber equipped with a freshwater recirculation system, for 3 weeks (Sébert et al., 2007). Significant increase in oocyte diameter (x1.11) and in plasma levels of 11-KT (x2.39), E2 (x1.99) and Vg (x12.47) were observed in females submitted to high hydrostatic pressure (HP). At the pituitary level, LHβ expression tended to increase, while FSHβ expression decreased, leading to an increase in LHβ/ FSHβ ratio. However, even if HP plays a specific and positive role in eel reproduction, additional environmental and internal factors are necessary to ensure complete sexual maturation.

### 7.5 Anthropic Factors Impairing Eel Silvering

The effects of contaminants in European eel was recently reviewed (Geeraerts and Belpaire, 2010).

During the growth phase of the eel, water and proteins are progressively replaced by fat (lipids and lipoproteins; Degani et al., 1986). An European silver eel of more than 500 g contains on average 30% of fat (Boëtius and Boëtius, 1980, 1985). In silver eels, up to 80% of available energy is stored as triglycerides in muscles (Boëtius and Boëtius, 1985; Sheridan, 1988). Lipids stored during the growth stage are catabolized to provide the energy necessary for migration, production of gametes and spawning. Some compounds such as organophosphorus pesticides, that are widely used in agriculture, can affect lipidogenesis in the eel. Thus, experimental exposures to diazinon (Ceron et al., 1996), fenitrothion (Sancho et al., 1997) and thiobencarb (Fernandez-Vega et al., 1999), as well as molinate (Sancho et al., 2000) inhibit acetylcholinerase activity in different tissues (brain, muscle, whole eye and plasma) of exposed eels. Eels exposed to endosulfan, diazinon or fenitrothion have significantly lower liver and muscle glycogen content, as well as fat contents than before exposure (Gimeno et al., 1995; Sancho et al., 1997, 1998). By studying the activity and expression level of several enzymes involved in liver lipolysis and lipogenesis, Pierron et al. (2007) demonstrated impairment of lipid storage by cadmium in yellow eels.

Corsi et al. (2005) reported, in the European eels from the Ortebello lagoon (Italy), presence of polychlorinated biphenyls (PCBs), lindane and DDTs in muscle tissue and organophosphate insecticides (OP) and carbamates (CBs) toxicity were also evidenced by reduced cholinesterase (ChE) activity.

Robinet and Feunteun (2002) hypothesized that cortisol production resulting from the repeated exposure to chemicals (demonstrated in other fish: Hontela, 1997), likely lead to lipolysis, which adversely affect energy accumulation in eels. This exaggerated lipolysis would be expected to delay silvering and related emigration until enough fat is stored by eels. However, several studies demonstrated that cortisol could stimulate both silvering and sexual maturation (see Section 2.1.3). Epstein et al. (1971) and Fontaine (1994) showed that high concentrations of plasma cortisol, lasting at least 7 days, triggers silvering and an early physiological adaptation to salinity. In addition, cortisol was shown to trigger gametogenesis (Olivereau, 1966), as well as synthesis and release of LH *in vivo* and *in vitro* (Huang et al., 1999).

As we and other studies demonstrated that androgens are the major factors controlling eel silvering, future experiments should aim at investigating the potential impact of antiandrogens on the process of eel silvering. In mammals, especially human, environmental anti-androgens have already been shown to affect the reproductive development (Kelce and Wilson, 1997) and timing of puberty (Den Hond and Schoeters, 2006; Toppari and Juul, 2010).

The unique life cycle of the eel, with a stage of growth in freshwater (yellow stage) when energy reserves (fat stored in muscle tissue) and lipophilic pollutants accumulated, followed by a long migration to the spawning areas when pollutants are released from the fat deposits, followed by a once-a-life spawning behaviour, makes the eel especially vulnerable to persistent pollutants (Larsson et al., 1991). Because of their lipid contents (up to 31%: Tesch, 1977; Boëtius and Boëtius, 1980; Geyer et al., 1994 compared to most freshwater species: 3–9% Geyer et al., 1994), eels are the favored targets for lipoliphilic contaminant accumulation. Contaminants, such as lipophilic xenobiotics, accumulate in the fat tissue of eels during their feeding stage (yellow eels). Yellow eels can stay for a prolonged period in freshwater during which they continuously bioaccumulate xenobiotics; they are mostly benthic and thus exposed to contaminants accumulated and adsorbed in sediments (Van der Oost et al., 1994). Since silver eels do not eat during their oceanic migration, contaminants previously accumulated can be remobilized and redistributed, thus triggering potential toxic events (Palsta et al., 2006). Moreover, as they do not reproduce during this stage, there is no loss of contaminants due to fat metabolisation and production/release of gametes. As long as the contaminants are stored in the fat reserves, toxic effects are minor. But, at the start of the migration, when the lipids are oxidized and the contaminants released, their levels in the blood plasma may increase up to toxic levels.

Other studies investigated the toxic effects after experimental maturation. Using an eel testicular organ culture system, Miura et al. (2005) showed that *para*-nonylphenol, an estrogenic endocrine disruptor, induced Sertoli cell hypertrophy and a decrease in germ cell number, in the presence of 11-KT (which induces spermatogenesis *in vitro*). Cadmium pre-exposure of experimentally matured silver eels strongly stimulated the pituitary-gonad-liver axis, leading to early and enhanced vitellogenesis, followed by oocyte atresia and death (Pierron et al., 2008). Van Ginneken et al. (2009) also tested the effects of PCBs on morphological and blood parameters as well as on energy metabolism in European eels that experienced a simulated partial migration of 800 km in Blazka swim tunnels. They reported increased spleen and liver weights and lower standard metabolic rate.

## Conclusions

All the external as well as internal modifications occurring during eel silvering have traditionally made scientists consider this event as a metamorphosis. However, as reviewed in this chapter, eel silvering may be primarily induced by the gonadotropic axis (Fig. 7). There is an overall activation of this axis, with increases of gonadotropin and sex steroid

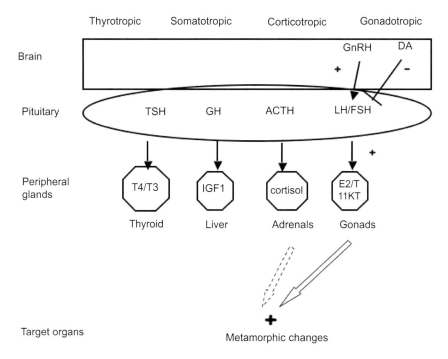

**Figure 7**. Neuroendocrine axes involved in the control of secondary metamorphosis (silvering) in eel. Silvering is triggered mainly by the gonatropic axis with some synergism with the corticotropic axis. TSH=thyrotropin; T4=thyroxine; T3=triiodothyronine ; GH=growth hormone; IGF=insulin-like growth factor; ACTH=corticotropin; LH=luteinizing hormone; FSH=follicle-stimulating hormone; E2=estradiol; T=testosterone; 11KT=11-ketotestosterone. DA=dopamine; GnRH=gonadotropin-releasing hormone.

levels. Moreover, exogenous sex steroids are able to induce peripheral morphological changes observed during this process. This let us regard eel silvering as a pubertal rather than a metamorphic event. The term "prepuberty" was first used by our group, as during eel silvering, puberty is blocked at an early stage and further sexual maturation only occurs during the reproductive migration. Other endocrine axes may participate in the control of eel silvering. This is the case of somatotropic and corticotropic axes acting in synergy with the gonadotropic axis (Fig. 7). Further studies should aim at investigating the neuroendocrine interactions controlling silvering, as well as the role of environmental and internal factors in the mechanisms leading to the activation of the gonadotropic axis.

# References

Aarestrup, K., F. Okland, M.M. Hansen, D. Righton, P. Gargan, M. Castonguay, L. Bernatchez, P. Howey, H. Sparholt, M.I. Pedersen and R.S. McKinley. 2009. Oceanic spawning migration of the European eel (*Anguilla anguilla*). Science 325: 1660.

Archer, S., A. Hope and J.C. Partridge. 1995. The molecular basis for the green-blue sensitivity shift in the rod visual pigments of the European eel. Proceedings of the Royal Society Series B 262: 289–295.

Aroua, S., M. Schmitz, S. Baloche, B. Vidal, K. Rousseau and S. Dufour. 2005. Endocrine evidence that silvering, a secondary metamorphosis in the eel, is a pubertal rather a metamorphic event. Neuroendocrinology 82: 221–232.

Baker, B.I. and P.M. Ingleton. 1975. Secretion of prolactin and growth hormone by teleost pituitaries *in vitro*. II. Effect of salt concentration during long-term organs culture. Journal of Comparative Physiology 100: 269–282.

Barbin, G.P. 1998 The role of olfaction in homing and estuarine migratory behavior of yellow-phase American eels. Canadian Journal of Fisheries and Aquatic Sciences 55: 564–575.

Barni, S., G. Bernocchi and G. Gerzeli. 1985. Morphohistochemical changes in hepatocytes during the life cycle of the European eel. Tissue and Cell 17: 97–109.

Beatty, D.D. 1975. Rhodopsin-porphyropsin changes in paired-pigment fishes. In: Vision in Fishes, M.A. Ali (Ed.). Plenum Press, New York, pp. 635–644.

Bertin, L. 1951. Les anguilles—Variation, croissance, euryhalinité, toxicité, hermaphrodisme juvenile et sexualité, migrations, métamorphoses. Payot, Paris.

Biro, P. 1992. Die geschichte des aals (*Anguilla anguilla* L.) im plattensee (Balaton). Österr. Fisch. 45: 197–207.

Björnsson, B.T. 1997. The biology of salmon growth hormone: From daylight to dominance. Fish Physiology and Biochemistry 17: 9–24.

Boëtius, J. and I. Boëtius. 1980. Experimental maturation of female silver eels, *Anguilla anguilla*. Estimates of fecundity and energy reserves for migration and spawning. Dana 1: 1–28.

Boëtius, I. and J. Boëtius. 1985. Lipid and protein content in *Anguilla anguilla* during growth and starvation. Dana 4: 1–17.

Boëtius, I. and L.O. Larsen. 1991. Effects of testosterone on eye size and spermiation in silver eels, *Anguilla anguilla*. General and Comparative Endocrinology 82: 238.

Boström, S.L. and R.G. Johansson. 1972. Enzyme activity patterns in white and red muscle of the eel (*Anguilla anguilla*) at different developmental stages. Comparative Biochemistry and Physiology B 42: 533–542.

Bowmaker, J.K., M. Semo, D.M. Hunt and G. Jeffery. 2008. Eel visual pigments revisited: the fate of retinal cones during metamorphosis. Visual Neuroscience 25: 1–7.

Bradley, E.A. 2001. Graves ophthalmopathy. Current Opinion in Ophthalmology 12: 347–351.

Braekevelt, C.R. 1985. Retinal fine structure in the European eel *Anguilla anguilla*. IV. Photoreceptors of the yellow eel stage. Anatomischer Anzeiger 158: 23–32.

Braekevelt, C.R. 1988. Retinal fine structure in the European eel *Anguilla anguilla*. VI. Photoreceptors of the sexually immature silver eel stage. Anatomischer Anzeiger 166: 23–31.

Callamand, O. and M. Fontaine. 1942. L'activité thyroïdienne de l'anguille au cours de son développement. Archives de Zoologie Expérimentale et Générale 82: 129–136.

Carlisle, D.B. and E.J. Denton. 1959. On the metamorphosis of the visual pigments of *Anguilla anguilla* (L.). Journal of the marine Biological Association of the United Kingdom 38: 97–102.

Castonguay, M., J.D. Dutil, C. Audet and R. Miller. 1990. Locomotor activity and concentration of thyroid hormones in migratory and sedentary juvenile American eels. American Fisheries Society 119: 946–956.

Ceron, J.J., M.D. Ferrando, E. Sancho, C. Gutierrez-Panizo and E. Andreu-Moliner. 1996. Effects of diazinon exposure on cholinesterase activity in different tissues of European eel (*Anguilla anguilla*). Ecotoxicology and Environmental Safety 35: 222–225.

Chow, S., H. Kurogi, N. Mochioka, S. Kaji, M. Okazaki and K. Tsukamoto. 2009. Discovery of mature freshwater eels in the open ocean. Fish Science 75: 257–259.

Comrie, M.M., C.P. Cutler and G. Cramb. 1999. Cloning and expression of guanylin from the European eel (*Anguilla anguilla*). Biochemical and Biophysical Research Communications 281: 1078–1085.

Corsi, I., M. Mariottini, A. Badesso, T. Caruso, N. Borghesi, S. Bonacci, A. Iacocca and S. Focardi. 2005. Contamination and sub-lethal toxicological effects of persistent organic pollutants in the European eel (*Anguilla anguilla*) in the Orbetello lagoon (Tuscany, Italy). Hydrobiologia 550: 237–249.

Cottrill, R.A., R.S. McKinley, G. van der Kraak, J-D. Dutil, K.B. Reid and K.J. McGrath 2001. Plasma non-esterified fatty acid profiles and 17b-oestradiol levels of juvenile immature and maturing adult American eels in the St Lawrence River. Journal of Fish Biology 59: 364–379.

Cottrill, P.B., W.L. Davies, M. Semo, J.K. Bowmaker, D.M. Hunt and G. Jeffery. 2009. Developmental dynamics of cone photoreceptors in the eel. BMC Developmental Biology 9: 71–79.

Cutler, C.P., I.L. Sanders, N. Hazon and G. Cramb. 1995a. Primary sequence, tissue specificity and expression of Na+,K+-ATPase α1 isoform in the European eel (*Anguilla anguilla*). Comparative Physiology and Biochemistry B 111: 567–573.

Cutler, C.P., I.L. Sanders, N. Hazon and G. Cramb. 1995b. Primary sequence, tissue specificity and expression of Na+,K+-ATPase β1 isoform in the European eel (*Anguilla anguilla*). Fish Physiology and Biochemistry 14: 423–429.

Dave, G., M. Johansson, A. Larsson, K. Lewander and U. Lidman. 1974. Metabolic and haematological studies on the yellow and silver shortfin phases of the European eel, *Anguilla anguilla* L. II. Fatty acid composition. Comparative Biochemistry and Physiology B 47: 583–591.

De Leo, G.A. and M. Gatto. 1995. A size and age-structured model of the European eel (*Anguilla anguilla* L.). Canadian Journal of Fisheries and Aquatic Sciences 52: 1351–1367.

De Silva, S.S., R.M. Gunasekera and R.O. Collins. 2002. Some morphometric and biochemical features of ready-to-migrate silver and pre-migratory yellow stages of the shortfin eel of south-eastern Australian waters. Journal of Fish Biology 61: 915–928.

Degani, G., H. Hahami and D. Levanon. 1986. The relationship of eel *Anguilla anguilla* (L.) body size, lipid, protein, glucose, ash, moisture composition and enzyme activity (aldolase). Comparative Biochemistry and Physiology A 84: 739–745.

Den Hond, E., G. Schoeters. 2006. Endocrine disrupters and human puberty. International Journal of Andrology 29: 264–271.

Denton, E.J., F.J. Warren. 1957. The photosensitive pigments in the retinae of deep-sea fish. Journal of the Marine Biological Association of the United Kingdom 36: 651–662.

Dufour, S. 1994. Neuroendocrinologie de la reproduction de l'anguille: de la recherche fondamentale aux problèmes appliqués. Bulletin Français de la Pêche et de la Pisciculture 335: 187–211.

Dufour, S. and Y.A. Fontaine. 1985. La migration de reproduction de l'anguille européenne (*Anguilla anguilla* L.): un rôle probable de la pression hydrostatique dans la stimulation de la fonction gonadotrope. Bulletin de la Société Zoologique de France 110: 291–299.

Dufour S. and K. Rousseau. 2007. Neuroendocrinology of fish metamorphosis and puberty: evolutionary and ecophysiological perspectives. Journal of Marine Science and Technology Special Issue 55–68.

Dufour, S., E. Burzawa-Gérard, N. Le Belle, M. Sbaihi and B. Vidal. 2003. Reproductive endocrinology of the European eel, *Anguilla anguilla*. In: Eel Biology, K. Aida, K. Tsukamoto and K. Yamauchi K (Eds.). Springer, Tokyo, pp. 107–117.

Durif, C., S. Dufour and P. Elie. 2005. The silvering process of *Anguilla anguilla*: a new classification from the yellow resident to the silver migrating stage. Journal of Fish Biology 66: 1025–1043.

Durif, C.M.F., S. Dufour and P. Elie. 2006. Impact of silvering stage, age, body size and condition on the reproductive potential of the European eel. Marine and Ecology Progress Series 327: 171–181.

Durif, C.M.F., F. Travade, J. Rives, P. Elie and C. Gosset. 2008. Relationship between locomotor activity, environmental factor, and timing of the spawning migration in the European eel, *Anguilla anguilla*. Aquatic Living Resources 21: 163–170.

Edeline, E., A. Bardonnet, V. Bolliet, S. Dufour and P. Elie. 2005. Endocrine control of *Anguilla anguilla* glass eel dispersal: Effect of thyroid hormones on locomotor activity and rheotactic behavior. Hormones and Behavior 48: 53–63.

Eggington, S. 1986. Metamorphosis of the American eel, Anguilla rostrata LeSueur: II. Structural reorganisation of the locomotory musculature. Journal of Experimental Zoology 238: 297–309.

Ellerby, D.J., I.L.Y. Spierts and J.D. Altringham. 2001. Slow muscle power output of yellow- and silver-phase European eels (*Anguilla anguilla* L.): changes in muscle performance prior to migration. The Journal of Experimental Biology 204: 1369–1379.

Epstein, F.H., M. Cynamon and W. McKay. 1971. Endocrine control of Na-K-ATPase and seawater adaptation in Anguilla rostrata. General and Comparative Endocrinology 16: 323–328.

Fernandez-Delgado, C., J.A. Hernando, Y. Ferrera and M. Bellido. 1989. Age and growth of yellow eels, *Anguilla anguilla*, in the estuary of the Guadalquivir river (south-west Spain). Journal of Fish Biology 34: 561–570.

Fernandez-Vega, C., E. Sancho, M.D. Ferrando and E. Andreu-Moliner. 1999. Thiobencarb toxicity and plasma AChE inhibition in the European eel. Journal of Environmental Science and Health B 34: 61–73.

Fontaine, M. 1953. La fonction hypophyso-thyroïdienne des poissons dans ses rapports avec leur morphologie et leur comportement. Journal du conseil international pour l'exploitation de la mer 19: 23–38.

Fontaine, M. 1985. Action de facteurs anormaux du milieu sur l'écophysiologie d'anticipation des poissons migrateurs amphihalins. Ichtyophysiologica Acta 9: 11–25.

Fontaine, M. and M. Olivereau. 1962. Nutrition et sexualité chez les poissons. Annales de la Nutrition et de l'Alimentation 16: A125–A152.

Fontaine, Y.A. 1994. L'argenture de l'anguille: métamorphose, anticipation, adaptation. Bulletin Français de la Pêche et de la Pisciculture 335: 171–186.

Fontaine, Y.A. and S. Dufour. 1991. The eels: from life cycle to reproductive endocrinology. Bulletin of the Institute of Zoology Academia Sinica 16: 237–248.

Fontaine, Y-A., E. Lopez, N. Delerue-Le Belle, E. Fontaine-Bertrand, F. Lallier and C. Salmon. 1976. Stimulation gonadotrope de l'ovaire chez l'anguille (*Anguilla anguilla* L.) hypophysectomisée. Journal of Physiology Paris 72: 871–892.

Fontaine, Y-A., S. Dufour, J. Alinat and M. Fontaine. 1985. A long immersion in deep-sea stimulates the pituitary gonatropic function of the female European eel (*Anguilla anguilla* L.). Comptes Rendus des Séances de l'Académie des Sciences 300: 83–87.

Fontaine, Y-A., M. Pisam, C. Le Moal and A. Rambourg. 1995. Silvering and gill "mitochondria-rich" cells in the eel, *Anguilla anguilla*. Cell and Tissue Research 281: 465–471.

Forrest, J.N. 1973. Na transport and Na-K-ATPase in gills during adaptation to seawater: effects of cortisol. American Journal of Physiology 224: 709–713.

Geeraerts, C. and C. Belpaire. 2010. The effects of contaminants in European eel: a review. Ecotoxicology 19: 239–266.

Geyer, H.J., I. Scheunert, R. Bruggemann, M. Matthies, C.E.W. Steinberg, V. Zitko, A. Kettrup, and W. Garrison. 1994. The relevance of aquatic organisms' lipid content to the toxicity of lipophilic chemicals: toxicity of lindane to different fish species. Ecotoxicology and Environmental Safety 28: 53–70.

Gimeno, L., M.D. Ferrando, S. Sanchez, L.O. Gimeno and E. Andreu. 1995. Pesticide effects on eel metabolism. Ecotoxicology and Environmental Safety 31: 153–157.

Han, Y-S., I-C. Liao, W-N. Tzeng, Y-S. Huang and J. Y-L. Yu. 2003. Serum estradiol–17β and testosterone levels during silvering in wild Japanese eel *Anguilla japonica*. Comparative Biochemistry and Physiology B 136: 913–920.

Han, Y.S., I.C. Liao, W.N. Tzeng and J.Y.L. Yu. 2004. Cloning of the cDNA for thyroid stimulating hormone ß subunit and changes in activity of the pituitary-thyroid axis during silvering of the Japanese eel, *Anguilla japonica*. Journal of Molecular Endocrinology 32: 179–194.

Haro, A. 2003. Downstream migration of silver-phase anguillid eels. In: Eel Biology, K. Aida, K. Tsukamoto and K. Yamauchi (Eds.). Springer, Tokyo, pp. 215–222.

Hontela, A. 1997. Endocrine and physiological responses of fish to xenobiotics: role of glucocorticosteroid hormones. Reviews in Toxicology 1: 1–46.

Hope, A.J., J.C. Partridge and P.K. Hayes. 1998. Switch in rod opsin gene expression in the European eel, *Anguilla anguilla* (L.). Proceedings of the Royal Society of London Series B 265: 869–874.

Huang, Y.S. 1998. Rôle des steroïdes sexuels et des hormones métaboliques dans le contrôle direct hypophysaire de l'hormone gonadotrope (GtH-II) chez l'anguille européenne, *Anguilla anguilla*. PhD thesis University Paris VI.

Huang, Y.S., K. Rousseau, N. Le Belle, B. Vidal, E. Burzawa-Gerard, J. Marchelidon and S. Dufour. 1998. Insulin-like growth factor-I stimulates gonadotrophin production from eel pituitary cells: a possible metabolic signal for induction of puberty. Journal of Endocrinology 159: 43–52.

Huang, Y.S., K. Rousseau, M. Sbaihi, N. Le Belle, M. Schmitz and S. Dufour. 1999. Cortisol selectively stimulates pituitary gonadotropin beta-subunit in primitive teleost, *Anguilla anguilla*. Endocrinology 140: 1228–1235.

Huertas, M., A.P. Scott, P.C. Hubbard, A.V.M. Canario and J. Cerda. 2006. Sexually mature European eels (*Anguilla anguilla* L.) stimulate gonadal development of neighbouring males: Possible involvement of chemical communication. General and Comparative Endocrinology 147: 304–313.

Hvidsten, N.A. 1985. Yield of silver eels and factors effecting downstream migration in the stream Imsa, Norway. Institute of Freshwater Research Drottningholm Report 62: 75–85.

Jegstrup, I.M. and P. Rosenkilde. 2003. Regulation of post-larval development in the European eel: thyroid hormone level, progress of pigmentation and changes in behaviour. Journal of Fish Biology 63: 168–175.

Kaneko, T., S. Hasegawa and S. Sasai. 2003. Chloride cells in the Japanese eel during their early stages and downstream migration. In: Eel Biology, K. Aida, K. Tsukamoto and K. Yamauchi (Eds.). Springer, Tokyo, pp. 457–468.

Kelce, W.R. and E.M. Wilson. 1997. Environmental antiandrogens: developmental effects, molecular mechanisms, and clinical implications. Journal of Molecular Medicine 75: 198–207.

Kleckner, R.C. 1980a. Swimbladder wall guanine enhancement related to migratory depth in silver phase Anguilla rostrata. Comparative Biochemistry and Physiology A 65: 351–354.

Kleckner, R.C. 1980b. Swimbladder volume maintenance related to initial oceanic migratory depth in silver-phase Anguilla rostrata. Science 208: 1481–1482.

Kleckner, R.C. and W.H. Krueger. 1981. Changes in swimbladder retial morphology in Anguilla rostrata during premigration metamorphosis. Journal of Fish Biology 18: 569–577.

Larsen L.O. and S. Dufour. 1993. Growth, reproduction and death in lampreys and eels. In: Fish Ecophysiology, J.C. Rankin and F.B. Jensen (Eds.). Chapman and Hall, London, pp. 72–104.

Larsson, P., S. Hamrin and L. Okla. 1990. Fat content as a factor inducing migratory behavior in the eel (*Anguilla anguilla* L.) to the Sargasso sea. Naturwissenschaften 77: 488–490.

Larsson, P., S. Hamrin and L. Okla. 1991. Factors determining the uptake of persistent pollutants in an eel population (*Anguilla anguilla* L.). Environmental Pollution 69: 39–50.

Lee, T.W. 1979. Dynamique des populations d'anguilles *Anguilla anguilla* (L.) des lagunes du basin d'Arcachon. PhD thesis, Université des Sciences et Techniques du Languedoc, Montpellier.

Lewander, K., G. Dave, M.L. Johansson, A. Larsson and U. Lidman. 1974. Metabolic and hematological studies on the yellow and silver phases of the European eel, *Anguilla anguilla* L. I. Carbohydrate, lipid, protein and inorganic ion metabolism. Comparative Biochemistry and Physiology B 47: 571–581.

Lokman, P.M., G.J. Vermeulen, J.G.D. Lambert and G. Young. 1998. Gonad histology and plasma steroid profiles in wild New Zealand freshwater eels (Anguilla dieffenbachii and *A. australis*) before and at the onset of the natural spawning migration. I. Females. Fish Physiology and Biochemistry 19: 325–338.

Lopez, E. and Y.A. Fontaine. 1990. Stimulation hormonale *in vivo* de l'ovaire d'anguille européenne au stade jaune. Reproduction Nutrition Development 30: 577–582.

Lowe, R.H. 1952. The influence of light and other factors on the seaward migration of the silver eel, *Anguilla anguilla*. Journal of Animal Ecology 21: 275–309.

Lythgoe, J.N. 1979. The ecology of vision. Clarendon Press, Oxford.

Marchelidon, J., N. Le Belle, A. Hardy, B. Vidal, M. Sbaihi, E. Burzawa-Gérard, M. Schmitz and S. Dufour. 1999. Etude des variations de paramètres anatomiques et endocriniens chez l'anguille européenne (*Anguilla anguilla*) femelle, sédentaire et d'avalaison: application à la caractérisation du stade argenté. Bulletin Français de la Pêche et de la Pisciculture 355: 349–368.

Martinez, A.S., C.P. Cutler, G.D. Wilson, C. Phillips, N. Hazon and G. Cramb. 2005. Cloning and expression of three aquaporin homologues from the European eel (*Anguilla anguilla*): effects of seawater acclimation and cortisol treatment on renal expression. Biology of the Cell 97: 615–627.

Mc Cleave, J.D., R.C. Kleckner and M. Castonguay. 1987. Reproductive sympatry of American and European eels and implications for migration and taxonomy. Proceedings of the Royal Society London 1: 86–297.

Miura, C., N. Takahashi, F. Michino and T. Miura. 2005. The effects of para-nonylphenol on Japanese eel (*Anguilla japonica*) spermatogenesis *in vitro*. Aquatic Toxicology 71: 133–141.

Okamura, A., Y. Yamada, N. Mikawa, S. Tanaka and H.P. Oka. 2002a. Exotic silver eels *Anguilla anguilla* in Japanese waters: seaward migration and environmental factors. Aquatic Living Resources 15: 335–341.

Okamura, A., Y. Yamada, S. Tanaka, N. Horie, T. Utoh, N. Mikawa, A; Akazawa and H.P. Oka. 2002b. Atmospheric depression as the final trigger for the seaward migration of the Japanese eel *Anguilla japonica*. Marine Ecology Progress Series 234: 281–288.

Olivereau, M. 1966. Effet d'un traitement par le cortisol sur la structure histologique de l'interrénal et de quelques tissus de l'anguille. Annales d'Endocrinologie Paris 27: 549–560.

Olivereau, M. and J. Olivereau. 1985. Effects of 17 α-methyltestosterone on the skin and gonads of freshwater male silver eels. General and Comparative Endocrinology 57: 64–71.

Palstra, A.P. and G.E.E.J.M. van den Thillart. 2010. Swimming physiology of European silver eels (*Anguilla anguilla* L.): energetic costs and effects on sexual maturation and reproduction. Fish Physiology and Biochemistry 36: 297–322.

Palstra, A.P., V.J.T. van Ginneken, A.J. Murk, and G.E.E.J.M. van den Thillart. 2006. Are dioxin-like contaminants responsible for the eel (*Anguilla anguilla*) drama? Naturwissenschaften 93: 145–148.

Palstra, A.P., D. Curiel, M. Fekkes, M. de Bakker, C. Székely, V. van Ginneken and G. van den Thillart. 2007. Swimming stimulates oocyte development in European eel (*Anguilla anguilla* L.). Aquaculture 270: 321–332.

Palstra, A.P., V. van Ginneken and G. van den Thillart. 2008a. Cost of transport and optimal swimming speeds in farmed and wild European eels (*Anguilla anguilla*). Comparative Biochemistry and Physiology A 151: 37–44.

Palstra, A.P., D. Schnabel, M.C. Nieveen, H.P. Spaink and G.E.E.J.M. van den Thillart. 2008b. Male silver eels mature by swimming. BMC Physiology 8: 14.

Palstra, A.P., V. van Ginneken and G. van den Thillart. 2009a. Effects of swimming on silvering and maturation. In: Spawning migration of the European eel, G. van den Thillart, S. Dufour and C. Rankin (Eds.). Springer, Netherlands, pp. 309–332.

Palstra, A.P., D. Schnabel, M.C. Nieveen, H.P. Spaink and G. van den Thillart. 2009b. Temporal expression of hepatic estrogen receptor 1, vitellogenin1 and vitellogenin2 in European silver eels. General and Comparative Endocrinology 166: 1–11.

Palstra, A.P., D. Schnabel, M.C. Nieveen, H.P. Spaink and G. van den Thillart. 2010. Swimming suppresses hepatic vitellogenesis in European female silver eels as shown by expression of the estrogen receptor 1, vitellogenin1 and vitellogenin2 in the liver. Reproductive Biology and Endocrinology 8: 27–36.

Pandi-Perumal, S.R., V. Srinivasan, G.J.M. Maestroni, D.P. Cardinali, B. Poeggeler and R. Hardeland. 2006. Melatonin. Nature's most versatile biological signal? FEBS Journal 273: 2813–2838.

Pankhurst, N.W. 1982a. Changes in the skin-scale complex with sexual maturation in the European eel, *Anguilla anguilla* (L). Journal of Fish Biology 21: 549–561.

Pankhurst, N.W. 1982b. Relation of visual changes to the onset of sexual maturation in the European eel, *Anguilla anguilla* (L). Journal of Fish Biology 21: 127–140.

Pankhurst, N.W. 1982c. Changes in body musculature with sexual maturation in the European eel, *Anguilla anguilla* (L). Journal of Fish Biology 21: 417–428.

Pankhurst, N.W. and J.N. Lythgoe. 1982. Structure and color of the tegument of the European eel *Anguilla anguilla* (L.). Journal of Fish Biology 21: 279–296.

Pankhurst, N.W. and J.N. Lythgoe. 1983. Changes in vision and olfaction during sexual maturation in the European eel *Anguilla anguilla* (L.). Journal of Fish Biology 23: 229–240.

Pankhurst, N.W. and P.W. Sorensen. 1984. Degeneration of the alimentary tract in sexually maturing European *Anguilla anguilla* (LeSueur). Canadian Journal of Zoology 62: 1143–1149.

Partridge, J.C., J. Shand, S.N. Archer, J.N. Lythgoe and W.A.H.M. Groningen-Luyben. 1989. Interspecific variation in the visual pigments of deep-sea fishes. Journal of Comparative Physiology A 164: 513–529.

Pierron, F., M. Baudrimont, A. Bossy, J.P. Bourdinaud, D. Brèthes, P. Elie and J.C. Massabuau. 2007. Impairment of lipid storage by cadmium in the European eel (*Anguilla anguilla*). Aquatic Toxicology 81: 304–311.

Pierron, F., M. Baudrimont, S. Dufour, P. Elie, A. Bossy, S. Baloche, N. Mesmer-Dudons N., P. Gonzalez, J.P. Bourdinaud and J.C. Massabuau. 2008. How cadmium could compromise the completion of the European eel's reproductive migration. Environmental Science and Technology 42: 4607–4612.

Poole, W.R. and J.D. Reynolds. 1996. Growth rate and age at migration of *Anguilla anguilla*. Journal of Fish Biology 48: 633–642.

Robinet, T. and E. Feunteun. 2002. Sublethal effects of exposure to chemical compounds: a cause for the decline in Atlantic eels? Ecotoxicology 11: 155–164.

Rohr, D.H., P.M. Lokman, P.S. Davie and G. Young. 2001. 11-ketotestosterone induces silvering-related changes in immature female short-finned eels, Anguilla australis. Comparative Biochemistry and Physiology A 130: 701–714.

Romeo, R.D. 2003. Puberty: a period of both organizational and activational effects of steroid hormones on neurobehavioural development. Journal of Neuroendocrinology 15: 1185–1192.

Rossi, R. and G. Colombo. 1976. Sex ratio, age and growth of silver eels in two brackish lagoons in the northern Adriatic Valli of Commacchio and Valli Nuova. Archivio di Oceanografia e Limnologia 18: 327–341.

Rousseau, K. and S. Dufour. 2004. Phylogenetic evolution of the neuroendocrine control of growth hormone: contribution from teleosts. Cybium 28: 181–198.

Rousseau K. and S. Dufour. 2008 Endocrinology of migratory fish life cycle in special environments: the role of metamorphoses. In: Fish Life in Special Environments, P. Sébert, D.W. Onyango, and B.G. Kapoor (Eds.). Science Publishers, Enfield (NH), USA, pp. 193–231.

Rousseau, K., Y.S. Huang, N. Le Belle, B. Vidal, J. Marchelidon, J. Epelbaum and S. Dufour. 1998. Long-term inhibitory effects of somatostatin and insulin-like growth factor 1 on growth hormone release by serum-free primary culture of pituitary cells from European eel (*Anguilla anguilla*). Neuroendocrinology 67: 301–309.

Rousseau, K., N. Le Belle, J. Marchelidon and S. Dufour. 1999. Evidence that corticotropin-releasing hormone acts as a growth hormone-releasing factor in a primitive teleost, the European eel (*Anguilla anguilla*). Journal of Neuroendocrinology 11: 385–392.

Rousseau, K., N. Le Belle, K. Pichavant, J. Marchelidon, B.K.C. Chow, G. Boeuf and S. Dufour. 2001. Pituitary growth hormone secretion in turbot, a phylogenetically recent teleost, is regulated by a species-specific pattern neuropeptides. Neuroendocrinology 74: 375–385.

Saglio, P., A.M. Escaffre and J.M. Blanc. 1988. Structural characteristics of the epidermal mucosa in yellow and silver European eel, *Anguilla anguilla* (L.). Journal of Fish Biology 32: 505–514.

Sancho, E., M.D. Ferrando and E. Andreu. 1997. Response and recovery of brain acetylcholinesterase activity in the European eel, *Anguilla anguilla*, exposed to fenitrothion. Ecotoxicology and Environmental Safety 38: 205–209.

Sancho, E., M.D. Ferrando and E. Andreu. 1998. *In vivo* inhibition of AChE activity in the European eel *Anguilla anguilla* exposed to tecnical grade fenitrothion. Comparative Biochemistry and Physiology C 120: 389–395.

Sancho, E., J.J. Ceron and M.D. Ferrando. 2000. Cholinesterase activity and hematological parameters as biomarkers of sublethal molinate exposure in *Anguilla anguilla*. Ecotoxicology and Environmental Safety 46: 81–86.

Sbaihi, M. 2001. Interaction des stéroïdes sexuels et du cortisol dans le contrôle de la reproduction et du métabolisme calcique chez un téléostéen migrateur, l'anguille (*Anguilla anguilla*). PhD thesis University Paris VI.

Sbaihi, M., M. Fouchereau-Peron, F. Meunier, P. Elie, I. Mayer, E. Burzawa-Gérard, B. Vidal and S. Dufour. 2001. Reproductive biology of the conger eel from the south coast of Brittany, France and comparison with the European eel. Journal of Fish Biology 59: 302–318.

Sbaihi, M., K. Rousseau, S. Baloche, F. Meunier M. Fouchereau-Peron and S. Dufour. 2009. Cortisol mobilizes mineral stores from vertebral skeleton in the European eel: an ancestral origin for glucocorticoid-induced osteoporosis? Journal of Endocrinology 201: 241–252.

Scaion, D., A. Vettier and P. Sébert. 2008. Pressure and temperature interactions on aerobic metabolism in migrating silver eels: results *in vitro*. Undersea Hyperbaric Medical Society 35: 27–33.

Schmidt, J. 1923. Breeding places and migration of the eel. Nature 111: 51–54.

Sébert, P. 2002. Fish at high pressure: a hundred history. Comparative Biochemistry and Physiology A 131: 575–585.

Sébert, P. 2003. Fish adaptations to pressure. In: Fish adaptation, A.L. Val and B.G. Kapoor BG (Eds.). Science Publishers, Enfield, pp. 73–95.

Sébert, P. and L. Barthélémy. 1985. Effects of high hydrostatic pressure per se, 101 atm on eel metabolism. Respiration Physiology 62: 349–357.

Sébert, P., B. Simon and L. Barthélémy. 1993. Hydrostatic pressure induces a state resembling histotoxic hypoxia in fish. Comparative Biochemistry and Physiology A 105: 255–258.

Sébert, M.E., A. Amérand, A. Vettier, F.A. Weltzien, C. Pasqualini, P. Sébert, S. Dufour . 2007. Effects of high hydrostatic pressure on the pituitary-gonad axis in the European eel, *Anguilla anguilla*. General and Comparative Endocrinology 153: 289–298.

Sébert, M.E., C. Legros, F.A. Weltzien, B. Malpaux, P. Chemineau, S. Dufour. 2008. Melatonin activates brain dopaminergic systems in the eel with an inhibitory impact on reproductive function. Journal of Neuroendocrinology 20: 917–929.

Sébert, P., D. Scaion, M. Belhomme. 2009. High hydrostatic pressure improves the swimming efficiency of European migrating silver eel. Respiratory Physiology and Neurobiology 165: 112–114.

Sheridan, M.A. 1988. Lipid dynamics in fish: aspects of absorption, transportation, deposition and mobilization. Comparative Biochemistry and Physiology B 90: 679–690.

Simon, B., P. Sébert and L. Barthélémy. 1989. Effects of long-term exposure to hydrostatic pressure per se (101 ATA) on eel metabolism. Canadian Journal of Physiology and Pharmacology 67: 1247–1251.

Sinha, V.P.R. and J.W. Jones. 1975. The European freshwater eel. Liverpool University Press, Liverpool.

Sorensen, P.W. and N.W. Pankhurst. 1988. Histological changes in the gonad, skin, intestine and olfactory epithelium of artificially-matured male American eels, Anguilla rostrata (LeSueur). Journal of Fish Biology 32: 297–307.

Suzuki, R., M. Kishida and T. Hirano. 1990. Growth hormone secretion during longterm incubation of the pituitary of the Japanese eel, *Anguilla japonica*. Fish Physiology and Biochemistry 8: 159–165.

Svedäng, H. and H. Wickström. 1997. Low fat contents in female silver eel: indications of insufficient energetic stores for migration and gonadal development. Journal of Fish Biology 50: 475–486.

Svedäng, H., E. Neuman and H. Wickström. 1996. Maturation patterns in female European eel: age and size at the silver eel age. Journal of Fish Biology 48: 342–351.

Tesch, F.W. 1977. The eel: biology and management of anguillid eels. Chapman and Hall, London.

Tesch, F.W. 1982. The sargasso Sea Eel Expedition 1979. Helgoländer Meeresunters 35: 263–277.

Tesch, F.W. 1989. Changes in swimming depth and direction of silver eels (*Anguilla anguilla* L.) from the continental shelf to the deep sea. Aquatic Living Resources 2: 9–20.

Tesch, F.W. 1991. Anguillidae. In: The Freshwater Fishes of Europe, H. Hoestlandt (Ed.). AULA-Verlag, Wiesbaden 388–437.

Tesch, F.W., H. Westerberg and L. Karlsson. 1991. Tracking studies on migrating silver eels in the Central Baltic. Meeresforschung 33: 183 –196.

Théron, M., F. Guerrero and P. Sébert. 2000. Improvement in the efficiency of oxidative phosphorylation in the freshwater eel acclimated to 10.1 MPa hydrostatic pressure. Journal of Experimental Biology 203: 3019–3023.

Thomson, A.J. and J.R. Sargent. 1977. Changes in the levels of chloride cells and (Na++K+)-dependent ATPase in the gills of yellow and silver eels adapting to seawater. Journal of Experimental Zoology 200: 33–40.

Todd, P.R. 1981a. Morphometric changes, gonad histology, and fecundity estimates in migrating New Zealand freshwater eels (*Anguilla* spp.). New Zealand Journal of Marine and Freshwater Research 15: 155–170.

Todd, P.R. 1981b. Timing and periodicity of migrating New Zealand freshwater eels (*Anguilla* spp.). New Zealand Journal of Marine and Freshwater Research 15: 225–235.

Toppari, J., A. Juul. 2010. Trends in puberty timing in humans and environmental modifiers. Molecular and Cellular Endocrinology 324: 39–44.

Tsukamoto, K. 1992. Discovery of the spawning area for Japanese eel. Nature 356: 789–791.

Tsukamoto, K. and T. Arai. 2001. Facultative catadromy of the eel *Anguilla japonica* between freshwater and seawater habitats. Marine Ecology Progress Series 220: 265–276.

Tsukamoto, K., I. Nakai and F.W. Tesch. 1998. Do all freshwater eels migrate? Nature 396: 635–636.

Tzeng, W.N., K.P. Severin and H. Wickström. 1997. Use of otolith microchemistry to investigate the environmental history of European eel *Anguilla anguilla*. Marine Ecology Progress Series 149: 73–81.

Tucker, D.W. 1959. A new solution to the Atlantic eel problem. Nature 183: 495–501.

van den Thillart, G., V. Van Ginneken, F. Korner, R. Heijmans, R. Van der Linden and A. Gluvers. 2004. Endurance swimming of European eel. Journal of Fish Biology 65: 312–318.

van der Oost, R., H. Heida, K. Satumalay, F-J. Van Schooten, F. Ariese and N.P.E. Vermeulen. 1994. Bioaccumulation biotransformation and DNA binding of pahs in feral eel (*Anguilla anguilla*) exposed to polluted sediments: a field survey. Environmental Toxicology and Chemistry 13: 859–870.

van Ginneken, V. and G. van den Thillart. 2000. Eel fat stores are enough to reach the Sargasso. Nature 403: 156–157.

van Ginneken, V., C. Durif, S.P. Balm, R. Boot, M.W.A. Verstegen, E. Antonissen, G. van den Thillart. (2007a) Silvering of European eel (*Anguilla anguilla* L.): seasonal changes of morphological and metabolic parameters. Animal Biology 57: 63–77.

van Ginneken, V., S. Dufour, M. Sbaihi, P. Balm, K. Noorlander, M. de Bakker, J. Doornbos, A. Palstra, E. Antonissen, I. Mayer and G. van den Thillart. 2007b. Does a 5500-km swim trial stimulate early sexual maturation in the European eel (*Anguilla anguilla* L.)? Comparative Biochemistry and Physiology A 147: 1095–1103.

van Ginneken, V., A. Palstra, P. Leonards, M. Nieveen, H. van den Berg, G. Glik and G., T. Spanings, P. Niemantsverdriet, G. van den Thillart, and A. Murk. 2009. PCBs and the energy cost of migration in the European eel (*Anguilla anguilla* L.). Aquatic Toxicology 92: 213–220.

Vettier, A., C. Szekely and P. Sébert. 2003. Are yellow eels from lake Balaton able to cope with high pressure encountered during migration to the Sargasso sea? The case of energy metabolism. Ann. Biol. 53: 329–338.

Vettier, A., A. Amérand, C. Cann-Moisan, and P. Sébert. 2005. Is the silvering process similar to the effects of pressure acclimatization on yellow eels? Respiratory Physiology and Neurobiology 145: 243–250.

Vettier, A., C. Labbé, A. Amérand, G. Da Costa, E. Le Rumeur, C. Moisan, and P. Sébert. 2006. Hydrostatic pressure effects on eel mitochondrial functioning and membrane fluidity. Undersea Hyperbaric Medical Society 33: 149–156.

Vidal, B., C. Pasqualini, N. Le Belle, M.C. Holland, M. Sbaihi, P. Vernier, Y. Zohar and S. Dufour. 2004. Dopamine inhibits luteinizing hormone synthesis and release in the juvenile European eel: a neuroendocrine lock for the onset of puberty. Biology of Reproduction 71: 1491–500.

Vollestad, L.A. 1988. Tagging experiments with yellow eel, *Anguilla anguilla* (L.) in brackish water in Norway. Sarsia 73: 157–161.

Vollestad, L.A. 1992. Geographic variation in age and length at metamorphosis of maturing European eel: environmental effects and phenotypic plasticity. Journal of Animal Ecology 61: 41–48.

Vollestad, L.A. and B. Johnson. 1986. Life-history characteristics of the European eel *Anguilla anguilla* in the Imsa River, Norway. Transactions of the American Fisheries Society 115: 864–871.

Vollestad, L.A., B. Jonsson, N.A. Hvidsten, T.F. Naesje, O. Haraldstad and J. Ruud-Hansen. 1986. Environmental factors regulating the seaward migration of European silver eels (*Anguilla anguilla*). Canadian Journal of Fisheries and Aquatic Sciences 43: 1909–1916.

Weltzien, F.A., C. Pasqualini, M-E. Sébert, B. Vidal, N. Le Belle, O. Kah, P. Vernier and S. Dufour. 2006. Androgen-dependent stimulation of brain dopaminergic systems in the female European eel (*Anguilla anguilla*). Endocrinology 147: 2964–2973.

Westerberg, H., I. Lagenfelt and H. Svedäng. 2007. Silver eel migration behaviour in the Baltic. ICES Journal of Marine Science 64: 1457–1462.

Westin, L. 1990. Orientation mechanisms in migrating European silver eel (*Anguilla anguilla*): temperature and olfaction. Marine Biology 106: 175 –179.

Wiersinga, W.M. and L. Bartalena. 2002. Epidemiology and prevention of Graves' ophthalmopathy. Thyroid 12: 855–860.

Wood, P. and J.G. Partridge. 1993. Opsin substitution induced in retinal rods of the eel, *Anguilla anguilla* (L.): a model for G-protein-linked receptors. Proceedings of the Royal Society of London Series B 254: 227–232.

Yada, T., A. Urano and T. Hirano. 1991. Growth hormone and prolactin gene expression and release in the pituitary of rainbow trout in serum-free culture. Endocrinology 129: 1183–1192.

Yamamoto, K. and K. Yamauchi. 1974. Sexual maturation of Japanese eel and production of eel larvae in the aquarium. Nature 251: 220–222.

Zachmann, A., J. Falcon, S.C. Knijff, V. Bolliet and M.A. Ali. 1992. Effects of photoperiod and temperature on rhythmic melatonin secretion from the pineal organ of the white sucker (Catostomus commersoni) *in vitro*. General and Comparative Endocrinology 86: 26–3.

Zara, V., L. Palmiei, A. Giudetti, A. Ferramosca, L. Capobianco and G.V. Gnoni. 2000. The mitochondrial tricarboxylate carrier: unexpected increased activity in starved silver eels. Biochemical and Biophysical Research Communications 276: 893–898.

Zhang, H., F. Futami, N. Horie, A. Okamura, T. Utoh, N. Mikawa, Y. Yamada, S. Tanaka and N. Okamoto. 2000. Molecular cloning of fresh water and deep-sea rod opsin genes from Japanese eel *Anguilla japonica* and expressional analyses during sexual maturation. FEBS Letters 469: 39–43.

# Index

# The Editors

**Sylvie Dufour** is Research Director at CNRS (National Centre for Scientific Research), and Head of Research Unit BOREA « Biology of Aquatic Organisms and Ecosystems » CNRS, Institute for Research and Development, University Pierre & Marie Curie, at the National Museum of Natural History, Paris. Her research activities concern Comparative and Evolutionary Neuroendocrinology: Origin and Evolution of Neuroendocrine Systems, Neuroendocrinology and Ecophysiology of Puberty and Metamorphosis.

**Karine Rousseau** is Associate Professor of Physiology at the National Museum of Natural History, Research Unit BOREA, Paris. Her research interests include Comparative Neuroendocrinology and Reproductive Physiology. She is currently studying the neuroendocrine systems involved in the control of metamorphosis and puberty in teleost fishes.

**B.G. Kapoor** was formerly professor and Head of Zoology in Jodhpur University (India). Dr Kapoor has co-edited 23 books published by Science Publishers, Enfield, NH, USA. The most recent ones are Fish Defenses (Volume 1), 2009 (with Giacomo Zaccone, José Meseguer Peñalver and Alfonsa García-Ayala); The Biology of Blennies, 2009 (with Robert A. Patzner, Emanuel J. Gonçalves and Philip A. Hastings); Development of Non-Teleost Fishes, 2009 (with Yvette W. Kunz and Carl A. Luer); Fish Defenses (Volume 2), 2009 (with Giacomo Zaccone, C. Perrière and A. Mathis); Fish Locomotion: An Eco-ethological Perspective (with Paolo Domenici), Biology of Subterranean Fishes, 2010 (with Eleonora Trajano and Maria Elina Bichuette); Biological Clock in Fish, 2010 (with Ewa Kulczykowska and Wlodzimierz Popek); The Biology of Gobies, 2011 (with Robert A. Patzner, James L. Van Tassell and Marcelo Kovačić). Dr Kapoor has also co-edited The Senses of Fish: Adaptations for the Reception of Natural Stimuli, 2004 (with G. von der Emde and J. Mogdans); and co-authored Ichthyology Handbook, 2004 (with B. Khanna), both from Springer, Heidelberg. He has also been a contributor in books from Academic Press, London (1969, 1975 and 2001). His E-mail is: bhagatgopal.kapoor@rediffmail.com

# Color Plate Section

## Chapter 2

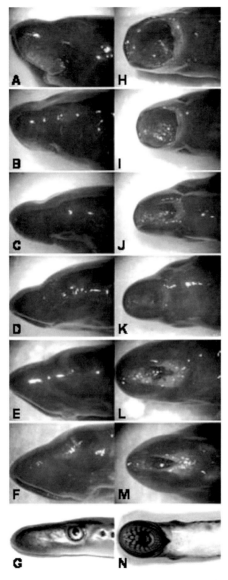

**Figure 1.** Anterior region in lateral (A- G) and ventral (H – N) views of larval *Petromyzon marinus* (A, H) induced to metamorphose following treatment with potassium perchlorate ($KClO_4$) (B-F and I-M) and juvenile *P. marinus* immediately following the completion of spontaneous metamorphosis (G, N). Approximate staging for $KClO_4$-induced metamorphosis is as follows: Stage 1 (B, I); Stage 2 (C, J); Stage 3 (D, K); Stage 4 (E, L); Stage 5/6 (F,M). Photos A-F and H-M kindly provided by Dr. John R. Gosche, University of Nevada.

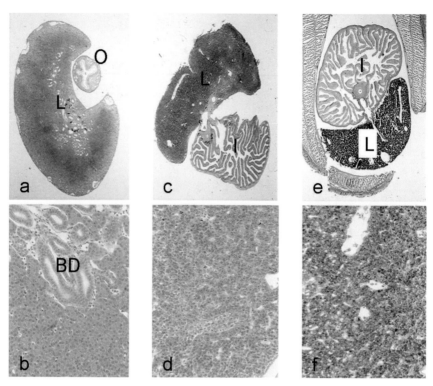

**Figure 15.** Light micrographs showing the staining for iron with Prussian Blue in the liver (L) of larva (a, b), stage 5 of metamorphosis (c,d) and in a juvenile (e,f) of *P. marinus*. Note the progressive increase of liver iron (blue stain) during metamorphosis into the iron-loaded state of the adult and the absence of this metal in the oesophagus (O) and intestine (I). There are bile ducts (BD) in the larva liver but none in the livers of metamorphosing animals or the adult. (a, c, and e. X25; b, d, and f. X250).

# Chapter 5

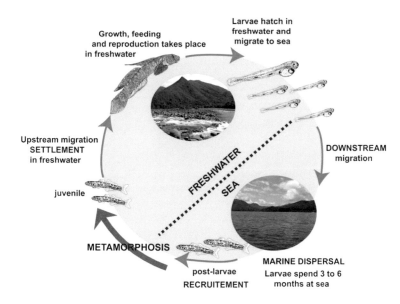

**Figure 1.** *S. lagocephalus* amphidromous life cycle.

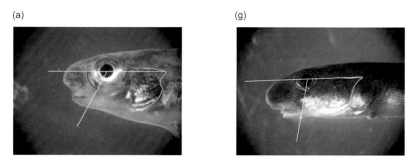

**Figure 2.** Photograph of *S. lagocephalus* (a) at day 0 (PL1) (arrival in estuary from the sea), corner of mouth angle is 63.0 ± 0.80°; (b) at day 37 (J1) (37 days spent in fluvarium since capture at the mouth river), corner of mouth angle is 80.4 ± 0.68°. Opercula is drawn on each photograph. Means are given ± SEM (n = 8 independant fish for each time point). (PL1: post-larval stage 1; J1: juvenile stage 1).

# Chapter 6

**Figure 2**. Atlantic salmon parr and smolt. Note the silvery color of smolt compared to the dark stripes of parr.

# Chapter 7

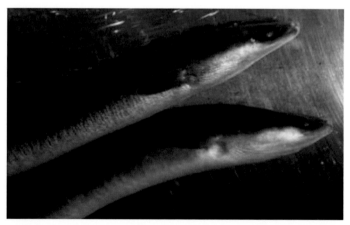

**Figure 2**. European eel yellow and silver stages. Note the silvery color of the ventral side of the silver eel as compared to the yellow eel. The eye diameter is enlarged in silver eels compared to yellow eels.

For Product Safety Concerns and Information please contact our EU representative GPSR@taylorandfrancis.com Taylor & Francis Verlag GmbH, Kaufingerstraße 24, 80331 München, Germany

Printed and bound by CPI Group (UK) Ltd, Croydon, CR0 4YY

01/05/2025

01858617-0001